The Physics of Fluids and Plasmas
An Introduction for Astrophysicists

A good working knowledge of fluid mechanics and plasma physics is essential for the modern astrophysicist. This graduate textbook provides a clear, pedagogical introduction to these core subjects. Assuming an undergraduate background in physics, this book develops fluid mechanics and plasma physics from first principles.

This book is unique because it presents neutral fluids and plasmas in a unified scheme, clearly indicating both their similarities and their differences. Also, both the macroscopic (continuum) and microscopic (particles) theories are developed, establishing the connections between them. Throughout, key examples from astrophysics are used, though no previous knowledge of astronomy is assumed. Exercises are included at the end of chapters to test the reader's understanding.

This textbook is aimed primarily at astrophysics graduate students. It will also be of interest to advanced students in physics and applied mathematics seeking a unified view of fluid mechanics and plasma physics, encompassing both the microscopic and macroscopic theories.

Arnab Rai Choudhuri is an Associate Professor of Physics at the Indian Institute of Science in Bangalore. After obtaining his Ph.D. at the University of Chicago in 1985, he spent two years at the High Altitude Observatory in Boulder and then joined the faculty of Indian Institute of Science in Bangalore. During his tenure in Bangalore, he has held visiting positions at the University of Chicago, the University of St Andrews and the Kiepenheuer Institut in Freiburg (as an Alexander von Humboldt Fellow). In the field of solar magnetohydrodynamics, he is one of the world's best-known scientists of his generation.

The Physics of Fluids and Plasmas

An Introduction for Astrophysicists

ARNAB RAI CHOUDHURI

CAMBRIDGE
UNIVERSITY PRESS

PUBLISHED BY THE PRESS SYNDICATE OF THE UNIVERSITY OF CAMBRIDGE
The Pitt Building, Trumpington Street, Cambridge CB2 1RP, United Kingdom

CAMBRIDGE UNIVERSITY PRESS
The Edinburgh Building, Cambridge CB2 2RU, United Kingdom
40 West 20th Street, New York, NY 10011-4211, USA
10 Stamford Road, Oakleigh, Melbourne 3166, Australia

First published 1998

Typeset in 10/13 pt Times   [TAG]

A catalogue record of this book is available from the British Library

Library of Congress Cataloguing in Publication data

Choudhuri, Arnab Rai, 1956–
The physics of fluids and plasmas : an introduction for
astrophysicists / Arnab Rai Choudhuri.
p.   cm.
Includes bibliographical references and index.
ISBN 0 521 55487 X. – ISBN 0 521 55543 4 (pbk.)
1. Fluid dynamics. 2. Plasma astrophysics. 3. Astrophysics
I. Title.
QB466.F58C46 1998
530.4′4–dc21   97-31092 CIP

ISBN 0 521 55487 X hardback
ISBN 0 521 55543 4 paperback

Transferred to digital printing 2004

To Gene Parker
from whom I learnt
much of what is
in this book

Contents

Preface

Hydrodynamics, magnetohydrodynamics, kinetic theory and plasma physics are becoming increasingly important tools for astrophysics research. Many graduate schools in astrophysics around the world nowadays offer courses to train graduate students in these areas. This was not the case even a few years ago—say around 1980—when it was rare for an astrophysics graduate school to teach these subjects, and the students who needed the knowledge of these subjects for their research were supposed to pick up the tricks of the trade on their own. With increasing applications of these subjects to astrophysics—especially to understand many phenomena discovered in the radio, X-ray or infrared wavelengths—the need is felt to impart a systematic training in these areas to all graduate students in astrophysics.

When I joined the faculty of the Astronomy Programme in Bangalore in 1987, I argued that a course covering these subjects should be introduced. My colleague and friend, Rajaram Nityananda, shared my enthusiasm for it, and we together managed to convince the syllabus committee of the need for it. From then onwards, this course has been taught regularly in our graduate programme, the responsibility of teaching it falling on my shoulders on several occasions. When I taught this course for the first time in 1988, I had to work very hard preparing lectures from different sources. I was lucky to have taken such a course myself as a graduate student in Chicago in 1981—taught by E. N. Parker—although it was somewhat unusual at that time for astronomy departments in the U.S.A. to offer such courses. The lecture notes which I had taken in that course and preserved carefully were a great help in preparing my course. I had to tell the students to use at least half-a-dozen basic books (such as Landau and Lifshitz's *Fluid Mechanics*, Chandrasekhar's *Hydrodynamic and Hydromagnetic Stability*, Cowling's *Magnetohydrodynamics*, Huang's *Statistical Mechanics*,

Chen's *Plasma Physics*) for different parts of the course. Ever since that time, I have been thinking of writing a comprehensive textbook based on my course. While I was trying to get some time off to write my book, the first textbook on the subject by Shu (1992) appeared. When I first saw the book, my heart sank. I thought that the book I had been contemplating had been written, and there would now be no point in my writing a book. However, on looking at Shu's book more closely, I realized that my point of view has been sufficiently different and there should be scope for more textbooks with different perspectives on such an important subject.

While writing such a textbook for astrophysics graduate students, two alternative approaches are possible. One is to take some astrophysical topics as central themes, and then develop fluid mechanics and plasma physics primarily as tools to be used. The other approach is to present fluid mechanics and plasma physics as logically coherent subjects, with some astrophysical examples to illustrate the applications of basic principles. I have followed the second approach, whereas the approach followed in Shu's book is closer to the first option. Fluid mechanics and plasma physics are venerable subjects with beautiful structures, and I personally believe that it is important for astrophysics students to appreciate the beauty of these structures rather than regarding these subjects only as tools for solving problems. Astrophysicists often have to deal with situations where it may not be obvious whether macroscopic continuum models work. Hence I have developed both the microscopic (particles) and the macroscopic (continuum) theories, establishing the connection between them.

I have attempted to present a unified discussion of neutral fluids and plasmas. One could think of writing a textbook in which theories of neutral fluids and plasmas are developed simultaneously. In such an approach, similar topics in fluid mechanics and plasma physics would be discussed together. I have followed the other approach of developing the theory of neutral fluids in Part 1 and the theory of plasmas in Part 2. Within each part, first I begin from microscopic theories and then develop the macroscopic continuum models from thereon. The main reason for presenting neutral fluids and plasmas separately is that the mathematical theory of neutral fluids is relatively simpler and I have tried to give a reasonably full account of it. On the other hand, the mathematical theory of plasmas is much more involved and it has often been necessary to leave some gaps in the arguments. Especially, the connection between microscopic and macroscopic theories of plasmas is an immensely complex subject which is still not on a rigorous and firm footing. I felt that readers would be in a better position to appreciate the complexities and subtleties of plasma physics if they already have a knowledge of the corresponding topics in the theory of neutral fluids. Throughout the book, however, I have emphasized both the similarities and the differences between the theories of neutral fluids and plasmas. The reader will find occasional discussions of stellar dynamics as well, pointing out how the techniques of stellar dynamics compare with the techniques used in the studies of neutral fluids and plasmas. A full treatment of stellar dynamics, however, is beyond the scope of this book.

Nowadays, it is often regarded as a great virtue of a textbook if its chapters are completely independent of each other. A reader should be able to read any chapter without reading any other chapter! I admit to committing the grievous sin of writing a book like an interwoven tapestry with connecting threads running from almost everywhere to everywhere. I know that only a foolish author of a technical book in this busy age would expect a reader to read his book from the first page to the last page. Once a reader develops some familiarity with this book, I do believe that it will be possible for him/her to find his/her way through the book to topics of personal interest without reading everything preceding it. I should, however, point out that I have taken particular pains to show as to how the various topics discussed in the book are connected to each other. These connections will be best appreciated by a reader who reads through major blocks of the text in the sequence in which they are presented. This book contains about 30% more material than what I can comfortably cover in a semester of fourteen weeks, lecturing three hours a week. So the instructor of a one-semester course should have ample opportunity of planning a suitably tailored course based on this textbook.

In a book of this size, it is not possible to start from the basics and cover the research frontiers properly. So I have tried mainly to emphasize the basics, although I hope to have avoided giving the impression that this is a closed classical subject in which everything has already been established for posterity. The reader should at least be able to form an imperfect impression of the research frontiers. In keeping with this philosophy of emphasizing the basics, instead of giving references to the most recent publications surveying the research frontiers, I have mainly given references to important classic papers in which new concepts emerged for the first time or new discoveries were reported for the first time. While urging the readers to go through Faraday's original papers, Maxwell (1891, Vol. I, p. xi) wrote: "It is of great advantage to the student of any subject to read the original memoirs on that subject, for science is always most completely assimilated when it is in the nascent state." I hope that my book would encourage the student to read the important landmark works in this field. While giving references in my publications, I, as a principle, try to look up the original materials. In the case of a few references of historical value, however, I had to rely on other authors who quoted these references, as they were not available in the libraries to which I had access.

This book is primarily meant for astrophysics graduate students. I do, however, hope that it will also be of interest to students of fluid mechanics and plasma physics as one of the rare textbooks treating neutral fluids and plasmas from a unified point of view, developing both the microscopic and macroscopic theories. Several years ago, I was attending a workshop on turbulence in Boulder. The workshop featured speakers on both fluid turbulence and plasma turbulence. The communication gap between fluid dynamicists and plasma physicists was very apparent, and it became particularly embarrassing when many fluid dynamicists were found absconding during the plasma turbulence sessions. A distinguished plasma physicist made some scathing remarks about the undesirability of the situation and suggested that somebody ought to write an elementary textbook stressing the unity of neutral fluids and plasmas, to help students develop a good attitude from the beginning. I leave it for others to judge if my book fits the bill. To make sure that the book is accessible to non-astrophysicists, astronomical jargon has been kept to a minimum and even the astrophysics examples discussed in the book should be understandable without any previous formal training in astrophysics.

Any comments or suggestions on this book may be sent to my e-mail address: arnab@physics.iisc.ernet.in

Arnab Rai Choudhuri
Bangalore

Acknowledgements

A large portion of this book was written when I was spending a sabbatical year in the very hospitable atmosphere of Kiepenheuer Institut für Sonnenphysik in Freiburg. I am grateful to the Alexander von Humboldt Foundation for making this stay possible with a Humbodlt Fellowship. Two of my friends with previous experience of writing books, Eric Priest and Michael Stix, gave many useful tips, including how to approach publishers. It was a delight to work with Adam Black of Cambridge University Press on the preparation of my book. When I started work on this book, I circulated a detailed plan of my book to many astrophysicists. Several persons provided me valuable suggestions and encouragement. They include S. Chandrasekhar, A. Feriz-Mas, R. Kippenhahn, A. Königl, L. Mestel, M. J. Rees, F. Shu, C. S. Shukre, L. Spitzer, R. J. Tayler, P. Ulmschneider, N. O. Weiss, and two anonymous referees to whom Cambridge University Press sent my plan. A few persons carefully read large portions of the manuscript, caught errors and suggested improvements. They are: my teachers, Gene Parker and Amal Kumar Raychaudhuri; my senior colleague, Roddam Narasimha; and my friend, Palash Pal. I am also grateful to the large number of students who have taken this course from me over the years. Their questions and comments helped in improving the presentation and avoiding mistakes. I inherited my love for learning and scholarship from my parents, who encouraged me to take up the academic profession at a time when it was not too fashionable. Finally, I express my thanks to my wife Mahua for taking up the major share of the responsibility of looking after our two little boys when I was busy writing this book.

This book is dedicated to E. N. Parker, whose course in Chicago served as the model for my course in Bangalore. Several portions of this book are, to a large extent, based on his lectures in that course.

I do not think that I have been able to capture the magic of his teaching, but I hope that this book is at least not unworthy of being dedicated to him.

Arnab Rai Choudhuri

Introduction

1

1.1 Fluids and plasmas in the astrophysical context

When a beginning student takes a brief look at an elementary textbook on fluid mechanics and at an elementary textbook on plasma physics, he or she probably forms the impression that these two subjects are very different from each other. Let us begin with some comments why we have decided to treat these two subjects together in this volume and why astrophysics students should learn about them.

We know that all substances are ultimately made up of atoms and molecules. Ordinary fluids like air or water are made up of molecules which are electrically neutral. By heating a gas to very high temperatures or by passing an electric discharge through it, we can break up a large number of molecules into positively charged ions and negatively charged electrons. Such a collection of ions and electrons is called a plasma, provided it satisfies certain conditions which we shall discuss later. Hence a plasma is nothing but a special kind of fluid in which the constituent particles are electrically charged.

When we watch a river flow, we normally do not think of interacting water molecules. Rather we perceive the river water as a continuous substance flowing smoothly as a result of the macroscopic forces acting on it. Engineers and meteorologists almost always deal with fluid flows which can be adequately studied by modelling the fluid as a continuum governed by a set of macroscopic equations. Usually most of the elementary fluid mechanics textbooks deal with these macroscopic equations without ever bothering about the molecular constitution of fluids. On the other hand, very often results of laboratory plasma experiments can be understood best in terms of forces acting on individual plasma particles and their motions. Hence elementary plasma physics textbooks often start from the dynamics of plasma particles.

Because of these very different approaches, elementary textbooks often hide the underlying unity in the sciences of fluids and plasmas.

It is intuitively obvious to us that fluids like water and air can be treated as macroscopic continuum systems. But astrophysicists often deal with systems like the solar wind or the interstellar medium having few particles per cm³ but extending over vast regions of space. It is not at once obvious if continuum fluid equations are applicable to such systems. Hence it is useful for astrophysicists to have some understanding of the microscopic basis of the continuum equations to know when they are applicable and when they break down. We shall try to understand in this book why and under what circumstances collections of particles can be modelled as continua. Since we shall develop both the particle and continuum aspects of the theory, it is useful to approach fluids and plasmas from a unified point of view, which is often obscured in elementary textbooks by stressing the continuum aspects of neutral fluids and particle aspects of plasmas.

Most objects in the astrophysical Universe are made up of ionized material which can be regarded as plasma. Hence it is no wonder that astrophysicists have to learn about plasmas to understand how the Universe works. Often, however, the ordinary fluid dynamics equations are adequate if electromagnetic interactions are not important in a problem. We have seen that a plasma is a special kind of fluid in which the constituent particles are charged. Hence the special character of plasmas becomes apparent only in circumstances in which electromagnetic interactions play important roles. When electromagnetic interactions are unimportant, plasmas behave very much like neutral fluids which obey simpler equations. Stellar structure and oscillations are examples of important astrophysical problems for which ordinary fluid equations are *almost* adequate, even though stars are made up of plasma. If the star has a strong magnetic field, it may be necessary to apply very small plasma corrections. One of the current research topics in the study of solar oscillations is to understand the *very small* effect of magnetic fields on these oscillations.

Since neutral fluid equations in a sense can be thought to constitute a special case of plasma equations in which the electromagnetic terms are set to zero, there may be some logical appeal in first developing the full plasma equations in complete glory and then considering the neutral fluids as a special case. For pedagogical reasons, however, we have decided to present things in the opposite order. The first half of the book is devoted to neutral fluids, which obey simpler equations than plasmas. Then, in the second half, we develop the theory of plasmas, which are governed by more complicated and more general equations. Within each half, we begin from microscopic or particle considerations and then develop the continuum models. It will be seen that the microscopic theory of neutral fluids is *not* exactly of the nature of a special case of the microscopic theory of plasmas with electromagnetic forces set to zero. The particles in a neutral fluid interact only when they collide, whereas the particles in a plasma interact through long-range electromagnetic interactions. This difference in the nature of interactions introduces some subtle differences in the microscopic theories.

Although we shall be considering astrophysical applications as examples throughout the text, we want to emphasize that what we present in this book is nothing but *standard* fluid mechanics and *standard* plasma physics. Astrophysical problems often necessitate the application of the basic theory to situations very different from any terrestrial situation, but the basic physics does not change. Although the material is presented in this book in a way which would be most suitable for somebody embarking on a career of astrophysics research, a careful reader of this book should be in a position to appreciate laboratory problems in fluid mechanics and plasma physics equally well.

1.2 Characteristics of dynamical theories

We would like to develop dynamical theories of fluids and plasmas. By *dynamical theory* we mean a physical theory with which the time evolution of a system can be studied. Classical mechanics, classical electrodynamics and quantum mechanics are some of the familiar examples of dynamical theories in physics. The structures of all these dynamical theories have certain common characteristics, which we would expect the dynamical theories of fluids and plasmas also to have. Let us begin by noting down these common characteristics.

First of all, we must have a way of describing the state of our system at one instant of time. For a mechanical system, this is done by specifying all the generalized position and momentum coordinates. The state of an electromagnetic field is given by $\mathbf{E}(\mathbf{x})$ and $\mathbf{B}(\mathbf{x})$ at all points at an instant of time. For a quantum system, the state is prescribed by the wave function $\psi(\mathbf{x})$. In other words, the state is always prescribed by giving the numerical values of a set of variables. The second requirement for a dynamical theory is that we should have a set of equations which tells us how these variables change with time. Once such a set of equations is given, if we know the values of all the variables prescribing the state of the system at one instant of time, we shall be able to calculate the values of all these variables at some future time. In other words, it is possible to calculate some future final state of the system from the initial state. In classical mechanics, Hamilton's equations give the time derivatives of position and momentum coordinates. Maxwell's equations contain the terms $\partial \mathbf{E}/\partial t$, $\partial \mathbf{B}/\partial t$ and hence provide the dynamical theory for the electromagnetic field. For a quantum system, time-dependent Schrödinger's equation tells us how $\psi(\mathbf{x})$ changes with time.

The mathematical theories for fluids and plasmas also should have similar structures with these two characteristics:

1 There should be a way to prescribe the state of the system with a set of variables.

2 There should be a set of equations giving the time derivatives of these variables.

We may begin by asking the question how the state of a fluid or a plasma can be prescribed at an instant of time. As we have already seen, there are different levels of looking at fluids and plasmas. At a certain level, they can be regarded as collections of particles. On another level, they can be treated as continua. We expect different dynamical theories at different levels having the two general characteristics listed above. The dynamical theories at different levels should also have some correspondence amongst them. In the next section, §1.3, we give a brief outline of the different levels at which we wish to look at fluids and plasmas, and the different dynamical theories that we wish to develop at these different levels. Section 1.3 should serve as a kind of guide map for this book.

Let us end this section by commenting that these two requirements for dynamical theories can be given geometrical representations by introducing a *phase space*. A phase space is an imaginary space having many dimensions such that each of the variables necessary to prescribe the state of the system corresponds to one dimension. Since continuous functions like $\psi(\mathbf{x})$ have to be specified at all the spatial points within a certain volume (i.e. at an infinite number of points), the corresponding phase space must have infinite dimensions, each dimension corresponding to the value of ψ at one point. It is easy to see that a state of the system corresponds to one point in the phase space. Since the dynamical equations tell us how the state changes with time, they make this point in phase space move with time and trace out a trajectory.

Table 1.1 *Different levels of theory for neutral fluids and plasmas*

Neutral fluids		
Level	Description of state	Dynamical equations
0: N quantum particles	$\psi(\mathbf{x}_1,\ldots,\mathbf{x}_N)$	Schrödinger's eqn.
1: N classical particles	$(\mathbf{x}_1,\ldots,\mathbf{x}_N,\mathbf{u}_1,\ldots,\mathbf{u}_N)$	Newton's laws
2: Distribution function	$f(\mathbf{x},\mathbf{u},t)$	Boltzmann eqn.
3: Continuum model	$\rho(\mathbf{x}),\,T(\mathbf{x}),\,\mathbf{v}(\mathbf{x})$	Hydrodynamic eqns.
Plasmas (Levels 0 and 1 same as above)		
Level	Description of state	Dynamical equations
2: Distribution function	$f(\mathbf{x},\mathbf{u},t)$	Vlasov eqn.
$2\frac{1}{2}$: Two-fluid model	See Chapter 11	
3: One-fluid model	$\rho(\mathbf{x}),\,T(\mathbf{x}),\,\mathbf{v}(\mathbf{x}),\,\mathbf{B}(\mathbf{x})$	MHD eqns.

1.3 Different levels of theory

Since fluids and plasmas are collections of particles, let us consider a collection of N particles and look at the different levels at which one may wish to develop dynamical theories for this system. These different levels are summarized in Table 1.1. At a very fundamental level, all microscopic particles obey quantum mechanics. Let us call it Level 0. The state of the system at this level is given by the N-particle wave function, which evolves in time according to Schrödinger's equation. In this book, however, we shall not discuss this level at all. At the next higher Level 1, the system can be modelled as a collection of N classical particles. Can we always pass on from Level 0 to Level 1? No, one often encounters collections of particles which are inherently quantum and a classical description is not adequate. The electron gas within a metal is an example of such a system from everyday life and the material inside a white dwarf star is an astrophysical example. Since we are not going to discuss Level 0 in this book, the dynamics of quantum gases remains outside the scope of this book.

Under what circumstances is a description at Level 1 possible for our system of N particles? Basically the wave packets for the different particles have to be widely separated so that quantum interference is not important. If p is the typical momentum of the particles, then the de Broglie wavelength is

$$\lambda = \frac{h}{p} \approx \frac{h}{\sqrt{m\kappa_{\mathrm{B}}T}},$$

where m is the mass of the particle, κ_{B} the Boltzmann constant and T the temperature (see, for example, Schiff 1968, p. 3; Mathews and Venkatesan 1976, §1.13). Since this is also a measure of the sizes of wave packets of individual particles, we have to compare this with the typical inter-particle distance, which is $n^{-1/3}$ if n is the particle number density per unit volume. Hence the condition for the non-overlapping of wave packets is

$$\frac{hn^{1/3}}{\sqrt{m\kappa_{\mathrm{B}}T}} \ll 1. \tag{1.1}$$

When this condition is satisfied, an individual wave packet evolves according to Schrödinger's equation in an isolated fashion and can be shown to move like a classical particle. This result is known as Ehrenfest's theorem and is derived from Schrödinger's equation in any textbook on quantum mechanics (see, for example, Schiff 1968, pp. 28–30; Mathews and Venkatesan 1976, §2.7). Hence (1.1) gives the condition that Level 1 can be *derived* from Level 0. We then have at Level 1 a system of N classical particles of which the state is prescribed by the position and velocity coordinates $(\mathbf{x}_1, \ldots, \mathbf{x}_N, \mathbf{u}_1, \ldots, \mathbf{u}_N)$. The time evolution of this system can be studied by Newton's laws of motion or by Hamilton's equations.

If N is large, then it is not realistic to solve the equations of motion for all the position and velocity coordinates. Hence, in the next higher Level 2, one introduces the distribution function $f(\mathbf{x}, \mathbf{u}, t)$ giving the particle number density in the six-dimensional $(\mathbf{x}, \mathbf{u})$ space at time t ($\mathbf{x}$ is the position coordinate of a particle and $\mathbf{u}$ is its velocity coordinate). A dynamical theory at this level requires an equation which tells us how $f(\mathbf{x}, \mathbf{u}, t)$ changes in time. The time derivative of $f(\mathbf{x}, \mathbf{u}, t)$ for a neutral fluid is given by the Boltzmann equation. The corresponding equation for plasmas is called the Vlasov equation. We shall see that this equation superficially resembles the Boltzmann equation, but has some subtle differences.

At the final Level 3, we model the systems as continua. Let us first consider how the state of a neutral fluid in the continuum model can be prescribed. We know that a single-component gas in thermodynamic equilibrium can be described by two thermodynamic variables. A moving fluid is not in thermodynamic equilibrium as a whole. But if we consider a small element of fluid and go to the frame in which it is at rest, then we can regard that element to be in *approximate* thermodynamic equilibrium in that frame. This idea and the exact meaning of the adjective *approximate* will be made clearer in Chapter 3, where we derive Level 3 from Level 2. Hence the state of that element of fluid can be prescribed by two thermodynamic variables and the velocity of that element with respect to some frame, say the laboratory frame of reference. Since we have to specify the state of each and every element of the fluid in this fashion, the state of the whole fluid is given by prescribing the two thermodynamic variables and the velocity at all points of the fluid. Taking density and temperature as examples of two thermodynamic variables, the specification of $\rho(\mathbf{x})$, $T(\mathbf{x})$ and $\mathbf{v}(\mathbf{x})$ at all points of the fluid at an instant of time gives the state of the fluid at that time. The usual macroscopic hydrodynamic equations tell us how all these variables vary in time and hence constitute a complete dynamical theory for neutral fluids at Level 3.

Since plasmas can have magnetic fields embedded in them, we have to take $\mathbf{B}(\mathbf{x})$ as an additional variable when considering the Level 3 for plasmas. We know that one takes the electric field $\mathbf{E} = 0$ inside conductors when solving electrostatics problems. Since plasmas are good conductors of electricity, electric fields in the local rest frames inside plasmas are also quickly shorted by currents and it is not necessary to take the electric field as an extra variable in the continuum model at Level 3. A state of the plasma at this level can be given by prescribing $\rho(\mathbf{x})$, $T(\mathbf{x})$, $\mathbf{v}(\mathbf{x})$ and $\mathbf{B}(\mathbf{x})$ at all points. We shall later derive a set of equations called the *magnetohydrodynamic* or MHD equations giving the time evolutions of these variables. They are more complicated than the ordinary fluid dynamics equations. But is it always justified to ignore the electric field? It turns out that one can have electric fields in plasmas over short distances existing for short times. To handle such situations, we introduce an intermediate Level $2\frac{1}{2}$ for plasmas. At this intermediate level, we regard plasmas as mixtures of two fluids having opposite electrical charges. The details of this two-fluid model will be discussed in Chapter 11. When we consider *slow* motions of plasmas under mechanical and magnetic stresses, MHD equations are adequate. Again the exact meaning of *slow* will be made clear later. Many astrophysical problems can be handled with MHD equations. Propagation of electromagnetic waves in plasmas, however, is a problem for which it is necessary to deal with the more complex two-fluid model at Level $2\frac{1}{2}$.

We have seen that the condition (1.1) has to be satisfied in order to pass from Level 0 to Level 1. Similarly some other conditions have to be met to derive Level 2 from Level 1 or Level 3 from Level 2. These conditions will be discussed in the appropriate places of the book. If a system of N particles satisfies these conditions, then it is possible to introduce the distribution function $f(\mathbf{x}, \mathbf{u}, t)$ or to model the system as a continuum.

Much of this book is devoted to studying the dynamics of neutral fluids and plasmas at Levels 2 and 3 (with the additional Level $2\frac{1}{2}$ for plasmas). To begin with, however, we need to understand how we can develop Level 2 from Level 1. For a proper appreciation of this subject, it is important to know some general results pertaining to phase spaces of dynamical systems. In view of the generality of these results, we have decided to discuss them in the next two sections of this introductory chapter and end the chapter with them.

We now end this section with a comment on predictability. It would seem that a dynamical theory satisfying the structural requirements described in §1.2 would be completely predictable. In other words, knowing the present state of the system, one would always be able to predict the future completely. Fluids and plasmas, however, can often have *turbulence*—a state of random and chaotic motions which appear unpredictable. Developing a proper theory of turbulence has remained one of the unsolved grand problems of physics for over a century. We shall discuss in Chapter 8 the question of how turbulence can arise in systems apparently governed by predictable equations. Even if a dynamical theory is predictable *in principle*, we shall see that there can be a loss of predictability *in practice*.

1.4 Ensembles in phase space. Liouville's theorem

Let us consider a dynamical system of which a state can be prescribed by the generalized position and momentum coordinates ($q_s, p_s; s = 1, \ldots, n$) and which evolves according to Hamilton's equations:

$$\dot{p}_s = -\frac{\partial H}{\partial q_s}, \tag{1.2}$$

$$\dot{q}_s = \frac{\partial H}{\partial p_s}, \tag{1.3}$$

where the Hamiltonian $H(q_s, p_s, t)$ can be a function of all the coordinates and time (see, for example, Goldstein 1980, Chapter 8; Raychaudhuri 1983, Chapters 8–9). If you do not have a deep understanding of Hamiltonian theory, you need not panic. We shall make use of Hamiltonian theory only rarely in this book and an acquaintance with the above two equations will suffice.

Considering Hamiltonian systems may seem somewhat restrictive, because not all dynamical theories can be put in the Hamiltonian form.

Readers familiar with the subject would know that it is not possible to make a Hamiltonian formulation of a dissipative system. However, dissipation in macroscopic systems usually means that the energy of some ordered macroscopic motion is being transferred into random molecular motions. When we look at a system at the microscopic level (say our Level 1) and include the molecular motions within the fold of the dynamical theory, usually a Hamiltonian formulation is possible. Our system at Level 1, a collection of N classical particles, certainly allows a Hamiltonian treatment.

For the statistical treatment of a system, it is often useful to introduce the concept of an *ensemble*. An ensemble means a set of many replicas of the same system, which are identical in all other respects apart from being in different states at an instant of time. Hence each member of the ensemble can be represented by a point in the phase space at an instant of time and their evolutions correspond to different trajectories in the phase space. If the ensemble points are distributed sufficiently densely and smoothly in the phase space, then it is meaningful to talk about the density of ensemble points at a location in the phase space. Let us denote this density by $\rho_{\mathrm{ens}}(q_s, p_s, t)$.

We now wish to prove Liouville's theorem, which is one of the fundamental theorems of statistical mechanics. Let us first state the theorem. Then we shall proceed to prove it. Let us consider one member of the ensemble and its trajectory $(q_s(t), p_s(t))$ in the phase space. We keep measuring the density $\rho_{\mathrm{ens}}(q_s(t), p_s(t), t)$ as a function of time varying as a parameter along this trajectory. Liouville's theorem states that the time derivative of this density as we move along the trajectory is zero, i.e.

$$\frac{D\rho_{\mathrm{ens}}}{Dt} = 0, \tag{1.4}$$

where D/Dt denotes the time derivative along the trajectory. If (q_s, p_s) and $(q_s + \delta q_s, p_s + \delta p_s)$ denote the states of the system at times t and $t + \delta t$ on this trajectory, then

$$\frac{D\rho_{\mathrm{ens}}}{Dt} = \lim_{\delta t \to 0} \frac{\rho_{\mathrm{ens}}(q_s + \delta q_s, p_s + \delta p_s, t + \delta t) - \rho_{\mathrm{ens}}(q_s, p_s, t)}{\delta t}. \tag{1.5}$$

Expansion in a Taylor series to linear terms in small quantities gives

$$\rho_{\mathrm{ens}}(q_s + \delta q_s, p_s + \delta p_s, t + \delta t) = \rho_{\mathrm{ens}}(q_s, p_s, t)$$
$$+ \sum_s \delta q_s \frac{\partial \rho_{\mathrm{ens}}}{\partial q_s} + \sum_s \delta p_s \frac{\partial \rho_{\mathrm{ens}}}{\partial p_s} + \delta t \frac{\partial \rho_{\mathrm{ens}}}{\partial t}.$$

Substituting the above expression in (1.5), we have

$$\frac{D\rho_{\text{ens}}}{Dt} = \frac{\partial\rho_{\text{ens}}}{\partial t} + \sum_s \dot{q}_s \frac{\partial\rho_{\text{ens}}}{\partial q_s} + \sum_s \dot{p}_s \frac{\partial\rho_{\text{ens}}}{\partial p_s}. \tag{1.6}$$

To establish Liouville's theorem, we now have to show that the R.H.S. of (1.6) is equal to zero.

As a next step, we derive another general result—the equation of continuity—which applies to any system that conserves mass. It can be applied to ordinary fluids in ordinary space or to ensemble point distributions in phase space (where the total number of ensemble points is conserved in time). If ρ be the density of the system in some space, then the mass $\int \rho dV$ within a volume can change only due to the mass flux across the surface bounding that volume, i.e.

$$\frac{\partial}{\partial t}\int \rho \, dV = -\int \rho\mathbf{v} \cdot d\mathbf{s}.$$

Here $\int \rho\mathbf{v} \cdot d\mathbf{s}$ is the outward mass flux through the bounding surface, the negative sign implying that an outward mass flux reduces the mass within the bounded volume. Transforming the surface integral to volume integral by Gauss's theorem, we get

$$\int \left[\frac{\partial\rho}{\partial t} + \nabla \cdot (\rho\mathbf{v}) \right] dV = 0.$$

Since this must be true for any arbitrary volume, we must have

$$\frac{\partial\rho}{\partial t} + \nabla \cdot (\rho\mathbf{v}) = 0, \tag{1.7}$$

which is the equation of continuity.

We shall have many occasions of applying the equation of continuity to continuum models of fluids and plasmas in ordinary space in the later chapters. Now we apply it to the ensemble point distribution in phase space by taking $\rho = \rho_{\text{ens}}$ and $\mathbf{v} = (\dot{q}_s, \dot{p}_s)$. Putting these in (1.7), we have

$$\frac{\partial\rho_{\text{ens}}}{\partial t} + \sum_s \frac{\partial}{\partial q_s}(\rho_{\text{ens}}\dot{q}_s) + \sum_s \frac{\partial}{\partial p_s}(\rho_{\text{ens}}\dot{p}_s) = 0,$$

i.e.

$$\frac{\partial\rho_{\text{ens}}}{\partial t} + \sum_s \dot{q}_s \frac{\partial\rho_{\text{ens}}}{\partial q_s} + \sum_s \dot{p}_s \frac{\partial\rho_{\text{ens}}}{\partial p_s} + \rho_{\text{ens}} \sum_s \left(\frac{\partial\dot{q}_s}{\partial q_s} + \frac{\partial\dot{p}_s}{\partial p_s} \right) = 0. \tag{1.8}$$

Using Hamilton's equations (1.2) and (1.3), we find that

$$\frac{\partial\dot{q}_s}{\partial q_s} + \frac{\partial\dot{p}_s}{\partial p_s} = \frac{\partial}{\partial q_s}\left(\frac{\partial H}{\partial p_s} \right) - \frac{\partial}{\partial p_s}\left(\frac{\partial H}{\partial q_s} \right) = 0. \tag{1.9}$$

From (1.8) and (1.9), it follows that the R.H.S. of (1.6) is zero. This completes the proof of Liouville's theorem.

We now point out an important corollary of Liouville's theorem. Suppose the ensemble points initially inside the phase space volume element $d^n q_s d^n p_s$ after some time fill the volume element $d^n q'_s d^n p'_s$. If ρ_{ens} and ρ'_{ens} are the corresponding densities, then it follows from the conservation of ensemble points that

$$\rho_{ens} \, d^n q_s \, d^n p_s = \rho'_{ens} \, d^n q'_s \, d^n p'_s. \tag{1.10}$$

Since Liouville's theorem implies

$$\rho_{ens} = \rho'_{ens},$$

we see at once from (1.10) that

$$d^n q_s \, d^n p_s = d^n q'_s \, d^n p'_s, \tag{1.11}$$

which is one of the important direct consequences of Liouville's theorem.

1.5 Collisionless Boltzmann equation

We again focus our attention on the system of N classical particles. We assume all the particles to be *similar*. This is an assumption which will be used throughout Chapters 2 and 3. This assumption helps us to keep the equations simpler, while conveying most of the important physics. It is straightforward to generalize our theoretical analysis to systems containing two or more types of particles. When discussing the kinetic theory of plasmas in later chapters, we shall see an example of this. The plasma will be regarded as made up of three types of particles: electrons, ions and neutral atoms.

A state of our system of N classical particles is prescribed by $6N$ position and velocity coordinates. We would refer to the corresponding $6N$-dimensional phase space as the Γ-space. We know that a state of the system is represented by a point in this Γ-space. Let us also introduce a six-dimensional space with the six dimensions corresponding to the position and velocity $(\mathbf{x}, \mathbf{u})$ of a particle. We call this the μ-space. Each of our N particles would be represented by a point in this μ-space at an instant of time. Hence we would require N points in the μ-space to represent a state of our system of N particles. We thus see that there is a correspondence between the representations in the Γ-space and the μ-space. The state of the system represented by one point in the Γ-space gets mapped into a configuration of N points in the μ-space. The time evolution of the system gives rise to a trajectory in the Γ-space. This trajectory gets mapped to N trajectories of the N points in the μ-space.

Let us now properly introduce the distribution function $f(\mathbf{x}, \mathbf{u}, t)$ in the μ-space. If δN be the number of points in a small volume δV of the μ-space, then $f(\mathbf{x}, \mathbf{u}, t)$ is defined as

$$f(\mathbf{x}, \mathbf{u}, t) = \lim_{\delta V \to 0^+} \frac{\delta N}{\delta V}, \qquad (1.12)$$

where the limit of δV has to be taken in a special way. We have to make it small compared to the overall spatial extension of the points in the μ-space, but still keep it large enough to have a sufficiently large number of points inside this volume. This special character of the limit is indicated by writing $\delta V \to 0^+$ in (1.12) rather than just $\delta V \to 0$. It is possible to introduce the distribution function $f(\mathbf{x}, \mathbf{u}, t)$ for a system only *if this special limit exists,* and then only we can pass from Level 1 to Level 2 in our Table 1.1.

In §1.4, we introduced a density of ensemble points in the phase space, which would be the Γ-space in the present context. We are now introducing the distribution function $f(\mathbf{x}, \mathbf{u}, t)$ which is a density of points in the μ-space. Just as we proved Liouville's theorem for the ensemble point density in the phase space (i.e. the Γ-space), is it possible to derive a similar result for $f(\mathbf{x}, \mathbf{u}, t)$ in μ-space? We used Hamilton's equations to prove Liouville's theorem. Hence an exactly similar proof of a similar result for $f(\mathbf{x}, \mathbf{u}, t)$ in the μ-space can be given only if the trajectories of the points in the μ-space can be obtained from a Hamiltonian $H(\mathbf{x}, \mathbf{u}, t)$ such that

$$\dot{\mathbf{u}} = -\nabla H, \qquad (1.13)$$

$$\dot{\mathbf{x}} = \nabla_{\mathbf{u}} H, \qquad (1.14)$$

where $\nabla_{\mathbf{u}}$ is the gradient in the velocity space given by

$$\nabla_{\mathbf{u}} = \hat{\mathbf{e}}_x \frac{\partial}{\partial u_x} + \hat{\mathbf{e}}_y \frac{\partial}{\partial u_y} + \hat{\mathbf{e}}_z \frac{\partial}{\partial u_z}. \qquad (1.15)$$

Note that the Hamiltonian in the Γ-space, from which Liouville's theorem is obtained, can be a function of $6N + 1$ variables ($6N$ coordinates plus time). On the other hand, the Hamiltonian appearing in (1.13) and (1.14) can be a function of seven scalar variables only (the components of $\mathbf{x}$ and $\mathbf{u}$ plus time t).

If the N particles do not interact with each other and move under the influence of some external potential $\phi(\mathbf{x})$ alone, then one can introduce a Hamiltonian

$$H(\mathbf{x}, \mathbf{u}, t) = \tfrac{1}{2} \mathbf{u}^2 + \phi(\mathbf{x}) \qquad (1.16)$$

appropriate for the μ-space. Substitution of (1.16) in (1.13) and (1.14) would give the equations of motion. However, we get into problems when we want to incorporate the mutual interactions amongst the particles themselves. Consider a particle with coordinates $(\mathbf{x}, \mathbf{u})$ interacting with a nearby particle $(\mathbf{x}', \mathbf{u}')$, i.e. $\mathbf{x}$ and $\mathbf{x}'$ are close. This interaction can usually be described by a potential of the form $\phi(\mathbf{x}, \mathbf{x}')$ and hence incorporating it in the Hamiltonian for the Γ-space (which can be a function of $6N + 1$ variables) would not be difficult. But it cannot be written in the form $\phi(\mathbf{x})$ and so cannot be incorporated in the Hamiltonian for the μ-space which appears in (1.13) and (1.14).

We conclude that a Hamiltonian formulation of the dynamics of N particles is always possible in the Γ-space, but the same is possible in the μ-space only if the mutual interactions amongst the particles can be neglected. If the mutual interactions are negligible, then we call the system *collisionless*. Only for a collisionless system, the dynamics in the μ-space is Hamiltonian and we can establish

$$\frac{Df}{Dt} = 0 \tag{1.17}$$

in exactly the same way in which we proved Liouville's theorem in §1.4. The total derivative D/Dt in (1.17) means the time derivative along the trajectory of a point in the μ-space. Just as the total derivative in the Γ-space could be put in the form (1.6), the total derivative in the μ-space can similarly be put in the form

$$\frac{Df}{Dt} = \frac{\partial f}{\partial t} + \dot{\mathbf{x}} \cdot \nabla f + \dot{\mathbf{u}} \cdot \nabla_{\mathbf{u}} f. \tag{1.18}$$

For collisionless systems, we then have from (1.17) and (1.18)

$$\frac{\partial f}{\partial t} + \dot{\mathbf{x}} \cdot \nabla f + \dot{\mathbf{u}} \cdot \nabla_{\mathbf{u}} f = 0. \tag{1.19}$$

This is the famous *collisionless Boltzmann equation*. Another way of writing this equation is

$$\frac{\partial f}{\partial t} + \dot{x}_i \frac{\partial f}{\partial x_i} + \dot{u}_i \frac{\partial f}{\partial u_i} = 0, \tag{1.20}$$

where we have used the summation convention that the index i repeated twice in a term implies summation over the coordinate axes x, y, z. Throughout this book, we shall use this summation convention *only* for summation over the coordinate axes and the Roman letters i, j, k, l, ... will be used for the index which is repeated twice. Any other kind of summation (i.e. summation not over the coordinate axes x, y, z) will be indicated explicitly with a summation sign. See Appendix A.3 for further discussion of the summation convention.

When the mutual interactions amongst the particles cannot be neglected, (1.19) has to be suitably modified. Treating inter-particle interactions for a neutral gas is relatively easy, because the particles can be assumed to interact *only when they collide*, i.e. when they are physically very close. For a plasma, however, the long-range nature of electromagnetic interactions implies that many particles will be interacting with each other even when they are not physically very close. We shall discuss the methods of handling inter-particle interactions both for neutral gases and plasmas in the appropriate places in later chapters.

Part I Neutral fluids

The analytical results obtained by means of this so-called "classical hydrodynamics" usually do not agree at all with the practical phenomena ... Hydrodynamics thus has little significance for the engineer because of the great mathematical knowledge required for it and the negligible possibility of applying its results. Therefore the engineers—such as Bernoulli, Hagen, Wiessbach, Darcy, Bazin, and Boussinesq—put their trust in a mass of empirical data collectively known as the "science of hydraulics", a branch of knowledge which grew more and more unlike hydrodynamics.

While the methods of classical hydrodynamics were of a specifically analytical character, those of hydraulics were mostly synthetic ... In classical hydrodynamics everything was sacrificed to logical construction; hydraulics on the other hand treated each problem as a separate case and lacked an underlying theory by which the various problems could be correlated. Theoretical hydrodynamics seemed to lose all contact with reality; simplifying assumptions were made which were not permissible even as approximations. Hydraulics disintegrated into a collection of unrelated problems; each individual problem was solved by assuming a formula containing some undetermined coefficients and then determining those so as to fit the facts as well as possible. Hydraulics seemed to become more and more a science of coefficients.

— L. Prandtl and O. G. Tietjens (1934a)

At an early stage in the development of the theory of turbulence the idea arose that turbulent motion consists of eddies of more or less definite range of sizes. This conception combined with the already existing ideas of the Kinetic Theory of Gases led Prandtl and me independently to introduce the length l which is often called a "Mischungsweg" and is analogous to the "mean free path" of the Kinetic Theory. The length l could only be defined in relation to the definite but quite erroneous conception that lumps of air behave like molecules of a gas, preserving their identity till some definite point in their path, when they mix with their surroundings and attain the same velocity and other properties as the mean value of the corresponding property in the neighbourhood. Such a conception must evidently be regarded as a very rough representation of the state of affairs ...

The difficulty of defining a "Mischungsweg" or scale of turbulence, without recourse to some definite hypothetical physical process which bears no relation to reality, ... led me, some years ago, to introduce the idea that the scale of turbulence and its statistical properties in general can be given an exact interpretation by considering the correlation between the velocities at various points of the field at one instant of time or between the velocity of a particle at one instant of time and that of the same particle at some definite time, ξ, later.

— G. I. Taylor (1935)

2 Boltzmann equation

2.1 Collisions in a dilute neutral gas

We have already derived the collisionless Boltzmann equation in Chapter 1. This is the equation satisfied by the distribution function $f(\mathbf{x}, \mathbf{u}, t)$ if we neglect the interactions amongst the particles. We now discuss how this equation has to be modified for a *dilute* neutral gas when collisions are taken into account.

The adjective *dilute* used above means that the total volume of the gas particles is negligible compared to the volume available to the gas, i.e.

$$na^3 \ll 1, \tag{2.1}$$

where n is the number density of the particles and a the radius of a particle. Since the particles in a neutral gas do not have long-range interactions like the particles in a plasma, these particles are assumed to interact only when they *collide*, i.e. when the separation between two particles is not much larger than $2a$. The term *collision* normally means the interaction between two such nearby particles. A particle moves freely in a straight line between two collisions. The average distance travelled by a particle between two collisions is known as the *mean free path*—a very important concept introduced by Clausius (1858). Exercise 2.2 asks the reader to show that the mean free path is given by

$$\lambda = \frac{1}{\sqrt{2}n\pi a^2}. \tag{2.2}$$

It should be straightforward to prove this after reading §2.2. One consequence of (2.1) and (2.2) is that $\lambda \gg a$. In other words, diluteness implies that the mean free path is much larger than the particle size

19

so that a typical particle trajectory appears as shown in Figure 2.1 and the particles move freely most of the time, a collision being of the nature of an instantaneous event. One can sometimes have situations where three or more particles come very close together and simultaneously interact with each other. If, however, the gas is *dilute*, then the probability of such multi-particle collisions is much smaller than the probability of binary collisions and it is sufficient to consider binary collisions only.

It is clear from (1.17) that the value of the distribution function $f(\mathbf{x}, \mathbf{u}, t)$ does not change as we move along the trajectory of a particle, provided collisions are neglected. Collisions, however, can produce changes in $f(\mathbf{x}, \mathbf{u}, t)$ due to two reasons:

1 Some particles originally having velocity $\mathbf{u}$ may have other velocities after collisions. This causes a decrease in $f(\mathbf{x}, \mathbf{u}, t)$.
2 Some particles originally having other velocities may have the velocity $\mathbf{u}$ after collisions, thereby causing an increase in $f(\mathbf{x}, \mathbf{u}, t)$.

We thus conclude that (1.17) has to be modified to the form

$$\frac{Df}{Dt} d^3x \, d^3u = -C_{\text{out}} + C_{\text{in}}, \qquad (2.3)$$

where C_{out} and C_{in} are the rates at which particles leave and enter the elementary volume $d^3x \, d^3u$ of the μ-space due to collisions. These rates are explicitly calculated in §2.2.

We now summarize some important characteristics of binary collisions which will be useful for future discussions. Suppose two particles with initial velocities $\mathbf{u}$ and $\mathbf{u}_1$ acquire velocities $\mathbf{u}'$ and $\mathbf{u}'_1$ after a collision. Since all the particles are assumed similar (i.e. have the same mass m), the conservation laws of momentum and energy imply

$$\mathbf{u} + \mathbf{u}_1 = \mathbf{u}' + \mathbf{u}'_1, \qquad (2.4)$$

$$\tfrac{1}{2}|\mathbf{u}|^2 + \tfrac{1}{2}|\mathbf{u}_1|^2 = \tfrac{1}{2}|\mathbf{u}'|^2 + \tfrac{1}{2}|\mathbf{u}'_1|^2. \qquad (2.5)$$

One would like to calculate the final velocities $\mathbf{u}'$ and $\mathbf{u}'_1$ from the

Figure 2.1 The typical trajectory of a particle in a dilute neutral gas.

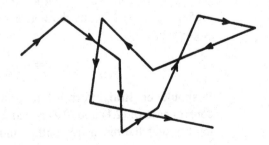

initial velocities. Since $\mathbf{u}'$ and $\mathbf{u}_1'$ have six scalar components, we need six scalar equations to solve them. Four of them are provided by (2.4) and (2.5). A fifth condition comes from the fact that collisions are coplanar if the force of interaction between the two particles is always radial, i.e. $\mathbf{u}'$ will have to lie in the plane of $\mathbf{u}$ and $\mathbf{u}_1$, forcing $\mathbf{u}_1'$ also to lie in the same plane by virtue of (2.4). We still need a sixth condition. This last condition should come from the nature of interaction between the particles. This is to be anticipated, as we do not expect the outcome of a collision to be independent of the nature of interaction and completely calculable from conservation laws. If the impact parameter of the collision is given, one can calculate the deflection produced by the collision from the interaction potential. This gives a sixth condition to determine the outcome of the collision. Textbooks of classical mechanics discuss methods of calculating such deflections from the interaction potential. Similar calculations can also be done by using quantum mechanics, where we can only predict the probability of deflection in a certain direction. We are interested here in a statistical treatment so that it is enough for us to know the probability of deflection in different directions. This can be handled by introducing the concept of a *differential scattering cross-section* discussed below. We shall not concern ourselves here with the question of calculating this cross-section from the interaction potential. We only discuss how the dynamics of our system can be studied when the scattering cross-section is given.

For the definition of the differential scattering cross-section in the context of classical mechanics, let us consider a beam of particles of number density n_1 and velocity $\mathbf{u}_1$ colliding with another beam of particles of number density n and velocity $\mathbf{u}$. A particle in the second beam experiences a flux $I = |\mathbf{u} - \mathbf{u}_1|n_1$ of particles from the first beam. We consider the number of collisions δn_c per unit volume per unit time which deflect particles from the second beam into a solid angle $d\Omega$. This number must be proportional to the number density n of particles in the second beam, proportional to the flux I these particles are exposed to and proportional to the solid angle $d\Omega$. Hence we write

$$\delta n_c = \sigma(\mathbf{u}, \mathbf{u}_1 | \mathbf{u}', \mathbf{u}_1') \cdot n \cdot |\mathbf{u} - \mathbf{u}_1| n_1 \cdot d\Omega, \qquad (2.6)$$

where the constant of proportionality $\sigma(\mathbf{u}, \mathbf{u}_1 | \mathbf{u}', \mathbf{u}_1')$ is the differential scattering cross-section. The conservation laws (2.4) and (2.5) along with the condition that we want the particles from the second beam to go into the solid angle $d\Omega$ would completely determine the final velocities $\mathbf{u}'$ and $\mathbf{u}_1'$. That is why we have written the differential scattering cross-section as an explicit function of these final velocities.

We can consider the reverse collision in which particles with initial velocities $\mathbf{u}'$ and $\mathbf{u}'_1$ acquire final velocities $\mathbf{u}$ and $\mathbf{u}_1$ after the collision. If molecular processes are assumed to be reversible, then we expect the reverse cross-section to equal the direct cross-section, i.e.

$$\sigma(\mathbf{u}', \mathbf{u}'_1 | \mathbf{u}, \mathbf{u}_1) = \sigma(\mathbf{u}, \mathbf{u}_1 | \mathbf{u}', \mathbf{u}'_1). \tag{2.7}$$

It should be noted that this condition of reversibility (2.7) is by no means self-evident. For a discussion of the subtleties associated with this reversibility condition, see Huang (1987, §3.2).

2.2 The collision integral

We now evaluate the term C_{out} appearing in (2.3). Let us consider the stream of particles having the tips of their velocity vectors within d^3u and the stream of particles having the tips of their velocity vectors within d^3u_1. The first particles make up a beam with number density $n = f(\mathbf{x}, \mathbf{u}, t) d^3u$ and velocity $\mathbf{u}$, whereas the second particles constitute a beam with number density $n = f(\mathbf{x}, \mathbf{u}_1, t) d^3u_1$ and velocity $\mathbf{u}_1$. Substituting for n and n_1 in (2.6), the collision rate between these two beams of particles is

$$\delta n_c = \sigma(\mathbf{u}, \mathbf{u}_1 | \mathbf{u}', \mathbf{u}'_1) |\mathbf{u} - \mathbf{u}_1| f(\mathbf{x}, \mathbf{u}, t) f(\mathbf{x}, \mathbf{u}_1, t) \, d\Omega \, d^3u \, d^3u_1, \tag{2.8}$$

Since C_{out} must be equal to the number of collisions per unit time within the volume $d^3x \, d^3u$, it is easy to see that it is obtained by multiplying δn_c by d^3x and then integrating over all Ω and $\mathbf{u}_1$. Hence

$$C_{\text{out}} = d^3x \, d^3u \int d^3u_1 \int d\Omega \, \sigma(\mathbf{u}, \mathbf{u}_1 | \mathbf{u}', \mathbf{u}'_1) |\mathbf{u} - \mathbf{u}_1| f(\mathbf{x}, \mathbf{u}, t) f(\mathbf{x}, \mathbf{u}_1, t). \tag{2.9}$$

To evaluate C_{in}, we consider the reverse collisions between particles with velocities in d^3u' and particles with velocities in $d^3u'_1$ such that their velocities after collisions lie within d^3u and d^3u_1 respectively. In analogy with (2.8), the number of such collisions per unit volume per unit time is

$$\delta n'_c = \sigma(\mathbf{u}', \mathbf{u}'_1 | \mathbf{u}, \mathbf{u}_1) |\mathbf{u}' - \mathbf{u}'_1| f(\mathbf{x}, \mathbf{u}', t) f(\mathbf{x}, \mathbf{u}'_1, t) \, d\Omega \, d^3u' \, d^3u'_1. \tag{2.10}$$

It is easy to see from (2.4) and (2.5) that

$$|\mathbf{u} - \mathbf{u}_1| = |\mathbf{u}' - \mathbf{u}'_1|. \tag{2.11}$$

Let us consider the phase space of two colliding particles. If the interaction between the two particles can be described by a Hamiltonian, then (1.11) holds for volumes in this two-particle phase space, which implies in the present context

$$d^3u\, d^3u_1 = d^3u'\, d^3u'_1. \tag{2.12}$$

Using (2.7), (2.11) and (2.12), the expression (2.10) for $\delta n'_c$ becomes

$$\delta n'_c = \sigma(\mathbf{u}, \mathbf{u}_1 | \mathbf{u}', \mathbf{u}'_1) |\mathbf{u} - \mathbf{u}_1| f(\mathbf{x}, \mathbf{u}', t) f(\mathbf{x}, \mathbf{u}'_1, t)\, d\Omega\, d^3u\, d^3u_1. \tag{2.13}$$

The term C_{in} is obtained by multiplying $\delta n'_c$ by d^3x and then integrating over Ω and $\mathbf{u}_1$, i.e.

$$C_{\text{in}} = d^3x\, d^3u \int d^3u_1 \int d\Omega\, \sigma(\mathbf{u}, \mathbf{u}_1 | \mathbf{u}', \mathbf{u}'_1) |\mathbf{u} - \mathbf{u}_1| f(\mathbf{x}, \mathbf{u}', t) f(\mathbf{x}, \mathbf{u}'_1, t). \tag{2.14}$$

Putting (2.9) and (2.14) in (2.3), we finally have

$$\frac{\partial f}{\partial t} + \mathbf{u}.\nabla f + \frac{\mathbf{F}}{m}.\nabla_{\mathbf{u}} f$$
$$= \int d^3u_1 \int d\Omega\, |\mathbf{u} - \mathbf{u}_1| \sigma(\Omega)(f' f'_1 - f f_1), \tag{2.15}$$

where we have substituted

$$f = f(\mathbf{x}, \mathbf{u}, t),$$

$$f_1 = f(\mathbf{x}, \mathbf{u}_1, t),$$

$$f' = f(\mathbf{x}, \mathbf{u}', t),$$

$$f'_1 = f(\mathbf{x}, \mathbf{u}'_1, t),$$

and

$$\mathbf{F} = m\dot{\mathbf{u}}$$

incorporates any force field (such as gravity) the particles may be subjected to. One does not include the inter-particle forces in $\mathbf{F}$ (at least in the case of neutral gases without long-range inter-particle forces), since their effects are already taken into account within the collision term on the R.H.S. of (2.15). We have also written the differential scattering cross-section as a function of the scattering angle Ω between $\mathbf{u}$ and $\mathbf{u}'$, since the differential scattering cross-section for a simple spherically symmetric interaction potential assumed here can be a function of the scattering angle alone (due to spherical symmetry). The full Boltzmann equation (2.15) with the collision integral for binary collisions is a nonlinear integro-differential equation for the distribution function $f(\mathbf{x}, \mathbf{u}, t)$ (Boltzmann 1872).

2.3 The Maxwellian distribution

We know that a uniform classical gas left to itself relaxes to a Maxwellian distribution (Maxwell 1860). Hence we expect the Maxwellian distribution to come out as an equilibrium solution of the full Boltzmann equation (2.15). This will be demonstrated now.

Let us consider a uniform gas such that the effect of any external force field is negligible, i.e. $\mathbf{F} = 0$. This does not literally mean that we have to take the gas to faraway space where gravitational fields are unimportant. If the gas is considered within a volume such that the differences in gravitational potential within that volume are negligible compared to the typical kinetic energies of the gas particles, then $\mathbf{F}$ can be neglected (because the motions of gas particles within that volume are not influenced much by gravity). In any case, a gas can be assumed *uniform* only within such limited volumes. A gas occupying a large volume in a gravitational field is always stratified and hence is not uniform.

For such a uniform gas in equilibrium, it is easy to see that the L.H.S. of (2.15) vanishes. Hence the R.H.S. should also be zero, which happens when

$$ff_1 = f'f_1'. \tag{2.16}$$

Since the distribution function for a uniform gas in equilibrium depends on the particle velocity alone, we need not explicitly indicate the dependences on $\mathbf{x}$ and t so that the logarithm of (2.16) would give

$$\log f(\mathbf{u}) + \log f(\mathbf{u}_1) = \log f(\mathbf{u}') + \log f(\mathbf{u}_1'), \tag{2.17}$$

where $\mathbf{u}$ and $\mathbf{u}_1$ are the velocities of two particles before a binary collision, and $\mathbf{u}'$ and $\mathbf{u}_1'$ are the velocities after the collision. If $\chi(\mathbf{u})$ is a quantity which is conserved during a binary collision, then we must also have

$$\chi(\mathbf{u}) + \chi(\mathbf{u}_1) = \chi(\mathbf{u}') + \chi(\mathbf{u}_1'). \tag{2.18}$$

Comparing (2.17) and (2.18), we conclude that the most general expression for the distribution function must be of the form

$$\log f(\mathbf{u}) = C_0 + \sum_r C_r \chi_r(\mathbf{u}), \tag{2.19}$$

where $\chi_r(\mathbf{u})$-s should include all the independently conserved quantities and C_r-s are constants. If conservations of energy and the three components of momenta are all the independent conservation laws (angular momentum conservation follows from linear momentum conservation), then we must have

$$\log f(\mathbf{u}) = C_0 + C_1 \mathbf{u}^2 + C_{2x} u_x + C_{2y} u_y + C_{2z} u_z.$$

One can rewrite this equation in a slightly different form

$$\log f(\mathbf{u}) = -B(\mathbf{u} - \mathbf{u}_0)^2 + \log A, \tag{2.20}$$

where A, B and $\mathbf{u}_0$ are constants. From (2.20) we finally get

$$f(\mathbf{u}) = A e^{-B(\mathbf{u} - \mathbf{u}_0)^2}. \tag{2.21}$$

If n is the particle number density per unit volume, then it easily follows from

$$n = \int d^3 u \, f(\mathbf{u})$$

that

$$A = \left(\frac{B}{\pi}\right)^{3/2} n. \tag{2.22}$$

To evaluate A and to perform many calculations described below, it is necessary to work out definite integrals of the type $\int e^{-\alpha x^2} dx$. See Appendix B for a discussion of such integrals.

The expression (2.21) would have the form of the Maxwellian distribution if we had $\mathbf{u}_0 = 0$. It is easy to see that the average velocity of the particles is given by

$$\langle \mathbf{u} \rangle = \frac{A}{n} \int d^3 u \, \mathbf{u} \, e^{-B(\mathbf{u} - \mathbf{u}_0)^2} = \frac{A}{n} \int d^3 u \, (\mathbf{u} + \mathbf{u}_0) e^{-B u^2} = \mathbf{u}_0, \tag{2.23}$$

i.e. a non-zero $\mathbf{u}_0$ implies an average streaming motion of the particles with respect to the frame of reference. If we go to a frame moving with velocity $\mathbf{u}_0$, then particles in that frame will have the usual Maxwellian distribution. If the particles in our system are the molecules in a gas, and if the temperature T and the Boltzmann constant κ_B are defined in the usual way, then the constant B should be

$$B = \frac{m}{2\kappa_B T}$$

so that the full expression of the distribution function is

$$f(\mathbf{u}) = n \left(\frac{m}{2\pi\kappa_B T}\right)^{3/2} \exp\left[-\frac{m(\mathbf{u} - \mathbf{u}_0)^2}{2\kappa_B T}\right]. \tag{2.24}$$

This distribution function is the equilibrium solution of the Boltzmann equation. If we start from a different initial distribution function, we expect it to evolve according to the Boltzmann equation and eventually relax to the equilibrium (i.e. Maxwellian) distribution, although we have not yet demonstrated that explicitly. The relaxation to the Maxwellian distribution is clearly an irreversible process, and we need to figure out how irreversibility arises if we want to understand the relaxation process. This is done through the celebrated H theorem of Boltzmann (1872).

2.4 Boltzmann's H theorem

Although microscopic processes at the molecular level are usually reversible, we know that most macroscopic processes are *not* reversible. This is the essence of the second law of thermodynamics. For example, when the distribution function of a gas relaxes to the Maxwellian distribution as a result of collisions, it is clearly an irreversible process. The reverse process of a Maxwellian distribution changing to a different distribution is extremely improbable. While deriving the Boltzmann equation, we have explicitly assumed the reversibility of microscopic processes in the form of the relation (2.7). Still the evolution of a distribution function according to the Boltzmann equation usually turns out to be irreversible. Hence the Boltzmann equation must hold the key to understanding how irreversibility at the macroscopic levels arises out of reversible microscopic processes. This is a profound subject, and a full discussion of it is beyond the scope of this book. We limit our discussion merely to the proof of the famous H theorem of Boltzmann (1872), which gives a tantalizing hint at the origins of macroscopic irreversibility. We shall not need this theorem for any further discussions in this book. We present it more for cultural reasons. Anybody learning about the Boltzmann equation should know what the H theorem is.

Let us construct the quantity H from the distribution function of a gas in the following way

$$H = \int d^3u\, f \log f. \tag{2.25}$$

Since we are integrating over the velocity space, H is a quantity per unit volume. The H theorem states the following: if the distribution function f appearing in the definition (2.25) of H evolves according to the Boltzmann equation, then H for a uniform gas in the absence of external forces can never increase with time, i.e.

$$\frac{dH}{dt} \leq 0. \tag{2.26}$$

We now prove this theorem.

Differentiating (2.25) with respect to time, we have

$$\frac{dH}{dt} = \int d^3u \frac{\partial f}{\partial t}(1 + \log f).$$

For a uniform gas with $\mathbf{F} = 0$, substitution from the Boltzmann equation (2.15) gives

$$\frac{dH}{dt} = \int d^3u \int d^3u_1 \int d\Omega\, \sigma(\Omega)|\mathbf{u} - \mathbf{u}_1|(f'f_1' - ff_1)(1 + \log f). \quad (2.27)$$

Since integrations have been carried out over both $\mathbf{u}$ and $\mathbf{u}_1$, we could interchange them within the integral. Let us replace the integral in (2.27) by half the original integral plus half the integral with $\mathbf{u}$ and $\mathbf{u}_1$ interchanged. This gives

$$\frac{dH}{dt} = \frac{1}{2} \int d^3u \int d^3u_1 \int d\Omega\, \sigma(\Omega)|\mathbf{u} - \mathbf{u}_1|(f'f_1' - ff_1)[2 + \log(ff_1)].$$
$$(2.28)$$

Since reverse collisions with velocity changes $\{\mathbf{u}', \mathbf{u}_1'\} \to \{\mathbf{u}, \mathbf{u}_1\}$ have the same cross-section as the direct collisions $\{\mathbf{u}, \mathbf{u}_1\} \to \{\mathbf{u}', \mathbf{u}_1'\}$, we must have an equation symmetrical to (2.28):

$$\frac{dH}{dt} = \frac{1}{2} \int d^3u' \int d^3u_1' \int d\Omega\, \sigma(\Omega)|\mathbf{u}' - \mathbf{u}_1'|(ff_1 - f'f_1')[2 + \log(f'f_1')].$$
$$(2.29)$$

Keeping (2.11) and (2.12) in mind, we add (2.29) to (2.28) and divide by 2. This gives

$$\frac{dH}{dt} = \frac{1}{4} \int d^3u \int d^3u_1 \int d\Omega\, \sigma(\Omega)|\mathbf{u} - \mathbf{u}_1|(f'f_1' - ff_1)\log\left(\frac{ff_1}{f'f_1'}\right).$$
$$(2.30)$$

For two real quantities α and β, one can easily show that

$$(\beta - \alpha)\log\left(\frac{\alpha}{\beta}\right) \le 0.$$

The H theorem (2.26) readily follows from (2.30) by virtue of this inequality.

It is easy to see that H for a gas with the equilibrium (i.e. Maxwellian) distribution function does not change. Hence, if we begin with a uniform gas having a non-equilibrium distribution function initially, H keeps decreasing until the gas relaxes to the equilibrium distribution when H attains a minimum value. Just as the entropy of an isolated system keeps increasing until it attains a maximum value, H keeps on decreasing to reach a minimum value. Although we began by assuming the reversibility of molecular interactions, we have eventually ended up with the quantity H of which the evolution is not symmetric in $+t$ and $-t$. The arrow of time has been picked up.

It should be noted that the H theorem is not the same thing as the principle of increase of entropy (which is one of the statements of the second law of thermodynamics). We have defined H only for a uniform gas, whereas the entropy can be defined for complicated systems. On the other hand, the usual thermodynamic definition of entropy is only for systems in thermodynamic equilibrium. Here we are considering the H function for gases not in thermodynamic equilibrium (i.e. having distribution functions different from the Maxwellian). Only in the special case of a dilute gas in thermodynamic equilibrium where H and the entropy can both be defined, there exists a simple relation between the two. The reader is asked to obtain it in Exercise 2.3.

We end the discussion of the H theorem by pointing out a profound paradox associated with it. Suppose, at one instant of time, we reverse the velocities of all the molecules in a gas. Then the gas molecules will retrace their previous paths. If the H function was previously decreasing, we now expect it to increase after the velocities have been reversed. Hence there exist distribution functions which evolve in such a way that the H function increases with time! We shall not discuss the resolution of this paradox here. The reader is encouraged to look up Chapman and Cowling (1970, §4.21) for a discussion of this and other related paradoxes.

2.5 The conservation equation

Let us consider a quantity $\chi(\mathbf{x}, \mathbf{u})$ which is conserved in binary collisions so that an equation like (2.18) holds. We can write it more compactly as

$$\chi + \chi_1 = \chi' + \chi_1', \tag{2.31}$$

where χ and χ_1 are values of this quantity which two particles have before a collision, whereas χ' and χ_1' are the corresponding values after the collision. The average value of χ per unit volume is expected to satisfy an equation which we now derive from the Boltzmann equation. This equation will be presented in this section merely as a formal result. We shall, however, see in the next chapter that this formal-looking equation plays a key role in deriving the hydrodynamic equations.

Let us multiply both sides of the Boltzmann equation (2.15) by χ and integrate over d^3u. The R.H.S. is

$$\text{R.H.S.} = \int d^3u \int d^3u_1 \int d\Omega \, \sigma(\Omega) |\mathbf{u} - \mathbf{u}_1| (f'f_1' - ff_1)\chi. \tag{2.32}$$

Since this involves integrations over $\mathbf{u}$ and $\mathbf{u}_1$, we can play the same tricks by which we got (2.30) from (2.27) in §2.4, i.e.

$$\text{R.H.S.} = \tfrac{1}{4} \int d^3u \int d^3u_1 \int d\Omega \, \sigma(\Omega)|\mathbf{u} - \mathbf{u}_1|(f'f_1' - ff_1)(\chi + \chi_1 - \chi' - \chi_1'),$$

(2.33)

which is zero by virtue of (2.31). We thus have

$$\int d^3u \, \chi \left(\frac{\partial f}{\partial t} + u_i \frac{\partial f}{\partial x_i} + \frac{F_i}{m} \frac{\partial f}{\partial u_i} \right) = 0.$$

(2.34)

We can manipulate the terms of (2.34) to put it in the following form:

$$\frac{\partial}{\partial t} \int d^3u \, \chi f + \frac{\partial}{\partial x_i} \int d^3u \, \chi u_i f - \int d^3u \, u_i f \frac{\partial \chi}{\partial x_i} + \frac{1}{m} \int d^3u \, \frac{\partial}{\partial u_i}(\chi F_i f)$$

$$- \frac{1}{m} \int d^3u \, \frac{\partial \chi}{\partial u_i} F_i f - \frac{1}{m} \int d^3u \, \chi \frac{\partial F_i}{\partial u_i} f = 0.$$

(2.35)

It is easy to show that the term

$$\int d^3u \, \frac{\partial}{\partial u_i}(\chi F_i f)$$

in the above equation is zero. This term is of the nature of a volume integral of a divergence in the velocity space, and hence can be converted, by Gauss's theorem, into a surface integral of $\chi F_i f$ integrated over a surface at infinity. For a well-behaved system, we expect $f \to 0$ at infinity in the velocity space. Hence this surface integral vanishes.

Let Q be any quantity associated with each particle. The average value of Q is defined by

$$\langle Q \rangle = \frac{1}{n} \int d^3u \, Qf,$$

(2.36)

where

$$n = \int d^3u \, f$$

is the number density per unit volume. It follows that

$$\int d^3u \, Qf = n\langle Q \rangle$$

is the total amount of the quantity Q per unit volume. Using the definition (2.36), the non-zero terms in (2.35) can be put in the form

$$\frac{\partial}{\partial t}(n\langle \chi \rangle) + \frac{\partial}{\partial x_i}(n\langle u_i \chi \rangle) - n \left\langle u_i \frac{\partial \chi}{\partial x_i} \right\rangle - \frac{n}{m} \left\langle F_i \frac{\partial \chi}{\partial u_i} \right\rangle$$

$$- \frac{n}{m} \left\langle \frac{\partial F_i}{\partial u_i} \chi \right\rangle = 0.$$

(2.37)

This equation tells us how the volume density $n\langle \chi \rangle$ of any quantity χ

conserved in binary collisions evolves in time. This equation will turn out to be of central importance in the next chapter.

Exercises

2.1 Consider particles obeying the Maxwellian distribution (2.24) with $u_0 = 0$. Show that the average *speed* of the particles is given by

$$\langle u \rangle = \sqrt{\frac{8\kappa_B T}{\pi m}}$$

and the root mean square speed is given by

$$u_{rms} = \sqrt{\frac{3\kappa_B T}{m}}.$$

2.2 Show that the number of collisions taking place per unit volume per unit time in a dilute gas is given by

$$N_{coll} = \int d^3u \int d^3u_1 \int d\Omega \, \sigma(\Omega) |\mathbf{u} - \mathbf{u}_1| f(\mathbf{u}) f(\mathbf{u}_1).$$

Now consider the molecules to be hard spheres of radius a so that the total scattering cross-section is πa^2. Evaluate the above integral for such molecules to obtain

$$N_{coll} = 4n^2 a^2 \left(\frac{\pi \kappa_B T}{m} \right)^{1/2}.$$

[Hint: first transform from $\mathbf{u}$ and $\mathbf{u}_1$ to the relative velocity $\mathbf{u}_{rel} = \mathbf{u} - \mathbf{u}_1$ and the centre-of-mass velocity $\mathbf{u}_{cm} = (\mathbf{u} + \mathbf{u}_1)/2$. Show that the Jacobian $\partial(\mathbf{u}, \mathbf{u}_1)/\partial(\mathbf{u}_{rel}, \mathbf{u}_{cm})$ is equal to 1 and then work out the transformed integral, which is straightforward.]

Finally show that the mean free path for such hard-sphere molecules is given by

$$\lambda = \frac{1}{\sqrt{2} n \pi a^2}.$$

2.3 For a gas obeying the Maxwellian distribution, show that the function H is equal to

$$H = n \left[\log n + \frac{3}{2} \log \left(\frac{m}{2\pi\kappa_B T} \right) - \frac{3}{2} \right].$$

Evaluate the entropy per unit volume S of an ideal monatomic gas from the usual thermodynamic definition and verify that

$$S = -\kappa_B H + \text{constant}.$$

3 March towards hydrodynamics

3.1 The moment equations

We have seen in the previous chapter that the evolution of a system of particles can be described by the Boltzmann equation, if binary collisions are the only interactions between the particles. We now want to find out when it becomes possible to model this system of particles as a continuum governed by macroscopic equations. The conservation equation (2.37) plays a key role in establishing a connection between the microscopic model based on the Boltzmann equation and the more macroscopic models. The symbol χ in (2.37) stands for any quantity conserved in binary collisions. In deriving (2.37), we began with the conserved quantity χ associated with each particle, i.e. χ was initially taken as a quantity at the microscopic level. We eventually ended up with an equation for $n\langle\chi\rangle$, which is the amount of χ per unit volume and is obviously a macroscopic quantity. Hence equation (2.37) provides the golden gateway for a passage from the world of microphysics to the world of macrophysics. Since mass, momentum and energy are conserved in the binary collisions between particles, we now substitute these quantities for χ in (2.37) and obtain a set of macroscopic equations. It may be noted that the translational kinetic energy is conserved in binary collisions only for monatomic gas particles. For more complex gas molecules, it is possible for the translational kinetic energy to be transformed into other forms like rotational energy. Hence some of the results obtained from the conservation of translational kinetic energy will be valid for monatomic gases only.

Putting $\chi = m$ in (2.37) and considering $\mathbf{F}$ to be independent of velocity, we have

$$\frac{\partial}{\partial t}(nm) + \frac{\partial}{\partial x_i}(nm\langle u_i\rangle) = 0, \tag{3.1}$$

where we have not put the averaging sign $\langle \ldots \rangle$ around m, because we are assuming all the particles to have the same mass m. Writing

$$\rho = mn \tag{3.2}$$

for the density and

$$\mathbf{v} = \langle \mathbf{u} \rangle \tag{3.3}$$

for the average flow velocity of the particles, (3.1) takes up the usual form of the continuity equation:

$$\frac{\partial \rho}{\partial t} + \frac{\partial}{\partial x_i}(\rho v_i) = 0, \tag{3.4}$$

which is the same as (1.7) derived by different arguments.

We next substitute $\chi = mu_j$ in (2.37), which gives

$$\frac{\partial}{\partial t}(nm\langle u_j \rangle) + \frac{\partial}{\partial x_i}(nm\langle u_i u_j \rangle) - nF_j = 0. \tag{3.5}$$

Noting that $\mathbf{u} - \mathbf{v}$ is the velocity of a particle with respect to the average flow, we define a tensor P_{ij} in terms of $\mathbf{u} - \mathbf{v}$ in the following way:

$$P_{ij} = nm\langle (u_i - v_i)(u_j - v_j) \rangle. \tag{3.6}$$

It is easy to see that

$$\langle (u_i - v_i)(u_j - v_j) \rangle = \langle u_i u_j \rangle + v_i v_j - \langle u_i \rangle v_j - v_i \langle u_j \rangle$$
$$= \langle u_i u_j \rangle - v_i v_j$$

on using (3.3) for the averages of u_i and u_j. Hence

$$P_{ij} = nm\langle u_i u_j \rangle - nmv_i v_j. \tag{3.7}$$

Using (3.7), we modify (3.5) to

$$\frac{\partial}{\partial t}(\rho v_j) + \frac{\partial}{\partial x_i}(\rho v_i v_j) = -\frac{\partial P_{ji}}{\partial x_i} + \frac{\rho}{m}F_j. \tag{3.8}$$

Finally, if the gas is monatomic and the translational kinetic energy is conserved in binary collisions, then we can substitute $\chi = \frac{1}{2}m|\mathbf{u} - \mathbf{v}|^2$ in (2.37). A few steps of straightforward algebra lead to

$$\frac{\partial}{\partial t}(\rho \epsilon) + \frac{\partial}{\partial x_i}(\rho \epsilon v_i) + \frac{\partial q_i}{\partial x_i} + P_{ij}\Lambda_{ij} = 0, \tag{3.9}$$

where

$$\epsilon = \frac{1}{2}\langle |\mathbf{u} - \mathbf{v}|^2 \rangle \tag{3.10}$$

is the internal energy per unit mass,

$$\mathbf{q} = \frac{1}{2}\rho\langle (\mathbf{u} - \mathbf{v})|\mathbf{u} - \mathbf{v}|^2 \rangle \tag{3.11}$$

is the energy flux, and

$$\Lambda_{ij} = \frac{1}{2}\left(\frac{\partial v_i}{\partial x_j} + \frac{\partial v_j}{\partial x_i}\right). \tag{3.12}$$

The derivation of (3.9) is being left as an exercise.

We now simplify (3.8) and (3.9) a little. The L.H.S. of (3.8) can be expanded to

$$\text{L.H.S.} = v_j\frac{\partial \rho}{\partial t} + \rho\frac{\partial v_j}{\partial t} + v_j\frac{\partial}{\partial x_i}(\rho v_i) + \rho v_i\frac{\partial v_j}{\partial x_i}.$$

It is easy to see that the first and the third terms jointly vanish by virtue of the equation of continuity (3.4). Keeping the remaining terms on the L.H.S., (3.8) can now be put in the form

$$\rho\left(\frac{\partial v_j}{\partial t} + v_i\frac{\partial v_j}{\partial x_i}\right) = -\frac{\partial P_{ji}}{\partial x_i} + \frac{\rho}{m}F_j. \tag{3.13}$$

We can play a similar trick with (3.9) by noting that

$$\frac{\partial}{\partial t}(\rho\epsilon) + \frac{\partial}{\partial x_i}(\rho\epsilon v_i) = \epsilon\frac{\partial \rho}{\partial t} + \rho\frac{\partial \epsilon}{\partial t} + \epsilon\frac{\partial}{\partial x_i}(\rho v_i) + \rho v_i\frac{\partial \epsilon}{\partial x_i}.$$

Again the first and the third terms on the R.H.S. vanish together due to the equation of continuity so that (3.9) becomes

$$\rho\left(\frac{\partial \epsilon}{\partial t} + v_i\frac{\partial \epsilon}{\partial x_i}\right) + \frac{\partial q_i}{\partial x_i} + P_{ij}\Lambda_{ij} = 0. \tag{3.14}$$

Readers familiar with the hydrodynamic equations will note that (3.4), (3.13) and (3.14) have some resemblances to those equations, although these are still *not* the hydrodynamic equations. Counting the three components of (3.13), we have five equations here. On the other hand, the number of variables is fourteen: ρ, three components of v_i, ϵ, six components of the symmetric tensor P_{ij}, three components of q_i. Hence, if some initial values of these variables are given, the equations (3.4), (3.13) and (3.14) are not adequate to calculate the future values of all these variables. In other words, these equations *do not constitute a dynamical theory*. If we have a system of particles evolving according to the Boltzmann equation, then the only thing we can say is that these fourteen variables defined for this system must satisfy (3.4), (3.13) and (3.14). We can get a dynamical theory out of these equations only if we are able to relate some of these variables in such a way that the number of independent variables becomes equal to the number of independent equations. We shall see in §3.3 and §3.4 how this can be done for a dilute gas, if certain conditions are satisfied. The equations (3.4), (3.13) and (3.14) are often called the moment equations of the Boltzmann equation, because they are obtained by multiplying the

Boltzmann equation by 0, 1 and 2 powers of the velocity and then integrating over the velocity space.

3.2 An excursion into stellar dynamics. Oort limit

Just as we have introduced the distribution function $f(\mathbf{x}, \mathbf{u}, t)$ for a system of molecules, one can introduce a distribution function similarly for a group of stars in a cluster or inside a galaxy. Will the Boltzmann equation (2.15) hold for the distribution function of the stars? The general answer is 'no'. Stars interact with each other through gravity, which is a long-range force, and hence the Boltzmann collision integral based on binary collisions is usually not very appropriate for a stellar system. There are, however, stellar systems in which collisions do not play an important role. The readers are referred to any book on stellar dynamics (such as Binney and Tremaine 1987) for a detailed discussion of collisional relaxation in a stellar system. A brief treatment of the subject can be found in §17.1.1. The rule of thumb is that a larger stellar system has longer collisional relaxation time. One cannot neglect collisions for a globular cluster (containing about 10^5 stars typically). For a galaxy with 10^{11} stars, however, the collisional relaxation time turns out to be much larger than the age of the Universe, and collisions can be neglected for such systems. Figures 3.1 and 3.2 show photographs of a spiral galaxy and a globular cluster, respectively.

For stellar systems in which collisions can be neglected, the collisionless Boltzmann equation (1.19) should hold. Since the moment equations (3.4), (3.8) and (3.9) would follow from the collisionless Boltzmann equation, they also must hold for such stellar systems. Jeans (1922) was the first scientist to appreciate this fact. He wrote down these moment equations in cylindrical coordinates in a way suitable for application to the stars in our Galaxy. Hence the moment equations in cylindrical coordinates are often referred to as Jeans equations in the astrophysical literature (see Exercise 3.3). We again refer the interested reader to Binney and Tremaine (1987) for a detailed discussion of the subject. To give a flavour of stellar dynamics, we shall consider just one famous example of the application of the moment equations to stars in our neighbourhood. The moment equations may appear formal at first sight. We want to convince the reader that these are actually equations which can be applied to the analysis of observational data, leading to far-reaching conclusions.

Oort (1932) used equation (3.5) to find out the average matter density near the solar neighbourhood of our Galaxy. Even if there is some matter in the solar neighbourhood which does not emit light and is not detected in direct observations, it will produce a gravitational field and hence will affect the motions of visible stars. Therefore, by analyzing the motions of visible stars, it is possible to estimate the total

Figure 3.1 A spiral galaxy: a stellar system in which collisions are unimportant. Photographed at Kavalur Observatory, India. Courtesy: Indian Institute of Astrophysics.

Figure 3.2 A globular cluster of stars: a stellar system in which collisions have been important. Photographed at Kavalur Observatory, India. Courtesy: Indian Institute of Astrophysics.

amount of matter in the solar neighbourhood. Let us take the z axis perpendicular to the plane of the Galaxy in the solar neighbourhood. We now make two assumptions:

1 The distribution of stars is in a steady state in a statistical sense so that

$$\frac{\partial}{\partial t} = 0$$

for any averaged quantity.

2 The variation of any quantity along the x or the y direction lying in the galactic plane is much smaller compared to the variation in the z direction. So we can put

$$\frac{\partial}{\partial x} = \frac{\partial}{\partial y} = 0.$$

With these two reasonable assumptions, the z component of (3.5) becomes

$$\frac{1}{n}\frac{d}{dz}(n\langle u_z^2\rangle) = g_z, \tag{3.15}$$

where we have introduced the gravitational field $\mathbf{g}$ by writing $\mathbf{F} = m\mathbf{g}$. Oort (1932) used the statistics of K giant stars to obtain the gravitational field at different heights from the galactic plane. K giants are very bright stars which can be observed to sufficiently large distances from the galactic plane and for which sufficiently good data existed in Oort's time about their number density and line-of-sight velocity at different heights from the galactic plane. Once g_z is obtained as a function of z, one can calculate the matter density producing this gravitational field from the Poisson equation for gravity $\nabla.\mathbf{g} = -4\pi G\rho_{\text{matter}}$, which here becomes

$$\frac{dg_z}{dz} = -4\pi G\rho_{\text{matter}}. \tag{3.16}$$

When the total matter density in the solar neighbourhood is estimated in this fashion, it turns out to be around

$$\rho_{\text{matter}} \approx 10 \times 10^{-24}\text{g cm}^{-3}.$$

On the other hand, if we calculate the density by estimating the amount of matter in the visible stars, then we find

$$\rho_{\text{star}} \approx 4 \times 10^{-24}\text{g cm}^{-3}.$$

Thus there must be unseen matter present in the solar neighbourhood in addition to the visible stars. This was a very important conclusion in 1932 when not much was known about the interstellar matter. This analysis also provides an upper limit for the amount of interstellar matter, since its density cannot exceed $(\rho_{\text{matter}} - \rho_{\text{star}})$. This is known as the *Oort limit*.

Let us end this section with some comments on mass determination in astronomy. From the light coming out of a system like a galaxy, we can estimate the number and the nature of the stars in the system, thus giving us a clue to the mass of the system. On the other hand, we can make a dynamical estimate of mass from the observed motions in the system by methods similar to what we have just discussed. Very often the dynamically estimated mass turns out to be larger than the mass estimated from light. In the case of Oort's analysis, the discrepancy was of the order of unity. In larger systems like clusters of galaxies, this discrepancy can be much larger. This means that our Universe contains a large amount of unseen matter, which produces gravitational fields, but emits no light. We still know precious little about the nature and distribution of this dark matter. This is the celebrated *dark matter problem* of modern astrophysics.

3.3 Zero-order approximation

We now return to the question of constructing a dynamical theory out of the equations (3.4), (3.13) and (3.14). This should be possible for a system of particles when the system behaves like a continuous fluid. To clarify what we may mean by fluidlike behaviour, let us consider a motion of my hand through the air. The air molecules hit by my hand gain a forward momentum and, when they collide with the molecules in front, a stress is conveyed to the forward layer of air molecules which have not yet come into contact with my hand. If, however, an astronaut sent to interstellar space were to move his hand in the interstellar medium (with about 1 particle per cm³), then the situation will be very different. Since collisions between molecules are infrequent there (by our standards!), the molecules hit by the hand will move forward and pass like projectiles through the other molecules in the interstellar medium which will remain unaffected. We tend to think of the first example of the response of air to my hand as fluidlike behaviour, whereas the second example presents a situation very different from our usual intuitive notion of a fluid. Hence collisions between molecules play an important role in establishing a fluidlike behaviour.

We have seen in §2.3 that collisions tend to set up a Maxwellian velocity distribution. Now we are suggesting that collisions play an important role in establishing fluidlike behaviour. Is there then any connection between a Maxwellian distribution at the microscopic level and a fluidlike behaviour at the macroscopic level? We shall see that the answer to this important question is 'yes'. If collisions are important enough in a system of particles to produce local Maxwellian distributions within different parts of the system, then we would expect the system to display fluidlike behaviour. We now demonstrate that it is indeed possible to obtain a dynamical theory out of the moment equations for such a system.

Let us assume that the distribution function at each point within a system of particles is given by

$$f^{(0)}(\mathbf{x}, \mathbf{u}, t) = n(\mathbf{x}, t) \left[\frac{m}{2\pi\kappa_B T(\mathbf{x}, t)} \right]^{3/2} \exp\left[-\frac{m\{\mathbf{u} - \mathbf{v}(\mathbf{x}, t)\}^2}{2\kappa_B T(\mathbf{x}, t)} \right], \quad (3.17)$$

which is the same as (2.24) with the number density n, the temperature T and the mean flow velocity $\mathbf{v}$ taken as functions of space and time as we expect in a real fluid. We now have to use (3.17) to calculate some of the quantities appearing in the moment equations.

We begin by calculating P_{ij} from (3.6). Keeping in mind that the averaging is defined through (2.36), we have

$$P_{ij} = mn \left(\frac{m}{2\pi\kappa_B T} \right)^{3/2} \int d^3U \, U_i U_j \exp\left(-\frac{mU^2}{2\kappa_B T} \right), \quad (3.18)$$

where

$$\mathbf{U} = \mathbf{u} - \mathbf{v}.$$

Since the integration is over the whole of velocity space, the integral vanishes when the integrand is odd. This is the case when i and j are unequal. Hence we have

$$P_{ij} = p\delta_{ij}, \quad (3.19)$$

where p can easily be shown from (3.18) to be

$$p = n\kappa_B T. \quad (3.20)$$

It will be made clear in the next chapter that p corresponds to *pressure* defined in the usual way. One sees at once that the definition (3.11) for the energy flux gives rise to an odd integrand and hence

$$\mathbf{q} = 0. \quad (3.21)$$

Using (3.17), we find from (3.10) that

$$\epsilon = \frac{3}{2} \frac{\kappa_B T}{m}. \quad (3.22)$$

It should be noted that this is the expression for internal energy for monatomic gases only. It further follows from (3.12) that

$$P_{ij}\Lambda_{ij} = \frac{1}{2}p\delta_{ij}\left(\frac{\partial v_i}{\partial x_j} + \frac{\partial v_j}{\partial x_i}\right) = p\nabla \cdot \mathbf{v}. \tag{3.23}$$

Substituting for P_{ij} from (3.19) in (3.13) and writing the equation in the vectorial form, we have

$$\frac{\partial \mathbf{v}}{\partial t} + (\mathbf{v} \cdot \nabla)\mathbf{v} = -\frac{1}{\rho}\nabla p + \frac{\mathbf{F}}{m}. \tag{3.24}$$

Using (3.21) and (3.23), it follows from (3.14) that

$$\rho\left(\frac{\partial \epsilon}{\partial t} + \mathbf{v} \cdot \nabla \epsilon\right) + p\nabla \cdot \mathbf{v} = 0. \tag{3.25}$$

We note that (3.24) and (3.25) along with the continuity equation in the vector form

$$\frac{\partial \rho}{\partial t} + \nabla \cdot (\rho \mathbf{v}) = 0 \tag{3.26}$$

make up a system of five independent scalar equations, because (3.24) has three scalar components. Apart from the three components of $\mathbf{v}$, the other variables appearing in these equations are ρ, p and ϵ (i.e. altogether six scalar variables). The external force $\mathbf{F}$ has to be prescribed and is not a dynamical variable. We see from (3.2), (3.20) and (3.22) that ρ, p and ϵ can be expressed in terms of only two variables n and T. Hence only two out of these three variables are independent, making the number of independent scalar variables in our equations to be equal to five. Thus the number of independent variables finally equals the number of independent equations, which means that we have ended up with a dynamical theory of macroscopic nature.

Although we have got a dynamical theory in terms of a set of macroscopic variables, some characteristics of real fluids are still lacking from our equations. For example, $\mathbf{q} = 0$ implies that there is no transport of energy from one part of the system to another part in the form of a flow of heat (i.e. the internal kinetic energy of the molecules). The tensor P_{ij} is found to be diagonal. We shall see later that a diagonal P_{ij} implies the absence of *viscosity*, which allows for the transport of momentum from one layer of the fluid to the other layers so that relative motions inside the fluid are damped out. In other words, we have got equations for a fluid which does not exhibit *transport phenomena* and hence does not behave like a real fluid.

What has gone wrong in our derivation? We had assumed local Maxwellian distributions at all points in the form of (3.17) without rigorous justification. Transport phenomena become very important in a fluid when there are strong gradients of temperature or velocity. This is exactly the situation which requires corrections to the local Maxwellian assumption. Let us consider a point P within a fluid with a strong temperature gradient such that the fluid on one side of P is much hotter than the fluid on the other side. Then molecules from the hotter side will stream in the neighbourhood of P with a velocity distribution corresponding to a higher temperature. Similarly molecules streaming in from the colder side will have a velocity distribution for a lower temperature. As a result of these molecular motions from different sides with different properties, the distribution function in the neighbourhood of P must have some departure from the Maxwellian distribution. It is necessary to handle this departure from the Maxwellian distribution in a systematic fashion in order to develop a theory for transport phenomena. This is done in §3.4.

3.4 Transport phenomena

Since we have to consider departures from the Maxwellian distribution in order to handle transport phenomena, let us write the distribution function as

$$f(\mathbf{x}, \mathbf{u}, t) = f^{(0)}(\mathbf{x}, \mathbf{u}, t) + g(\mathbf{x}, \mathbf{u}, t), \tag{3.27}$$

where $f^{(0)}(\mathbf{x}, \mathbf{u}, t)$ is the Maxwellian distribution given by (3.17) and the remaining part $g(\mathbf{x}, \mathbf{u}, t)$ corresponds to the departure from the Maxwellian which we assume to be small. Substituting (3.27) in the Boltzmann equation, we find that the collision integral (i.e. the R.H.S. of (2.15)) is given by

$$\text{R.H.S.} \approx \int d^3 u_1 \int d\Omega \, |\mathbf{u} - \mathbf{u}_1| \sigma(\Omega)(f^{(0)\prime} g_1' + f_1^{(0)\prime} g' - f^{(0)} g_1 - f_1^{(0)} g), \tag{3.28}$$

where we have neglected the quadratic terms in the small quantity g. An order-of-magnitude estimate of the collision integral can be obtained from one of the terms, say the last term, which gives

$$-\int d^3 u_1 \int d\Omega \, |\mathbf{u} - \mathbf{u}_1| \sigma(\Omega) f^{(0)}(\mathbf{x}, \mathbf{u}_1, t) g(\mathbf{x}, \mathbf{u}, t) \approx -g(\mathbf{x}, \mathbf{u}, t) \cdot n\sigma_{\text{tot}} \bar{u}_{\text{rel}},$$

where σ_{tot} is the total collision cross-section and $\bar{u}_{\text{rel}}$ is the average relative velocity between the particles. We note that $n\sigma_{\text{tot}}$ is approximately the inverse of the mean free path λ (given by (2.2) if the particles are taken to be rigid spheres) and $n\sigma_{\text{tot}} \bar{u}_{\text{rel}}$ is the inverse of the collision time τ. Hence we can approximate the collision integral in the Boltzmann equation by $-(g/\tau)$ so that (2.15) becomes

$$\left(\frac{\partial}{\partial t} + \mathbf{u} \cdot \nabla + \frac{\mathbf{F}}{m} \cdot \nabla_u\right) f = -\frac{f - f^{(0)}}{\tau}. \tag{3.29}$$

If an arbitrary distribution function f relaxes to the Maxwellian distribution $f^{(0)}$ in the collision time τ, then the rate of change of f is roughly given by the R.H.S. of (3.29). We present a treatment of transport phenomena based on this approximate equation (3.29), often called the BGK equation—after Bhatnagar, Gross and Krook (1954) who popularized its use. The treatment of transport phenomena using the BGK equation was given by Liepmann, Narasimha and Chahine (1962). It becomes a formidable problem if one keeps all the terms in the collision integral (3.28) and treats it without further approximations. Techniques for attacking this full problem were developed by Chapman (1916) and Enskog (1917). Readers desirous of learning more about this extremely complicated subject should delve into the classic monograph of Chapman and Cowling (1970).

The approximate equation (3.29) illuminates much of the basic physics in a relatively less painful way. We first use (3.29) to make an estimate of the departure of the distribution function from a pure Maxwellian. If a system has a strong gradient, then the term $\mathbf{u}.\nabla$ on the L.H.S. of (3.29) is what gives rise to the departures from a Maxwellian distribution. To have a rough estimate, we balance this term with the R.H.S. to get

$$\frac{|u|f^{(0)}}{L} \approx \frac{|g|}{\tau}, \tag{3.30}$$

where $|u|$ is the typical molecular velocity and L is the typical length scale over which properties of the system change appreciably. We see from (3.30) that

$$\frac{|g|}{f^{(0)}} \approx \frac{\lambda}{L}. \tag{3.31}$$

Hence the departures from a pure Maxwellian distribution will be small if the mean free path is small compared to the typical length scale. Using

$$\alpha = \frac{\lambda}{L}$$

as a small expansion parameter, we can write the distribution function as a series

$$f = f^{(0)} + \alpha f^{(1)} + \alpha^2 f^{(2)} + \cdots, \tag{3.32}$$

where the terms $f^{(0)}$, $f^{(1)}$, $f^{(2)}$, ... are of the same order. This is known as the *Chapman–Enskog expansion*. By substituting (3.32) in (3.29), one can evaluate the successive terms. Here we shall only consider the evaluation of the first-order term.

To calculate the first-order corrections, we approximate f by $f^{(0)}$ on the L.H.S. of (3.29), which gives

$$g = -\tau \left(\frac{\partial}{\partial t} + u_i \frac{\partial}{\partial x_i} + \frac{F_i}{m} \frac{\partial}{\partial u_i} \right) f^{(0)}. \tag{3.33}$$

It is easy to see from (3.17) that $f^{(0)}$ depends on t and x_i through $n(\mathbf{x}, t)$, $T(\mathbf{x}, t)$ and $\mathbf{v}(\mathbf{x}, t)$. We therefore have

$$\frac{\partial f^{(0)}}{\partial t} = \frac{\partial n}{\partial t} \frac{\partial f^{(0)}}{\partial n} + \frac{\partial T}{\partial t} \frac{\partial f^{(0)}}{\partial T} + \frac{\partial v_i}{\partial t} \frac{\partial f^{(0)}}{\partial v_i}.$$

A similar expression holds for $\partial f^{(0)}/\partial x_i$. Substituting (3.17) in such expressions, we first have to evaluate $\partial f^{(0)}/\partial t$ and $\partial f^{(0)}/\partial x_i$. Then, putting them in (3.33), we find after a few steps of algebra

$$g = -\tau \left[\frac{1}{T} \frac{\partial T}{\partial x_i} U_i \left(\frac{m}{2\kappa_B T} U^2 - \frac{5}{2} \right) + \frac{m}{\kappa_B T} \Lambda_{ij} \left(U_i U_j - \frac{1}{3} \delta_{ij} U^2 \right) \right] f^{(0)}. \tag{3.34}$$

The detailed steps of algebra are left as Exercise 3.4 and can be found in books like Huang (1987, §5.5). We have finally found the expression giving the departure from the Maxwellian distribution. Since this departure results from the non-uniformities of the system as we have already pointed out, it is no wonder that g depends linearly on the gradients of temperature T and of flow velocity v_i (through Λ_{ij} as defined in (3.12)). The linear dependence of g on the collision time τ is also easy to understand. A larger τ will imply that particles will be able to stream in from faraway regions of the non-uniform system without suffering collisions and the local distribution function will depart from a Maxwellian more substantially. On the other hand, if collisions are efficient in a system, then τ will be small making g small and the distribution function will be close to a local Maxwellian.

In §3.3, we had calculated quantities like P_{ij}, $\mathbf{q}$ and ϵ from a purely Maxwellian distribution $f^{(0)}$. Our task now is to take the modified distribution $f^{(0)} + g$ and calculate these quantities from (3.6), (3.10) and (3.11) (taking (2.36) as the definition of averaging). We readily see that

$$\mathbf{q} = \frac{\rho}{2n} \int d^3 U \, \mathbf{U} U^2 g.$$

We again note that if a part of the integrand is odd, it does not contribute to the integral. Hence only the term involving $\partial T/\partial x_i$ in the expression (3.34) for g contributes, giving us

$$\mathbf{q} = -K\nabla T, \tag{3.35}$$

where

$$K = \frac{\tau m}{6T} \int d^3U \, U^4 \left(\frac{m}{2\kappa_B T} U^2 - \frac{5}{2} \right) f^{(0)} = \frac{5}{2} \tau n \frac{\kappa_B^2 T}{m} \tag{3.36}$$

on working out the integrals in accordance with the rules outlined in Appendix B. We see from the definition (3.11) that $\mathbf{q}$ is of the nature of a flux of internal energy, i.e. a flux of heat. Hence it is tempting to regard K appearing in (3.35) as the coefficient of thermal conductivity. We shall introduce the coefficient of thermal conductivity through the usual macroscopic definition in the next chapter and shall show there that K is indeed that coefficient.

It is easy to see that P_{ij} now becomes

$$P_{ij} = p\delta_{ij} + \pi_{ij}, \tag{3.37}$$

where

$$\pi_{ij} = m \int d^3U \, U_i U_j g.$$

Only the term involving Λ_{ij} in (3.34) contributes to this integral. So we have

$$\pi_{ij} = -\frac{\tau m^2}{\kappa_B T} \Lambda_{kl} \int d^3U \, U_i U_j \left(U_k U_l - \frac{1}{3}\delta_{kl} U^2 \right) f^{(0)}. \tag{3.38}$$

One notes from this expression that π_{ij} is a traceless symmetric tensor, i.e. $\pi_{ii} = 0$ (summation over i). Since π_{ij} depends linearly on Λ_{kl}, it has got to have the traceless form

$$\pi_{ij} = -2\mu \left(\Lambda_{ij} - \frac{1}{3}\delta_{ij}\nabla \cdot \mathbf{v} \right), \tag{3.39}$$

where we note that the trace of Λ_{ij} is nothing but $\nabla.\mathbf{v}$. To obtain the coefficient μ, we evaluate one of the components of π_{ij} from (3.38), say π_{12}, which is

$$\pi_{12} = \frac{\tau m^2}{\kappa_B T} \Lambda_{kl} \int d^3U \, U_1 U_2 \left(U_k U_l - \frac{1}{3}\delta_{kl} U^2 \right) f^{(0)}$$

$$= -2\frac{\tau m^2}{\kappa_B T} \Lambda_{12} \int d^3U \, U_1^2 U_2^2 f^{(0)},$$

because the integrand makes non-vanishing contributions only when

k and l are equal to 1 and 2 (or 2 and 1). Comparing this expression with (3.39), we conclude

$$\mu = \frac{\tau m^2}{\kappa_B T} \int d^3 U\, U_1^2 U_2^2 f^{(0)} = \tau n \kappa_B T. \qquad (3.40)$$

We thus see that P_{ij} has a non-zero off-diagonal part π_{ij} given by (3.39) with μ obtained from (3.40). We shall show in Chapter 5 that a non-diagonal P_{ij} implies that it is possible to transport momentum from a fast-moving layer of fluid to the adjoining slow-moving layer. We shall also introduce the coefficient of viscosity in Chapter 5 in the usual way and find that μ appearing here turns out to be just this coefficient.

To summarize, the distribution function is close to the Maxwellian if the mean free path is small compared to the length scale of the system. By calculating the small departure from the Maxwellian in the first-order, we are able to evaluate the transport coefficients K and μ. We shall, however, have to wait for later chapters where the coefficients of thermal conductivity and viscosity are introduced in the usual way, and we find K and μ are indeed these coefficients.

3.5 Comparison with experiments

Assuming that μ and K obtained in the previous section are indeed the viscosity and the thermal conductivity, do our expressions for these quantities agree with experimental data? We can take the collision time τ to be the mean free path λ divided by the mean speed $\langle u \rangle$. Using the results of Exercises 2.1 and 2.2, we get

$$\tau = \frac{\lambda}{\langle u \rangle} = \frac{1}{4na^2} \left(\frac{m}{\pi \kappa_B T} \right)^{1/2} \qquad (3.41)$$

if the molecules are assumed to be rigid spheres. On substituting this expression of τ in (3.40), we get

$$\mu = \frac{1}{4a^2} \left(\frac{m \kappa_B T}{\pi} \right)^{1/2}. \qquad (3.42)$$

Chapman and Cowling (1970, Chapter 12) present expressions for the viscosity of dilute gases obtained by the rigorous Chapman–Enskog method. If the molecules are assumed to be rigid spheres, then

$$\mu = \frac{5}{64a^2} \left(\frac{m \kappa_B T}{\pi} \right)^{1/2}. \qquad (3.43)$$

Considering the fact that (3.42) was obtained by the crude method of approximating the four terms in (3.28) by a single term, we are pleased that it is not too far off from the rigorous expression (3.43). Reif (1965, Chapter 14) derives the rigorous expression (3.43) by a somewhat simpler method keeping all the four terms of (3.28), but without developing the full technique of the Chapman–Enskog procedure.

One of the first things to note in (3.42) or (3.43) is that the viscosity is independent of the density of the gas. This remarkable consequence of kinetic theory, which was first pointed out by Maxwell (1860), appears surprising at first sight, because we might have intuitively expected a denser gas to be more viscous. But this independence of density has been confirmed by experiments. A denser gas has more molecules to transport various physical quantities. But, since the mean free path becomes shorter, each individual molecule is less efficient as an agent of transport. Due to the combination of these two opposing effects, the transport phenomena in a gas turn out to be independent of density. We also see from (3.42) and (3.43) that the viscosity is proportional to $\sqrt{T}$. Experiments show that the viscosity of a gas increases with temperature faster than $\sqrt{T}$. This is due to the fact that the molecules in a real gas are not rigid spheres. At higher temperatures, the molecules have more kinetic energies and come closer during collisions. Therefore the effective radius a decreases with temperature (see Exercise 3.5). A look at (3.42) or (3.43) then makes it clear why the viscosity rises more rapidly than $\sqrt{T}$. We want to point out that this increase of viscosity with temperature is a property of gases alone. The viscosity of a liquid decreases with temperature. Anybody who has ever heated oil knows that hot oil is less viscous than cold oil and flows more easily.

From (3.36) and (3.40), we find that

$$K = \frac{5}{2}\mu\frac{\kappa_B}{m}. \tag{3.44}$$

From the expression (3.22) for the internal energy of a monatomic gas, we see that the specific heat per unit mass of a monatomic gas is given by

$$c_V = \frac{3}{2}\frac{\kappa_B}{m}. \tag{3.45}$$

It follows from (3.44) and (3.45) that

$$\frac{K}{\mu c_V} = \frac{5}{3}. \tag{3.46}$$

This numerical factor turns out to be 5/2 instead of 5/3 in the more rigorous calculations presented in Chapman and Cowling (1970,

Chapter 13). Again the result based on our approximate theory compares favourably. One can experimentally find $K/\mu c_V$ for the inert gases which are monatomic. The experimental values for different inert gases are: helium -2.45; neon -2.52; argon -2.48; krypton -2.54; xenon -2.58 (Chapman and Cowling 1970, §13.2).

To sum up, the results of transport phenomena obtained by kinetic theory are in general agreement with experimental data. We give the values of viscosity and thermal conductivity for some common fluids in Appendix D.

3.6 Hydrodynamics at last

In §3.4 we obtained expressions for $\mathbf{q}$ and P_{ij} involving the transport coefficients. We can now substitute them in (3.13) and (3.14) to obtain the hydrodynamic equations incorporating transport phenomena. We see from (3.37) and (3.39) that

$$\frac{\partial P_{ji}}{\partial x_i} = \frac{\partial p}{\partial x_j} - \mu \left[\nabla^2 v_j + \frac{1}{3} \frac{\partial}{\partial x_j} (\nabla \cdot \mathbf{v}) \right], \qquad (3.47)$$

if we treat μ as a constant while differentiating with respect to the spatial variables. This is a reasonable assumption in most circumstances. Substituting this in (3.13), we have

$$\rho \left(\frac{\partial v_j}{\partial t} + v_i \frac{\partial v_j}{\partial x_i} \right) = -\frac{\partial p}{\partial x_j} + \mu \left[\nabla^2 v_j + \frac{1}{3} \frac{\partial}{\partial x_j} (\nabla \cdot \mathbf{v}) \right] + \frac{\rho}{m} F_j. \qquad (3.48)$$

From (3.12), (3.37) and (3.39), we also have

$$P_{ij}\Lambda_{ij} = p\nabla \cdot \mathbf{v} - 2\mu \left[\Lambda_{ij}\Lambda_{ij} - \frac{1}{3}(\nabla \cdot \mathbf{v})^2 \right]. \qquad (3.49)$$

On substituting (3.35) and (3.49) in (3.14),

$$\rho \left(\frac{\partial \epsilon}{\partial t} + \mathbf{v} \cdot \nabla \epsilon \right) - \nabla \cdot (K\nabla T) + p\nabla \cdot \mathbf{v} - 2\mu \left[\Lambda_{ij}\Lambda_{ij} - \frac{1}{3}(\nabla \cdot \mathbf{v})^2 \right] = 0. \qquad (3.50)$$

These basic equations (3.48) and (3.50) can be simplified further in most circumstances. The viscosity term involving $\nabla \cdot \mathbf{v}$ in (3.48) is usually small and can be neglected. The terms involving μ in (3.50) correspond to heat production due to the viscous damping of fluid motions. This is also unimportant in most problems. We also make a change of notation from now. While discussing the dynamics of a system of particles, we had defined $\mathbf{F}$ as the force acting on a particle. Now that we shall be discussing continuum models of fluids, it makes more sense to talk about force acting per unit mass. The force per unit mass is also denoted in many hydrodynamics textbooks by **F**. We can conform to this notation if we now write **F** where we were previously writing **F**$/m$ and henceforth take **F** to mean force per unit mass. Adding the equation of continuity to (3.48) and (3.50) simplified suitably, we finally write down the full set of hydrodynamic equations:

$$\frac{\partial \rho}{\partial t} + \nabla \cdot (\rho \mathbf{v}) = 0, \tag{3.51}$$

$$\frac{\partial \mathbf{v}}{\partial t} + (\mathbf{v} \cdot \nabla)\mathbf{v} = -\frac{1}{\rho}\nabla p + \mathbf{F} + \frac{\mu}{\rho}\nabla^2 \mathbf{v}, \tag{3.52}$$

$$\rho \left(\frac{\partial \epsilon}{\partial t} + \mathbf{v} \cdot \nabla \epsilon \right) - \nabla \cdot (K \nabla T) + p \nabla \cdot \mathbf{v} = 0, \tag{3.53}$$

where we have neglected the various small terms and written **F** for **F**$/m$. Equation (3.52) is known as the *Navier–Stokes equation*, whereas the last equation (3.53) is essentially a statement of the conservation of energy.

It is not difficult to convince ourselves that (3.51–3.53) constitute a dynamical theory. If the coefficients μ and K are specified along with the applied force **F**, then the variables appearing in the equations other than the three components of **v** are ρ, p, T and ϵ. We again see from (3.2), (3.20) and (3.22) that these variables are interrelated in such a way that only two of them are independent (it is to be noted that (3.22) remains unchanged even when the the first-order corrections are incorporated). Hence the number of equations is the same as the number of independent dynamical variables. Chapters 4–9 are devoted to studying solutions of (3.51–3.53) under different circumstances. Let us recapitulate the simplifying assumptions that we made in this section to obtain these equations in the present form: (i) spatial variation of viscosity μ is neglected; (ii) heat production due to the viscous damping of motions is neglected; (iii) the viscosity term involving $\nabla \cdot \mathbf{v}$ in (3.47) is neglected. These assumptions hold for all the problems which we study in this book. In the exceptional circumstance in which one of these assumptions may not hold, one can figure out the extra terms to be included in the equations. Since these assumptions are valid under most circumstances we are likely to encounter, it makes more sense to write down the equations in their present form, rather than always writing the extra terms which may be of use only in some exceptional situation.

Since astrophysicists often have to deal with systems like the interstellar medium or the solar wind with very few particles per cm^3, it was our aim to find out the conditions under which the hydrodynamic model holds. We have seen that if the mean free path is small compared to the length scale, then the distribution function remains close to a Maxwellian and a hydrodynamic description becomes possible. We point out that astrophysical gases are often in the plasma state and contain magnetic fields, which also may play a crucial role in establishing a fluidlike behaviour. This will be discussed later when we consider plasmas.

After obtaining the conditions necessary for the hydrodynamic equations to apply to dilute gases, let us consider the other limit of dense fluids. Do the same hydrodynamic equations hold for dense fluids? Since we have derived the hydrodynamic equations starting from the Boltzmann equation and the Boltzmann equation holds only for sufficiently dilute gases where it is enough to consider binary collisions alone, the derivation presented above does not hold for dense fluids. One can, however, give macroscopic derivations by assuming the fluid to be a continuum and arrive at *exactly the same hydrodynamic equations*. This is discussed in the next chapter. The crucial assumption we had to make in deriving the hydrodynamic equations from the Boltzmann equation was that the binary collisions conserve momentum and energy, which led us to the moment equations. If more complicated multi-particle collisions in dense fluids also conserve momentum and energy, it is no surprise that eventually we end up with the same hydrodynamic equations for dense fluids from the macroscopic considerations. Since we intuitively feel that modelling a dense fluid as continuum is not something too bad, the macroscopic continuum derivation of hydrodynamic equations presented in the next chapter should convince anybody about the validity of these equations for dense fluids. The continuum derivations, however, do not make it apparent that these equations may be applicable for very dilute gases with few particles per cm^3. The purpose of this chapter has been to establish that the hydrodynamic equations can be applied to sufficiently dilute gases if the mean free path is small compared to the length scale.

3.7 Concluding remarks

We have seen that a system of particles can be treated like a continuum fluid if frequent collisions keep the distribution functions in local regions close to the Maxwellian. In other words, local regions should have the properties of a system in thermodynamic equilibrium, even though the different parts of the overall system may not be in thermodynamic equilibrium with each other. The smaller the mean free paths compared to the length scales, the assumption of *local thermodynamic equilibrium* the better holds and the continuum description makes more sense.

An astrophysicist may be interested in studying three types of particle systems:

(i)   particles without long-range interactions;
(ii) charged particles interacting through electromagnetic forces;
(iii) particles interacting through gravitational forces.

Even particles in a neutral gas, of course, exert gravitational forces on each other. However, the typical gravitational potential energy between two gas particles is so negligible compared to the thermal kinetic energy that we are usually justified in putting a neutral gas in category (i) above. On the other hand, for stars in a cluster or in a galaxy, the gravitational potential energy is usually comparable to the kinetic energy of the random motions—a result known as the *virial theorem* (see §17.1.1). This result follows from the fact that it is often the random motions which sustain the stellar system in a dynamic steady state against gravitational collapse. Such a stellar system obviously has to be put in category (iii). A plasma is the simplest example of category (ii).

Fluid models hold for systems in category (i) if the mean free path is small compared to the length scale. Since systems in categories (ii) and (iii) have long-range interactions, many particles interact with each other all the time and the concept of a mean free path cannot be introduced in a very clean way. Hence the Chapman–Enskog expansion (3.32) cannot be carried out easily, as it is difficult to define the small parameter α. In the case of a plasma, however, the strong electrostatic attractions between positively and negatively charged particles ensure that volumes of plasma with statistically significant number of particles are nearly neutral. Hence a plasma has some characteristics in common with a neutral gas. Although a charged particle in a plasma in principle produces a long-range electromagnetic field, its effect is usually screened off by particles of the opposite charge within a short distance called the *Debye length* (to be discussed in Chapter 11). It is possible for local regions of a plasma to relax to local thermodynamic equilibria such that a continuum description can be introduced. Establishing a continuum fluid model for a plasma rigorously, however, is a much more difficult problem than it is a for a neutral fluid, since the Chapman–Enskog expansion cannot be done for a plasma in a straightforward way. Later chapters will address the question of introducing continuum models for plasmas.

It may seem at first sight that a stellar system will be easier to study than a plasma, because there is gravitational charge (i.e. mass) of only one sign compared to the electric charges of two opposite signs. The reality, however, is the other way round. As we have pointed out, opposite charges in a plasma help in screening the long-range interactions, making the local thermodynamic equilibrium and consequently continuum models possible. The gravitational forces in a stellar system cannot be screened similarly and hence they are of the nature of long-range interactions in a much more profound way. The most dramatic consequence of this is that thermodynamic equilibrium *is not possible* for self-gravitating stellar systems, making them very different from neutral fluids and plasmas.

A deep unifying theme which underlies many astrophysical results is that self-gravity is incompatible with thermodynamic equilibrium. Consider what happens when collisions try to relax a self-gravitating stellar system. One expects the gravitational potential well to be deepest at the centre of the system. When collisions try to establish a thermodynamic equilibrium, more particles are put in the central well, thereby deepening the well further and requiring more particles there for the establishment of a thermodynamic equilibrium. This may lead to a runaway situation. The lack of thermodynamic equilibrium makes collisional stellar dynamics very different from the dynamical theories in this book and we shall not discuss it here, leaving the interested reader to texts like Binney and Tremaine (1987). We have given a flavour of collisionless stellar dynamics in §3.2, pointing out that the moment equations hold for stellar systems in which collisions are unimportant. It should not, however, be assumed that a primordial particle distribution function remains frozen in collisionless stellar systems due to the lack of collisional relaxation. For example, although the stars in our Galaxy make up a collisionless system (i.e. a system for which the collisional relaxation time is much larger than the age of the Universe), the velocity distribution of the stars in the solar neighbourhood seems to follow the so-called *Schwarzschild velocity ellipsoid* (see Mihalas and Binney 1981), which is like an anisotropic Maxwellian distribution. If a collisionless stellar system has come into the present steady state after a gravitational contraction from a more extended initial state, then the changing gravitational fields during the contraction phase could have acted somewhat like the forces of collision, thereby producing a relaxed velocity distribution of the particles—a process first suggested by Lynden-Bell (1967) and christened *violent relaxation*. A collisionless stellar system can also have some waves and instabilities similar to the waves and instabilities in fluids which we shall study later. Binney and Tremaine (1987) present discussions of some such analogies between fluid systems and stellar systems. In spite of these common characteristics, the lack of a proper thermodynamic equilibrium makes a stellar system quite different from neutral fluids and plasmas, and we shall not discuss stellar dynamics in this book. Only when we discuss collisions in plasmas in Chapter 11, we shall make a few comments on collisional stellar dynamics.

Exercises

3.1 Carry out all the algebra to obtain (3.9).

3.2 Consider an *isothermal* gas in a gravitational field, the gravitational potential being $\Phi(\mathbf{x})$. Assuming that the distribution function is of the form $S(\mathbf{x})f_M(\mathbf{u})$ where $f_M(\mathbf{u})$ is the Maxwellian distribution

$$f_M(\mathbf{u}) = \left(\frac{m}{2\pi\kappa_B T}\right)^{3/2} \exp\left(-\frac{mu^2}{2\kappa_B T}\right),$$

determine $S(\mathbf{x})$ from the Boltzmann transport equation.

3.3 In cylindrical coordinates, the distribution function f is a function of the coordinates r, θ, z and the velocity components $\Pi = \dot{r}$, $\Theta = r\dot{\theta}$, $Z = \dot{z}$. If the particles move in a gravitational potential $\Phi(\mathbf{x})$, show that the collisionless Boltzmann equation in cylindrical geometry is

$$\frac{\partial f}{\partial t} + \Pi\frac{\partial f}{\partial r} + \frac{\Theta}{r}\frac{\partial f}{\partial \theta} + Z\frac{\partial f}{\partial z} + \left(\frac{\Theta^2}{r} - \frac{\partial \Phi}{\partial r}\right)\frac{\partial f}{\partial \Pi}$$

$$-\left(\frac{\Pi\Theta}{r} + \frac{1}{r}\frac{\partial \Phi}{\partial \theta}\right)\frac{\partial f}{\partial \Theta} - \frac{\partial \Phi}{\partial z}\frac{\partial f}{\partial Z} = 0.$$

Now consider the special case of a steady, axisymmetric situation so that you can put $\partial/\partial t = \partial/\partial \theta = 0$. Multiply this simplified equation by Π and Z respectively and integrate over the velocity space. Show that this leads to equations

$$\frac{\partial}{\partial r}(n\langle\Pi^2\rangle) + \frac{\partial}{\partial z}(n\langle\Pi Z\rangle) + \frac{n}{r}[\langle\Pi^2\rangle - \langle\Theta^2\rangle] = -n\frac{\partial \Phi}{\partial r},$$

$$\frac{\partial}{\partial r}(n\langle\Pi Z\rangle) + \frac{\partial}{\partial z}(n\langle Z^2\rangle) + \frac{n}{r}\langle\Pi Z\rangle = -n\frac{\partial \Phi}{\partial z},$$

where the averages are defined in the usual way and n is the number density of particles. These are two of the Jeans equations referred to in §3.2.

If the terms involving $\langle \Pi Z \rangle$ are neglected in the last equation, note that it reduces to (3.15) from which the Oort limit was obtained. Do you think it justified to neglect these terms?

3.4   Work out all the algebra for the derivations given in §3.4.

3.5   The viscosity of helium at temperatures -102.6, 0, 183.7 and $815\,^\circ\mathrm{C}$ has the values 1.392×10^{-4}, 1.887×10^{-4}, 2.681×10^{-4} and $4.703 \times 10^{-4}\,\mathrm{g\,cm^{-1}\,s^{-1}}$ respectively. Calculate the effective radii of the atoms at these temperatures using (3.43).

3.6   Consider a uniform gas of electrons with mass m and charge $-e$ in the presence of a fixed lattice of ions such that the system has overall charge-neutrality and the electrons with the number density n have the distribution $n f_\mathrm{M}(\mathbf{u})$, where $f_\mathrm{M}(\mathbf{u})$ is the usual Maxwellian as given in Exercise 3.2. Now a weak uniform electric field $\mathbf{E}$ is switched on. Using the BGK equation (3.29), calculate the departure from the Maxwellian in the presence of this electric field and find out the electrical conductivity σ defined by

$$-ne\langle \mathbf{u} \rangle = \sigma \mathbf{E}.$$

4 Properties of ideal fluids

4.1 Macroscopic derivation of hydrodynamic equations

In the previous chapter, we derived the hydrodynamic equations (3.51–3.53) from the molecular dynamics of a dilute gas for which the mean free path is small compared to the length scale. We now show that exactly the same equations follow from macroscopic considerations by treating the fluid as a continuum. Hence the hydrodynamic equations derived for dilute gases in Chapter 3 should also hold for dense fluids which can be regarded as continua at the macroscopic level. The macroscopic derivations will make the reader feel more at home with the hydrodynamic equations, since these derivations are based on notions close to our everyday experience.

Since we shall be discussing how various fluid dynamical variables evolve with time, let us begin by drawing attention to the two different kinds of time derivatives: *Eulerian* and *Lagrangian*. The *Eulerian* derivative denoted by $\partial/\partial t$ implies differentiation with respect to time at a fixed point. On the other hand, one can think of moving with a fluid element with the fluid velocity $\mathbf{v}$ and time-differentiating some quantity associated with this moving fluid element. This type of time derivative is called *Lagrangian* and is denoted by d/dt. If $\mathbf{x}$ and $\mathbf{x}+\mathbf{v}\delta t$ are the positions of a fluid element at times t and $t + \delta t$, then the Lagrangian time derivative of some quantity $Q(\mathbf{x}, t)$ is defined as

$$\frac{dQ}{dt} = \lim_{\delta t \to 0} \frac{Q(\mathbf{x} + \mathbf{v}\delta t, t + \delta t) - Q(\mathbf{x}, t)}{\delta t}. \tag{4.1}$$

Keeping the first-order terms in the Taylor expansion, we have

$$Q(\mathbf{x} + \mathbf{v}\delta t, t + \delta t) = Q(\mathbf{x}, t) + \delta t \frac{\partial Q}{\partial t} + \delta t \, \mathbf{v} \cdot \nabla Q.$$

Putting this in (4.1), we have the very useful relation between the Lagrangian and the Eulerian derivatives:

$$\frac{dQ}{dt} = \frac{\partial Q}{\partial t} + \mathbf{v} \cdot \nabla Q. \tag{4.2}$$

This equation is exactly similar to (1.6) for the time derivative in phase space. We were concerned there with a $6N$-dimensional phase space, whereas here we are now dealing with the three-dimensional physical space.

We already derived the equation of continuity (1.7) in §1.4 by continuum arguments and then again have seen in §3.1 that it follows from the Boltzmann equation. Hence we need not derive it here again. It is usually written in terms of the Eulerian derivative of ρ. Noting that

$$\nabla \cdot (\rho \mathbf{v}) = \mathbf{v} \cdot \nabla \rho + \rho \nabla \cdot \mathbf{v}$$

from (A.4) and using the relation (4.2), we can easily write down the equation of continuity using the Lagrangian derivative

$$\frac{d\rho}{dt} + \rho \nabla \cdot \mathbf{v} = 0. \tag{4.3}$$

This form is often useful.

4.1.1 The equation of motion

To find the equation for the velocity, we consider a fluid element of volume δV. The mass of this fluid element is $\rho \, \delta V$ and its acceleration is given by the Lagrangian derivative $(d\mathbf{v}/dt)$. Hence it follows from Newton's second law of motion that

$$\rho \, \delta V \frac{d\mathbf{v}}{dt} = \delta \mathbf{F}_{\text{body}} + \delta \mathbf{F}_{\text{surface}}, \tag{4.4}$$

where we have split the force acting on the fluid element into two parts: the body force $\delta \mathbf{F}_{\text{body}}$ and the surface force $\delta \mathbf{F}_{\text{surface}}$. A body force is something which acts at all points within the body of a fluid. Gravity is an example of such a force. It is customary to denote the body force per unit mass as $\mathbf{F}$ so that

$$\delta \mathbf{F}_{\text{body}} = \rho \, \delta V \, \mathbf{F}. \tag{4.5}$$

The surface force on a fluid element is the force acting on it across the surface bounding the fluid element. Let $d\mathbf{S}$ be an element of area on the bounding surface. The surface force $d\mathbf{F}_{\text{surface}}$ acting across this area is assumed proportional to this area. Since $d\mathbf{S}$ and $d\mathbf{F}_{\text{surface}}$ are both vectors, a proportional relation between them implies that they must be related through a second-rank tensor which we write as P_{ij}, i.e.

$$(d\mathbf{F}_{\text{surface}})_i = -P_{ij}dS_j. \tag{4.6}$$

The total surface force acting on a volume of fluid is then given by a surface integral

$$(\mathbf{F}_{\text{surface}})_i = -\oint P_{ij}dS_j.$$

Using (A.18), we transform this surface integral into a volume integral, i.e.

$$(\mathbf{F}_{\text{surface}})_i = -\int \frac{\partial P_{ij}}{\partial x_j}dV.$$

Hence the surface force acting on a small fluid element of volume δV is

$$(\delta \mathbf{F}_{\text{surface}})_i = -\frac{\partial P_{ij}}{\partial x_j}\delta V. \tag{4.7}$$

Substituting (4.5) and (4.7) into (4.4), we finally have

$$\rho\frac{dv_i}{dt} = \rho F_i - \frac{\partial P_{ij}}{\partial x_j}. \tag{4.8}$$

This is the same as (3.13) if we keep in mind that we are now writing $\mathbf{F}$ where we were writing $\mathbf{F}/m$ previously.

For a fluid in static equilibrium, it is an experimentally established fact that the force acting across an element of area inside the fluid or on its boundary is always perpendicular to that element of area. In fact, this is often taken as the definition of a fluid. A fluid is defined as a substance in which motions are induced whenever there is a part of the surface force not perpendicular to the surface (i.e. a shear force). This is in contrast to elastic solids within which shear forces can be present in static equilibrium. Mathematically, for a static fluid, we write

$$P_{ij} = p\delta_{ij}, \tag{4.9}$$

which substituted in (4.6) gives

$$d\mathbf{F}_{\text{surface}} = -p\,d\mathbf{S}. \tag{4.10}$$

It should be clearly apparent that p introduced here is the *pressure*, which is defined as the force acting per unit area. The negative sign in (4.10) signifies that the pressure force acting across the bounding surface of a fluid volume is always inward directed, whereas the vector area $d\mathbf{S}$ is taken by convention as outward directed.

Although (4.9) holds for a fluid at rest, it is generally no longer valid when there are motions inside the fluid. For example, consider two layers of a fluid having different velocities on the two sides of a surface as shown in Figure 4.1. If the surface force was given by (4.10), then the force across the surface of separation could only be in the vertical direction. We, however, expect a horizontal tangential shear force to act across the surface of separation and to transport momentum from the faster-moving layer to the slower layer such that the faster layer slows down and the slower layer speeds up. We shall see in the next chapter how such tangential stresses can be handled by introducing the coefficient of viscosity. In this chapter, we restrict our discussion to the simplifying assumption that (4.9) holds even for fluids in motion. Fluids for which the condition (4.9) always holds are known as *ideal fluids*. Apart from liquid helium at very low temperatures, we do not know of any other fluid which can be regarded as ideal. Still, in many fluid problems in which viscosity does not play an important role, the ideal fluid equations give sufficiently good results, while keeping the calculations simpler. Hence it is worthwhile to spend some time understanding the properties of ideal fluids before we launch into a study of viscous fluids in the next chapter. Substituting (4.9) into (4.8) and using (4.2), we have the equation

$$\frac{\partial \mathbf{v}}{\partial t} + (\mathbf{v} \cdot \nabla)\mathbf{v} = -\frac{1}{\rho}\nabla p + \mathbf{F}. \tag{4.11}$$

This is known as the *Euler equation* (Euler 1755, 1759). We note that the Navier–Stokes equation (3.52) reduces to the Euler equation (4.11) if the coefficient of viscosity μ is set to zero. The macroscopic basis of the Navier–Stokes equation will be discussed in the next chapter.

Although scientists like Newton and Bernoulli considered isolated problems involving fluids, it was Euler (1755, 1759) who laid down

Figure 4.1 A fluid with two layers moving with different velocities.

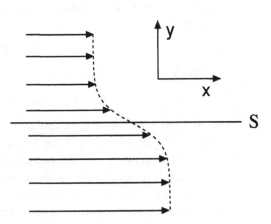

the foundations of hydrodynamics through a systematic investigation of its basic equations. Hence it is only appropriate that the basic equation of motion is named after Euler—often regarded as the father of modern hydrodynamics.

4.1.2 The equation of energy

Finally we come to the derivation of (3.53). This equation follows from the first law of thermodynamics, which is essentially a statement of the conservation of energy. If we have a fluid in thermodynamic equilibrium to which some amount of heat dQ has been added, then the first law of thermodynamics says that

$$dQ = dU + p\,dV, \tag{4.12}$$

where dU is the change in internal energy and $p\,dV$ is the work done by the system. All thermodynamic variables can broadly be classified into two categories: *extensive* and *intensive*. The variables which are added up when we combine several thermodynamic systems are called extensive. For example, when we combine two systems with internal energies U_1 and U_2, the internal energy of the resultant system is $U_1 + U_2$. Hence internal energy is an extensive variable. On the other hand, variables like pressure and temperature cannot be added like this. They are called intensive variables. It is to be noted that each term in (4.12) is linear in some extensive variable. This ensures that (4.12) is consistent. An equation with different powers of extensive variables in different terms would not make sense.

We now wish to adapt (4.12) to a continuous fluid in which there may be motions. The different parts of the fluid will in general not be in thermodynamic equilibrium and hence (4.12), which holds for systems in thermodynamic equilibrium, cannot be applied to a large part of the fluid. However, a sufficiently *small* fluid element (i.e. small compared to the length scales) can be regarded in local thermodynamic equilibrium as we have discussed in Chapter 3. We now write down (4.12) for such a small fluid element of mass δm. Handling intensive variables like pressure p is not a problem, as one can use the values of these variables within this small fluid element. In the case of extensive variables, however, it is necessary to define them *per unit mass*. If Q and U per unit mass are denoted by q and ϵ, then the heat added to the system and its change of internal energy are $dQ = \delta m\,dq$ and $dU = \delta m\,d\epsilon$. Since volume per unit mass is $1/\rho$, we have to replace dV in the last term of (4.12) by $\delta m\,d(1/\rho)$. Putting these in (4.12) and dividing by dt, we get

$$\frac{dq}{dt} = \frac{d\epsilon}{dt} + p\frac{d}{dt}\left(\frac{1}{\rho}\right).$$

Since

$$p\frac{d}{dt}\left(\frac{1}{\rho}\right) = -\frac{p}{\rho^2}\frac{d\rho}{dt} = \frac{p}{\rho}\nabla\cdot\mathbf{v}$$

from (4.3), we have

$$\rho\frac{d\epsilon}{dt} + p\nabla\cdot\mathbf{v} = -\mathscr{L}, \tag{4.13}$$

where

$$-\mathscr{L} = \rho\frac{dq}{dt}$$

is the rate of heat gain per unit volume ($\mathscr{L}$ is the heat loss rate).

If there are heat flows within a fluid, then some regions may gain heat. Let us estimate this in order to calculate $\mathscr{L}$. The heat flux $\mathscr{F}$ inside a continuous substance can usually be assumed to be proportional to the temperature gradient, i.e.

$$\mathscr{F} = -K\nabla T, \tag{4.14}$$

where K is called the coefficient of thermal conductivity and the negative sign implies that the heat flow is in the direction of decreasing temperature. The heat loss rate from a volume of fluid is equal to the heat flux integrated over the bounding surface, which is

$$\oint \mathscr{F}\cdot d\mathbf{S} = \int \nabla\cdot\mathscr{F}\, dV.$$

Hence the heat loss rate per unit volume must be given by $\nabla.\mathscr{F}$ so that

$$\mathscr{L} = \nabla\cdot\mathscr{F} = -\nabla\cdot(K\nabla T). \tag{4.15}$$

Substituting this for $\mathscr{L}$ in (4.13), we finally end up with the energy equation (3.53). We also confirm that K appearing in (3.53) coincides with the coefficient of thermal conductivity defined in the usual way. We have already indicated in §3.6 that there can be heat gain due to the viscous dissipation of motions, which is usually small and is not included in $\mathscr{L}$ here. In many astrophysical problems, however, heat gains and losses due to radiation are very important and *have to be included* in $\mathscr{L}$. The detailed discussion of radiative gain and loss processes is beyond the scope of this book.

4.2 The vorticity equation. Incompressible and barotropic fluids

From (A.6), it follows that

$$(\mathbf{v} \cdot \nabla)\mathbf{v} = \tfrac{1}{2}\nabla(\mathbf{v} \cdot \mathbf{v}) - \mathbf{v} \times (\nabla \times \mathbf{v}). \qquad (4.16)$$

Substituting this in the Euler equation (4.11),

$$\frac{\partial \mathbf{v}}{\partial t} + \frac{1}{2}\nabla(\mathbf{v} \cdot \mathbf{v}) - \mathbf{v} \times (\nabla \times \mathbf{v}) = -\frac{1}{\rho}\nabla p + \mathbf{F}.$$

We now take the curl of this equation. Since body forces like gravity are usually conservative, we assume $\mathbf{F}$ to drop out when we take the curl so that we are left with

$$\frac{\partial \omega}{\partial t} = \nabla \times (\mathbf{v} \times \omega) + \frac{1}{\rho^2}\nabla \rho \times \nabla p, \qquad (4.17)$$

where

$$\omega = \nabla \times \mathbf{v} \qquad (4.18)$$

is called the *vorticity*.

We now show that hydrodynamic calculations become considerably simplified when the fluid is incompressible. It is fairly obvious that incompressibility is a reasonably good assumption for liquids like water. What may not be so obvious is that even gases like air can be *regarded as incompressible* in many fluid dynamical problems! When air is compressed in some localized region, the excess air there tries to spread around quickly. We shall see in Chapter 6 that this spreading takes place at the sound speed. Hence only objects moving at speeds comparable to or faster than the sound speed (such as jet planes) are capable of producing appreciable compressions in air. When an object moves slowly, the air in front of it can get rid of the compression at a speed faster than the speed at which the object is trying to build up the compression. Air can therefore be regarded as incompressible as long as all the motions inside have velocities small compared to the sound speed. When I move my hand through the air, the flow of air around my hand can be described adequately with the equations of incompressible hydrodynamics.

Since the density ρ for an incompressible fluid does not change, we see from (4.3) that

$$\nabla \cdot \mathbf{v} = 0. \qquad (4.19)$$

This can be taken as the condition of incompressibility. If $\nabla \cdot \mathbf{v}$ remains negligible for a flow pattern, then we are justified in regarding the fluid as incompressible. We note that (4.17) reduces to the following simpler equation for incompressible fluids

$$\frac{\partial \omega}{\partial t} = \nabla \times (\mathbf{v} \times \omega). \tag{4.20}$$

We see that (4.20) involves two fluid variables, $\mathbf{v}$ and ω, which are certainly not independent. If $\mathbf{v}$ is given, then we can easily find ω from (4.18). On the other hand, if ω is given, then $\mathbf{v}$ can be solved from (4.18) and (4.19). It is a well-known result of mathematical physics that a vector field can be solved if its divergence and curl are given (see, for example, Panofsky and Phillips 1962, §1–1). Since (4.20) gives the evolution of ω and since $\mathbf{v}$ can be found from ω, it should be clear that equation (4.20) along with (4.18) and (4.19) provides the complete dynamical theory of incompressible fluids.

It thus turns out that the energy equation (3.53) is redundant while treating incompressible fluids. Since many elementary textbooks on fluid mechanics (especially textbooks for engineering students) often deal mainly with flows that can be regarded as incompressible, the energy equation is paid much less attention compared to the other basic hydrodynamic equations. In astrophysical problems, however, we often have to deal either with systems having appreciable density variations or with systems having within them motions with speeds comparable to the speed of sound. Hence the energy equation has to be taken into account in most astrophysical fluid dynamics problems. In §4.3, we shall discuss the hydrostatic equilibrium of some systems with variable density. We, however, defer to Chapter 6 the study of fluids within which there are motions with speeds comparable to the sound speed. The study of such fluids is usually referred to as *gas dynamics*.

In certain problems, it becomes possible to assume a functional relation between p and ρ, i.e. we can write

$$p = p(\rho). \tag{4.21}$$

Such a relation is called a *barotropic relation* and fluids satisfying such a relation are known as *barotropic fluids*. It is easy to see that the vorticity equation reduces to (4.20) for barotropic fluids as well. If p is no longer an independent variable, then ρ and $\mathbf{v}$ are the only independent variables appearing in the equation of continuity and the Euler equation, and it is obvious that these equations constitute a full dynamical theory. The barotropic relation (4.21) is in a sense a substitute for the energy equation and it is not consistent to include an additional energy equation for barotropic fluids.

4.3 Hydrodynamic equations in conservative forms

We note that the equation of continuity (3.51) has the form

$$\frac{\partial}{\partial t}(\text{density}) + \mathbf{div}(\text{its flux}) = 0. \tag{4.22}$$

The significance of this equation is that the the amount of material within a certain volume can change only as a result of the fluxes of that material across the bounding surface.

The other hydrodynamic equations can also be put in forms similar to (4.22). With the help of the equation of continuity, the Euler equation can easily be put in the form

$$\frac{\partial}{\partial t}(\rho v_i) + \frac{\partial}{\partial x_j}(\rho v_i v_j) = -\frac{\partial p}{\partial x_i} + \rho F_i.$$

In fact, we proceeded in exactly the opposite way in §3.1 when (3.8) was transformed into (3.13). In the absence of external forces, i.e. when $\mathbf{F} = 0$, we write

$$\frac{\partial}{\partial t}(\rho v_i) + \frac{\partial T_{ij}}{\partial x_j} = 0, \tag{4.23}$$

a form similar to (4.22), where

$$T_{ij} = p\delta_{ij} + \rho v_i v_j \tag{4.24}$$

can be regarded as the flux of momentum. This interpretation becomes clear when we consider the amount of momentum $\int \rho v_i \, dV$ within a volume. Its rate of change according to (4.23) is given by

$$\frac{\partial}{\partial t}\int \rho v_i \, dV = -\int \frac{\partial T_{ij}}{\partial x_j} dV = -\oint T_{ij} dS_j,$$

where the volume integral has been transformed into a surface integral by (A.18). It is clear from this equation that the amount of momentum within a volume changes only as a result of the momentum flux T_{ij} across the bounding surface (when no external force acts). This result leads to an obvious corollary: if we consider a large volume of fluid with no momentum flux across the faraway bounding surfaces, then the total amount of momentum of this fluid is conserved.

To obtain an energy equation in the same form, we note that energy per unit volume is given by $\rho\epsilon + \frac{1}{2}\rho v^2$, which is the sum of internal and kinetic energies. The time derivative of this can be obtained by combining all the hydrodynamic equations (3.51–3.53). In the absence of the external force $\mathbf{F}$, if we neglect the viscosity terms, then it is found that

$$\frac{\partial}{\partial t}\left(\rho\epsilon + \frac{1}{2}\rho v^2\right) = -\nabla \cdot \left[\rho\mathbf{v}\left(\frac{1}{2}v^2 + w\right) - K\,\nabla T\right], \tag{4.25}$$

where

$$w = \epsilon + \frac{p}{\rho} \qquad (4.26)$$

is the enthalpy per unit mass. We see that (4.25) is the energy equation having the form of (4.22). The derivation of (4.25), which is being left here as an exercise problem (Exercise 4.1), can be found in Landau and Lifshitz (1987, §6). It is clear that $[\rho v(\frac{1}{2}v^2 + w) - K\,\nabla T]$ should be interpreted as the energy flux in the fluid.

The equations (4.23) and (4.25) are referred to as the conservative forms of hydrodynamic equations. They essentially state the conservation laws of momentum and energy. Sometimes, especially in numerical work, these equations turn out to be more convenient than the hydrodynamic equations in their usual form.

4.4 Hydrostatics. Modelling the solar corona

Now that we have derived all the hydrodynamic equations constituting a dynamical theory, we would like to study some solutions of these equations. We begin by considering what is probably the simplest possibility, a fluid at rest. By putting $\mathbf{v} = 0$ and setting all the time derivatives equal to zero, we obtain from (3.51–3.53) that

$$\nabla p = \rho \mathbf{F}, \qquad (4.27)$$

$$\nabla \cdot (K\,\nabla T) = 0. \qquad (4.28)$$

We have already noted in §4.2 that the energy equation is redundant for incompressible fluids so that (4.27) alone is necessary to study the static equilibrium of incompressible fluids.

Let us consider the one-dimensional problem of a fluid in static equilibrium in a uniform gravitational field

$$\mathbf{F} = -g\hat{\mathbf{e}}_z,$$

where z is taken in the vertically upward direction. Then the z component of (4.27) is

$$\frac{dp}{dz} = -\rho g. \qquad (4.29)$$

For an incompressible fluid with constant density ρ, the solution is

$$p = p_0 - \rho g z, \qquad (4.30)$$

where p_0 is the pressure at the level $z = 0$. We thus establish the well-known result that pressure in an incompressible fluid at rest increases linearly with depth. Although (4.28) is needed for compressible fluids in general, we now consider the simple case of an isothermal ideal gas, for which the condition T = constant replaces (4.28). From (3.2) and (3.20), we have the relation

$$p = \frac{\kappa_B}{m} \rho T \tag{4.31}$$

for the ideal gas. Substituting this in (4.29) and remembering that T is constant, we have

$$\frac{\kappa_B T}{m} \frac{d\rho}{dz} = -\rho g,$$

of which the solution is

$$\rho = \rho_0 \exp\left(-\frac{mgz}{\kappa_B T}\right), \tag{4.32}$$

where ρ_0 is the density at $z = 0$.

The hydrostatic equation (4.27) is one of the fundamental equations in the study of stellar structure. Since conduction is a less important process for the transport of energy compared to radiative transport (or convection if convection does take place) in a main-sequence star, (4.28) has to be replaced by the appropriate equations for the other energy transport processes. These equations, which are presented in any textbook on stellar structure, will not be discussed here. We note that if (4.28) is replaced by the polytropic relation $p = A\rho^\gamma$, then the hydrostatic problem in the spherical geometry leads to the famous Lane–Emden equation (Exercise 4.3). As an example of a hydrostatic problem in astrophysics, we discuss the problem of constructing the static model of the solar corona, which led to a paradox when it was first studied by Chapman (1957) and Parker (1958).

Figure 4.2 is a photograph of a total solar eclipse showing that the Sun has a corona of hot tenuous gas extending well beyond the solar surface. Although the corona appears fairly non-spherical, one can try to construct a first approximate model of the corona by assuming spherical symmetry so that quantities like density and pressure can be regarded as functions of radius r alone. It is known that the corona has a temperature much higher than the surface temperature of the Sun. The possible reasons for this high temperature will be discussed in §15.5. Here we assume that the heat is produced in the lower layers of the corona so that outer regions of the corona can be modelled by taking a boundary condition that $T = T_0$ at some radius $r = r_0$ near the base of the corona. This suffices to pose the problem even though we may not know what is heating the corona. This is somewhat like calculating the temperature distribution in a metal rod with one end heated in a furnace. If the temperature of the furnace is given as a boundary condition, then the problem can be solved without knowing whether the furnace is heated by charcoal, gas or electricity.

Since the corona has very little mass, the gravitational field in the corona can be regarded as an inverse-square field created by the mass of the Sun $M_\odot$. In spherical geometry, equations (4.27) and (4.28) become

$$\frac{dp}{dr} = -\frac{GM_\odot}{r^2}\frac{m}{\kappa_B}\frac{p}{T}, \tag{4.33}$$

$$\frac{d}{dr}\left(Kr^2\frac{dT}{dr}\right) = 0, \tag{4.34}$$

where we have eliminated ρ from (4.33) by making use of (4.31). It follows from the kinetic theory of plasmas that the thermal conductivity K of a plasma goes as the 5/2 power of temperature (see §13.5). Hence (4.34) gives

$$r^2 T^{5/2}\frac{dT}{dr} = \text{constant},$$

of which the solution is

$$T = T_0\left(\frac{r_0}{r}\right)^{2/7} \tag{4.35}$$

satisfying the boundary conditions that $T = T_0$ at $r = r_0$ and $T = 0$

Figure 4.2 The solar corona during a total solar eclipse photographed from India. Courtesy: Jagdev Singh, Indian Institute of Astrophysics.

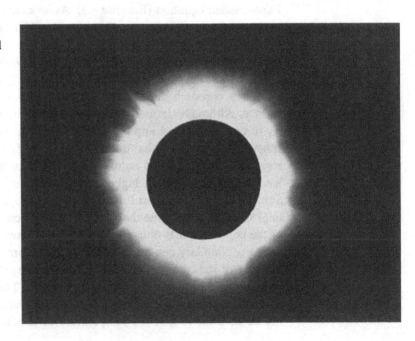

at infinity. Substituting for T in (4.33) from (4.35), we get

$$\frac{dp}{p} = -\frac{GM_\odot m}{\kappa_{\mathrm{B}} T_0 r_0^{2/7}} \frac{dr}{r^{12/7}},$$

of which the solution satisfying $p = p_0$ at $r = r_0$ is

$$p = p_0 \exp\left[\frac{7GM_\odot m}{5\kappa_{\mathrm{B}} T_0 r_0} \left\{\left(\frac{r_0}{r}\right)^{5/7} - 1\right\}\right]. \qquad (4.36)$$

The surprising thing to note is that the pressure has a non-zero asymptotic value as r goes to infinity. It is not possible to obtain a solution of the problem such that both p and T are zero at infinity. The asymptotic value of p at infinity is much larger than the typical value of the pressure of the interstellar medium.

What is the significance of this non-zero pressure at infinity? Parker (1958) concluded correctly that the hot solar corona could be in static equilibrium only if some appropriate pressure is applied at infinity to stop it from expanding. Since there is nothing to contain the corona by applying the necessary pressure, Parker (1958) suggested that the outer parts of the corona must be expanding in the form of a *solar wind*. The solar wind was detected from spacecraft observations just a few years after Parker's bold prediction. Parker (1958) worked out a detailed hydrodynamic model of the solar wind as well. This model will be discussed in §6.8.

4.5 Bernoulli's principle for steady flows

After considering hydrostatics, we now look at the next simplest hydrodynamic problem, which is the study of steady flows (i.e. flows which are independent of time). For any flow, one can introduce the concept of a *streamline*, which is a curve such that the fluid velocity **v** is tangential to it at every point. It is exactly like a line of force in an electric or a magnetic field. For steady flows, however, a streamline has another physical significance: it is the path which a fluid element traces out in time.

Since the body force **F** acting on a fluid (such as gravity) is usually conservative, we write

$$\mathbf{F} = -\nabla \Phi.$$

Using (4.16), it follows from the Euler equation (4.11) that steady flows satisfy

$$\nabla\left(\frac{1}{2}v^2\right) - \mathbf{v} \times (\nabla \times \mathbf{v}) = -\frac{1}{\rho}\nabla p - \nabla \Phi.$$

We now take a line integral of this equation along a streamline. If $d\mathbf{l}$ is a line element of the streamline, then we have

$$\int d\mathbf{l} \cdot \left[\nabla \left(\frac{1}{2}v^2 \right) - \mathbf{v} \times (\nabla \times \mathbf{v}) + \frac{1}{\rho}\nabla p + \nabla \Phi \right] = 0. \qquad (4.37)$$

Since $d\mathbf{l}$ and $\mathbf{v}$ are in the same direction, it is easy to see that $d\mathbf{l} \cdot \mathbf{v} \times (\nabla \times \mathbf{v})$ is zero. The remaining terms in (4.37) readily admit an integral

$$\frac{1}{2}v^2 + \int \frac{dp}{\rho} + \Phi = \text{constant} \qquad (4.38)$$

along a streamline, where the integral $\int dp/\rho$ has to be evaluated from a reference point on the streamline to the point where all the other quantities are considered. This result is known as *Bernoulli's principle* (Bernoulli 1738).

We often use this result to situations from everyday life involving liquids on the Earth's surface. For an incompressible fluid, the integral $\int dp/\rho$ in (4.38) can simply be replaced by p/ρ, and we write $\Phi = gh$, where h the height from some horizontal reference level. Then (4.38) becomes

$$\frac{1}{2}v^2 + \frac{p}{\rho} + gh = \text{constant} \qquad (4.39)$$

along a streamline in the incompressible fluid.

To illustrate the application of (4.39), let us consider a tank of water with an outlet at the bottom turned upward as shown in Figure 4.3. We can apply (4.39) to find the velocity of water coming through the outlet. We consider the streamline indicated in Figure 4.3 going from the upper surface of the tank to the outlet, and apply (4.39) to the points at the top and at the outlet. The pressure at both of these points is the atmospheric pressure p_0. If the outflow rate is small, then the level of water falls very slowly and we can take the velocity at the top point to be zero. Then, measuring the height h from the level

Figure 4.3 Water coming out of an upturned outlet in a tank.

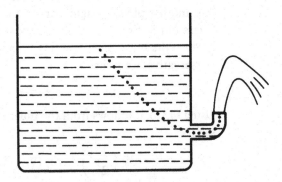

where the outlet is, we find from (4.39) that

$$\frac{p_0}{\rho} + gH = \frac{1}{2}v_{\text{out}}^2 + \frac{p_0}{\rho},$$

where H is the height of the water level above the level of the outlet and v_{out} is the velocity of the water flowing through the outlet. We then have

$$v_{\text{out}} = \sqrt{2gH}. \tag{4.40}$$

It follows that the velocity of outflowing water is the same irrespective of the direction in which the outlet is turned. If it is turned upward, then the outflowing water rises with the initial upward velocity $\sqrt{2gH}$. It is easy to see that the water should rise through the height H, i.e. it should reach the height where the inside water level is. In reality, however, water is always found to rise to a height slightly less. This is due to the fact that (4.40) is obtained by neglecting viscosity. Viscosity causes some dissipation of motion and hence the actual outflow velocity turns out to be slightly less than what is predicted from the ideal fluid equations.

Bernoulli's principle implies that the pressure is less at a point on a streamline if the velocity is more there. Hence, if a fluid is made to move through some constricted area where the velocity increases, then the pressure drops there. In the *Feynman Lectures* (Feynman, Leighton and Sands 1964, §40–3), a striking illustration of this is given in the following words: "Have you ever held two pieces of paper close together and tried to blow them apart? Try it! They come *together*. The reason, of course, is that the air has a higher speed going through the constricted space between the sheets than it does when it gets outside. The pressure between the sheets is *lower* than atmospheric pressure, so they come together rather than separating."

4.6 Kelvin's vorticity theorem

We have seen that the vorticity ω in an incompressible or barotropic fluid satisfies (4.20). A remarkable theorem follows from this equation. We first state this theorem before proving it.

Consider a surface S_1 inside a fluid at time t_1. The flux of vorticity linked with this surface is $\int_{S_1} \omega \cdot d\mathbf{S}$. At some future time t_2, the fluid elements which made up the surface S_1 at time t_1 will make up a different surface S_2. The vorticity flux linked with this surface S_2 at time t_2 will be $\int_{S_2} \omega \cdot d\mathbf{S}$. Kelvin's theorem (Helmholtz 1858; Kelvin

1869) states that

$$\int_{S_1} \omega \cdot d\mathbf{S} = \int_{S_2} \omega \cdot d\mathbf{S}.$$

We write this more compactly in the form

$$\frac{d}{dt} \int_S \omega \cdot d\mathbf{S} = 0, \tag{4.41}$$

where the Lagrangian derivative d/dt implies that we are considering the variation of the vorticity flux $\int_S \omega \cdot d\mathbf{S}$ linked with the surface S as we follow the surface S with the motion of the fluid elements constituting it.

We now have to start from (4.20) and prove the important result (4.41). Several authors, such as Landau and Lifshitz (1987, §8), give a proof using the fact that ω is the curl of **v**. It is, however, possible to give a proof without using this relation between ω and **v**. In other words, we prove the following more general theorem: if any vector field **Q** in a fluid satisfies the equation

$$\frac{\partial \mathbf{Q}}{\partial t} = \nabla \times (\mathbf{v} \times \mathbf{Q}), \tag{4.42}$$

then

$$\frac{d}{dt} \int_S \mathbf{Q} \cdot d\mathbf{S} = 0. \tag{4.43}$$

We shall see later that there are other quantities apart from vorticity which satisfy (4.42) and hence this general theorem holds for them.

To proceed with the proof now, we note that the flux $\int_S \mathbf{Q} \cdot d\mathbf{S}$ linked with the surface S can change with time due to two reasons: (i) intrinsic

Figure 4.4
Displacement of a
surface element due
to fluid motions.

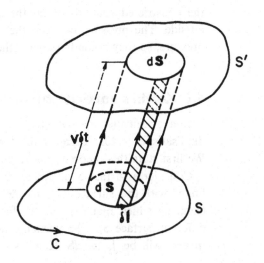

variation in $\mathbf{Q}$, and (ii) motion of the surface S. Mathematically we write

$$\frac{d}{dt}\int_S \mathbf{Q}\cdot d\mathbf{S} = \int_S \frac{\partial\mathbf{Q}}{\partial t}\cdot d\mathbf{S} + \int_S \mathbf{Q}\cdot\frac{d}{dt}(d\mathbf{S}). \qquad (4.44)$$

Figure 4.4 shows an element of area which has changed from $d\mathbf{S}$ at time t to $d\mathbf{S}'$ at time $t' = t + \delta t$. We see that $d\mathbf{S}$ and $d\mathbf{S}'$ make up the two ends of a cylinder. The vector area of a side strip of this cylinder is $-\delta t \mathbf{v} \times \delta\mathbf{l}$, where $\delta\mathbf{l}$ is a length element from the curve encircling the surface $d\mathbf{S}$ as shown in Figure 4.4. Since the vector area $\oint d\mathbf{S}$ for a closed surface is zero, the surfaces of this cylinder satisfy the equation

$$d\mathbf{S}' - d\mathbf{S} - \delta t\oint \mathbf{v}\times\delta\mathbf{l} = 0,$$

where the line integral is taken around the surface element $d\mathbf{S}$. It then follows that

$$\frac{d}{dt}(d\mathbf{S}) = \lim_{\delta t\to 0}\frac{d\mathbf{S}' - d\mathbf{S}}{\delta t} = \oint \mathbf{v}\times\delta\mathbf{l}.$$

The last term of (4.44) now becomes

$$\int_S \mathbf{Q}\cdot\frac{d}{dt}(d\mathbf{S}) = \int\oint \mathbf{Q}\cdot(\mathbf{v}\times\delta\mathbf{l}) = \int\oint(\mathbf{Q}\times\mathbf{v})\cdot\delta\mathbf{l}.$$

Here the double integral $\int\oint$ means that we first take a line integral around surface elements like $d\mathbf{S}$ and then sum up such line integrals for the many surface elements which would make up the surface S. It is easy to see that this ultimately gives a line integral along the curve C encircling the whole surface S, because the contributions from the line integrals in the interior cancel out when we sum over all surface elements. Hence we have

$$\int_S \mathbf{Q}\cdot\frac{d}{dt}(d\mathbf{S}) = \oint_C(\mathbf{Q}\times\mathbf{v})\cdot\delta\mathbf{l} = \int_S[\nabla\times(\mathbf{Q}\times\mathbf{v})]\cdot d\mathbf{S}$$

by Stokes's theorem. We then have from (4.44)

$$\frac{d}{dt}\int_S \mathbf{Q}\cdot d\mathbf{S} = \int_S d\mathbf{S}\cdot\left[\frac{\partial\mathbf{Q}}{\partial t} - \nabla\times(\mathbf{v}\times\mathbf{Q})\right]. \qquad (4.45)$$

We now see that (4.43) follows from (4.42) and (4.45). This completes our proof.

When $\mathbf{Q}$ stands for the vorticity ω, Kelvin's theorem follows as a special case of this general theorem, implying that the flux of vorticity is conserved and the vortex lines move with the fluid. For example, consider the case that there are some vortices in the neighbourhood of A in a fluid (Figure 4.5). If there is a general flow taking the fluid from the neighbourhood of A to the neighbourhood of B after some time, then Kelvin's theorem indicates that the vortices also move with the fluid to the neighbourhood of *B*. The evidence for this is found by observing that vortices in rivers are often carried with the general flows of the rivers. Lord Kelvin, who proved this theorem, was so intrigued by the permanence of vortices that he even attempted to build a theory of matter by suggesting that atoms are vortices in the ether! One of the strong predictions of Kelvin's theorem is that an ideal fluid having no vortices at an instant of time cannot subsequently develop vortices. This could have been inferred from (4.20) itself. We shall see in the next chapter that viscosity makes the generation and decay of vortices possible. A fluid without viscosity would not behave the way we expect real fluids to behave. When water is splashed against a solid surface, lots of swirling motions are produced. If a fluid were *really ideal* and vortices could not be generated, then such swirling motions would not take place. In the *Feynman Lectures*, the very apt phrases 'dry water' and 'wet water' are used to refer to ideal fluids and viscous fluids respectively—ascribing the credit of coining these phrases to John von Neumann (Feynman, Leighton and Sands 1964, §40–2).

4.7 Potential flows. Flow past a cylinder

Since an ideal incompressible fluid with zero vorticity at an instant will never generate vorticity, it is mathematically consistent to consider a fluid with

$$\nabla \times \mathbf{v} = 0 \tag{4.46}$$

everywhere and at all times. Such fluid flows are called *irrotational flows*. They are often called *potential flows* also, because the velocity for an irrotational flow can be written as the gradient of a potential, i.e.

$$\mathbf{v} = -\nabla\phi, \tag{4.47}$$

where ϕ is called the *velocity potential*. The incompressibility condition $\nabla \cdot \mathbf{v} = 0$ then implies

$$\nabla^2\phi = 0, \tag{4.48}$$

Figure 4.5 A vortex being carried in fluid motions.

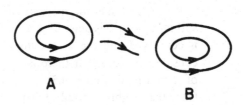

A   B

which is the well-known Laplace equation. Since the normal component of the fluid velocity has to be zero at a solid surface at rest, the appropriate boundary condition is

$$\frac{\partial \phi}{\partial n} = 0 \tag{4.49}$$

on a static solid boundary. This is often referred to as a Neumann-type boundary condition, and it is well known that the Laplace equation with this type of boundary condition constitutes a well-posed problem of which the solution is unique (see, for example, Panofsky and Phillips 1962, §3–1; Jackson 1975, §1.9). Elegant mathematical solutions have been found for many potential flows inside or around solid boundaries. Some of the older textbooks of fluid mechanics devote large portions to discussions of such potential flow problems. The excessive attention paid to this subject in some of these textbooks is *not* because of the fact that this is of central importance in understanding the physics of fluids, but rather because it is one of the fluid mechanics topics that provides ample opportunity for displaying mathematical cleverness. Since a display of mathematical virtuosity is not our main concern here, we shall just consider one example of a potential flow problem to give an idea of the subject. We assume that the reader is already familiar with the techniques of solving boundary value problems and indicate only the main steps.

Consider a uniform flow past a right circular cylinder, which is placed with its axis perpendicular to the direction of the flow. By 'uniform flow' we mean that at large distances from the cylinder the flow has a uniform velocity $-U\hat{\mathbf{e}}_x$. We want to find out the velocity field around the cylinder. This is clearly a two-dimensional problem. We can choose the centre of the cylinder as the origin of the coordinate system. If the radius of the cylinder is $r = a$, then we have to solve the Laplace equation (4.48) in the region $a \leq r \leq \infty$. The boundary condition (4.49) applied to the surface of the cylinder is

$$\frac{\partial \phi}{\partial r} = 0 \quad \text{at} \quad r = a. \tag{4.50}$$

The boundary condition at infinity can be taken to be

$$\phi = Ur \cos \theta \quad \text{at} \quad r \to \infty, \tag{4.51}$$

which would give the velocity

$$\mathbf{v}_\infty = -U \cos \theta \, \hat{\mathbf{e}}_r + U \sin \theta \, \hat{\mathbf{e}}_\theta = -U \, \hat{\mathbf{e}}_x. \tag{4.52}$$

The general solution of the Laplace equation in cylindrical coordinates can be written in the form of the Fourier series

$$\phi = (A_0 + B_0 \ln r)(C_0 + D_0\theta) + \sum_{n=1}^{\infty} \left(A_n r^n + \frac{B_n}{r^n} \right) (C_n \cos n\theta + D_n \sin n\theta),$$

(4.53)

where A_n, B_n, C_n and D_n are constants (see, for example, Panofsky and Phillips 1962, §4–9; Jackson 1975, §2.11). A few easy steps show that the boundary conditions (4.50) and (4.51) can be satisfied only if two terms of the series (4.53) are taken in the following combination

$$\phi = U \cos \theta \left(r + \frac{a^2}{r} \right).$$

(4.54)

The flow velocity around the cylinder is now easily found by taking the negative gradient of ϕ, which gives

$$\mathbf{v} = -U\, \hat{\mathbf{e}}_x + U \frac{a^2}{r^2} (\cos \theta\, \hat{\mathbf{e}}_r + \sin \theta\, \hat{\mathbf{e}}_\theta).$$

(4.55)

The streamlines of this flow are shown in Figure 4.6.

By a simple Galilean transformation, we can now consider the problem of the motion of a cylinder with velocity $U\hat{\mathbf{e}}_x$ through a fluid at rest in regions away from the cylinder. In the coordinate frame moving with the cylinder, the velocity field around the cylinder is given by (4.55). Hence, if we take the origin of the coordinate system at an instantaneous position of the centre of the cylinder, then the flow pattern around the cylinder at that instant is given by

$$\mathbf{v}' = \mathbf{v} + U\, \hat{\mathbf{e}}_x = U \frac{a^2}{r^2} (\cos \theta\, \hat{\mathbf{e}}_r + \sin \theta\, \hat{\mathbf{e}}_\theta).$$

(4.56)

We now want to calculate the kinetic energy associated with the fluid motion around the cylinder. The kinetic energy in a fluid layer of unit thickness parallel to the cylinder axis is given by

$$K_{\text{fluid}} = \tfrac{1}{2}\rho \int_a^{\infty} (\mathbf{v}')^2 2\pi r\, dr.$$

Figure 4.6 Flow of an ideal fluid past a cylinder.

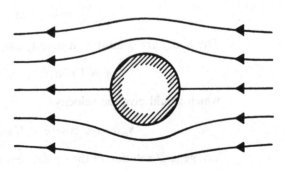

Substituting for $\mathbf{v}'$ from (4.56), this integral is found to be

$$K_{\text{fluid}} = \tfrac{1}{2}\pi\rho a^2 U^2 = \tfrac{1}{2}M'U^2,$$

where

$$M' = \pi a^2 \rho$$

is the mass of fluid displaced by the unit length of the cylinder. If M is the mass of the cylinder per unit length, then the total kinetic energy associated with unit length of the cylinder (i.e. kinetic energy of the cylinder plus the surrounding fluid) is

$$K_{\text{total}} = \tfrac{1}{2}(M + M')U^2. \tag{4.57}$$

We can now deduce the equation of motion of a cylinder moving through an ideal fluid. If $\mathbf{F}$ be the force acting per unit length of the cylinder moving with velocity $\mathbf{U}$, then the rate of work done $\mathbf{F}\cdot\mathbf{U}$ has to be equal to the rate of change of total kinetic energy, i.e.

$$\mathbf{F}\cdot\mathbf{U} = \frac{dK_{\text{total}}}{dt} = (M + M')\mathbf{U}\cdot\frac{d\mathbf{U}}{dt},$$

from which

$$\mathbf{F} = (M + M')\frac{d\mathbf{U}}{dt}. \tag{4.58}$$

Since this equation has the form of Newton's second law of motion with $M+M'$ appearing instead of just the mass of the cylinder, $M+M'$ is often referred to as the *effective mass*. One surprising conclusion following from (4.58) is that there is no force exerted on a cylinder moving through a fluid with a uniform speed $\mathbf{U}$. This result is certainly at variance with our everyday experience that a fluid exerts a drag force on any object moving through it. This contradiction between the theoretical conclusion and the everyday fact is known as *d'Alembert's paradox*. This paradox arises because we have neglected viscosity in our equations. If a body moving through a fluid were to be slowed by a drag force, the kinetic energy lost in the slowing process has to go somewhere. In a viscous fluid, it can go to heat. Since this is not possible in an ideal fluid, we come up with the result that an ideal fluid cannot exert drag on objects moving through it. This paradox warns us that the neglect of viscosity can sometimes give rise to qualitatively wrong conclusions, even though superficially it may appear that the viscosity is not important in the problem. The reason behind this will be discussed further in the next chapter.

We close by finding the acceleration of a cylinder falling in a gravitational field through a large volume of fluid (such that the edge effects can be neglected and we can apply results for motions through fluids extending to infinity). According to Archimedes's principle (Exercise 4.2), the gravitational force acting per unit length of the cylinder is

$$F = (M - M')g,$$

where g is the acceleration due to gravity. It then readily follows from (4.58) that

$$\frac{dU}{dt} = \frac{M - M'}{M + M'}g. \tag{4.59}$$

This equation implies that an object like a cylinder dropped in the ocean keeps on accelerating—again a conclusion resulting from the neglect of viscosity. In reality, viscosity limits the terminal velocity to an asymptotic value.

4.8 Stream function

For an incompressible fluid satisfying $\nabla \cdot \mathbf{v} = 0$, the velocity can be represented by the curl of a vector field (just like the vector potential for a magnetic field). For two-dimensional problems without z-dependence, this becomes particularly useful if we write

$$\mathbf{v} = -\nabla \times [\psi(x, y)\hat{\mathbf{e}}_z], \tag{4.60}$$

from which it readily follows that

$$v_x = -\frac{\partial \psi}{\partial y}, \qquad v_y = \frac{\partial \psi}{\partial x}. \tag{4.61}$$

The function $\psi(x, y)$ is called the *stream function*. We shall now show that it is constant on streamlines. Since a streamline is defined as a curve which has $\mathbf{v}$ in the tangential direction, its equation is

$$\frac{dx}{v_x} = \frac{dy}{v_y}, \tag{4.62}$$

i.e.

$$v_y \, dx - v_x \, dy = 0.$$

Substituting from (4.61), we have

$$\frac{\partial \psi}{\partial x}dx + \frac{\partial \psi}{\partial y}dy = 0, \tag{4.63}$$

which means that ψ is constant along streamlines.

If the two-dimensional flow is irrotational in addition to being incompressible, then we can introduce the velocity potential as well through (4.47). There exist some beautiful mathematical relations between the stream function and the velocity potential for a flow which is both incompressible and irrotational. The components of velocity in terms of the velocity potential are

$$v_x = -\frac{\partial \phi}{\partial x}, \qquad v_y = -\frac{\partial \phi}{\partial y}. \tag{4.64}$$

It is easy to see from (4.61) and (4.64) that

$$\nabla \phi \cdot \nabla \psi = 0,$$

which means that the velocity potential and the stream function are orthogonal functions. We also note from (4.61) and (4.64) that

$$\frac{\partial \phi}{\partial x} = \frac{\partial \psi}{\partial y}, \qquad \frac{\partial \phi}{\partial y} = -\frac{\partial \psi}{\partial x}. \tag{4.65}$$

These are the Cauchy–Riemann conditions for the analyticity of the function $\phi(x, y) + i\psi(x, y)$ (see, for example, Arfken 1985, §6.2; Copson 1935, §3.4). In other words, in a two-dimensional incompressible and irrotational fluid flow problem, it is possible to introduce an analytic function of the complex variable $z = x + iy$ of which the real and imaginary parts correspond to the velocity potential and the stream function:

$$f(z) = f(x + iy) = \phi(x, y) + i\psi(x, y). \tag{4.66}$$

One can also introduce the *complex velocity* defined as

$$w = -\frac{df}{dz}. \tag{4.67}$$

It is straightforward to show that

$$w = v_x - iv_y. \tag{4.68}$$

In a particular case, of course, the question is how to find the appropriate analytic function $f(z)$. To give an example, the problem of uniform flow past a cylinder discussed in §4.7 has the solution

$$f(z) = U \left(z + \frac{a^2}{z} \right).$$

It is easy to see that the real part of this is the the velocity potential given in (4.54). The imaginary part gives the stream function

$$\psi = U \sin \theta \left(r - \frac{a^2}{r} \right).$$

The streamlines plotted in Figure 4.6 are nothing but the curves on which this function is constant.

Exercises

4.1　Derive the energy conservation equation (4.25) starting from the basic equations of hydrodynamics.

4.2　Suppose an object heavier than water is fully immersed in water within which pressure varies as given by (4.30). Show that the net force exerted on the object by the surrounding water is $-M'g$, where M' is the mass of water displaced by the object. This is the celebrated *Archimedes's principle*.

4.3　Consider a spherically symmetric mass of gas in hydrostatic equilibrium under *its own* gravity. Show that the pressure $p(r)$ and density $\rho(r)$ inside satisfy the equation

$$\frac{1}{r^2}\frac{d}{dr}\left(\frac{r^2}{\rho}\frac{dp}{dr}\right) = -4\pi G\rho, \qquad (4.69)$$

where r is the radial distance measured from the centre.

Now assume the pressure and the density to be related by

$$p = K\rho^{(1+1/n)}$$

and substitute

$$\rho = \rho_c\theta^n, \qquad r = a\xi,$$

where ρ_c is the central density and

$$a = \left[\frac{(n+1)K\rho_c^{(1/n-1)}}{4\pi G}\right]^{1/2}.$$

Show that (4.69) becomes

$$\frac{1}{\xi^2}\frac{d}{d\xi}\left(\xi^2\frac{d\theta}{d\xi}\right) = -\theta^n.$$

This is the *Lane–Emden equation* (Homer Lane 1869; Emden 1907), which is solved with the boundary conditions

$$\theta(0) = 1, \qquad \frac{d\theta}{d\xi}\bigg|_{\xi=0} = 0$$

to obtain the structures of self-gravitating polytropic spheres.

4.4　Suppose water is coming out of a hole of cross-sectional area S_h on the side of a water tank. Show that the asymptotic cross-sectional area of the water jet is $S_j = \frac{1}{2}S_h$. [Hint: use the expression for the momentum flux.]

4.5　Consider two parallel vortex lines of equal strength with (i) the same sign and (ii) opposite signs. Find out how these vortex lines move in each other's velocity fields.

4.6   Consider a sphere of mass M moving through an incompressible and irrotational fluid extending to infinity. Find the velocity distribution around the sphere and show that the effective mass of the sphere is $M + \frac{1}{2}M'$, where M' is the mass of the displaced fluid.

4.7   Suppose a cylinder moving through an incompressible fluid with velocity U has a circulation around it such that the velocity potential (4.54) has an additional term

$$\frac{\kappa\theta}{2\pi}.$$

Note that this keeps the flow outside the cylinder irrotational. After finding the velocity distribution around, apply Bernoulli's principle to derive the pressure at points next to the surface of the cylinder. By integrating the pressure, show that the net force exerted on the cylinder by the surrounding fluid is $\rho U\kappa$ in a direction perpendicular to U. This is known as the *Magnus effect* (Magnus 1853). As $\kappa \to 0$, we are led to d'Alembert's paradox discussed in §4.7.

5 Viscous flows

5.1 Tangential stress in a Newtonian fluid

We have already seen that the surface forces inside a fluid can be handled by introducing the second-rank tensor P_{ij} appearing in (4.6). As in §3.4, we write

$$P_{ij} = p\delta_{ij} + \pi_{ij}. \tag{5.1}$$

Chapter 4 has been devoted to the study of *ideal fluids* in which π_{ij} is assumed to be zero so that the surface force on an element of surface is always normal. We pointed out that the ideal fluid equations do not reproduce some of the properties of real fluids. Now we want to find out an expression for π_{ij} from macroscopic considerations. We shall see that it will be in agreement with what we found in §3.4 from microscopic theory.

To motivate our discussion, let us again look at Figure 4.1. We know from everyday experience that the internal friction in a fluid (which we call *viscosity*) opposes relative motions amongst different layers of a fluid. Since the fluid above the separating surface S is moving slower than the fluid below, we expect the fluid above to be subjected to a shear force trying to accelerate it in the x direction such that the velocity difference above and below S is reduced. This force has to act across the surface S of which the normal is in the y direction. From the expression (4.6) for a surface force, we see that there can be a force in the x direction acting across a surface with the normal in the y direction only if π_{xy} is non-zero. The shear force is expected to be larger for a larger velocity gradient dv_x/dy. Newton postulated that the shear force is proportional to the velocity gradient, i.e. we write

$$\pi_{xy} = -\mu \frac{dv_x}{dy}, \tag{5.2}$$

where μ is the coefficient of viscosity and the minus sign ensures that the direction of the force is correct. Fluids obeying the proportionality relation between the shear stress and the velocity gradient are known as *Newtonian fluids*. It is found experimentally that most fluids are very close to being Newtonian. Although (5.2) correctly gives the shear stress in a Newtonian fluid for the particularly simple situation of Figure 4.1, we have to find a more general expression for π_{ij} to handle more complicated situations.

The following simple consideration shows that (5.2) could not be a general expression for the shear stress. Suppose a fluid mass is undergoing rigid rotation around the z axis passing through the point O. Figure 5.1 shows the variation of velocity on points on the y axis. Certainly the velocity gradient $\partial v_x / \partial y$ is non-zero. But the shear stress π_{xy} has to be zero, as there are no relative motions inside the fluid. Hence (5.2) could not be true in this case. We have to subtract the component of rotation from the velocity gradient to get the correct expression for the shear stress. We write

$$\frac{\partial v_i}{\partial x_j} = \frac{1}{2}\left(\frac{\partial v_i}{\partial x_j} + \frac{\partial v_j}{\partial x_j}\right) + \frac{1}{2}\left(\frac{\partial v_i}{\partial x_j} - \frac{\partial v_j}{\partial x_i}\right). \tag{5.3}$$

One can give several different arguments to show that the first part of the expression

$$\Lambda_{ij} = \frac{1}{2}\left(\frac{\partial v_i}{\partial x_j} + \frac{\partial v_j}{\partial x_i}\right) \tag{5.4}$$

corresponds to pure shear and the second part, which is a component of the vorticity, to rotation. Perhaps the simplest way to convince one

Figure 5.1 Velocity vectors due to a rigid rotation around the z axis.

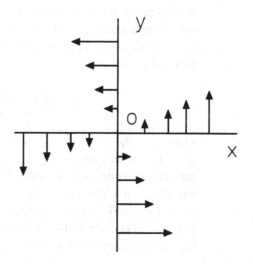

that the part Λ_{ij} does not include rotation is to show that this part vanishes for pure rigid rotation. If a fluid mass is undergoing rigid rotation with the angular velocity $\mathbf{\Omega}$, then the fluid velocity $\mathbf{v}$ at a distance $\mathbf{x}$ from the rotation axis is given by $\mathbf{v} = \mathbf{\Omega} \times \mathbf{x}$. Using (A.23), this can be written in the tensorial notation as

$$v_i = \epsilon_{ikl}\Omega_k x_l.$$

It is easy to see that this expression substituted in (5.4) gives $\Lambda_{ij} = 0$. Hence, if we want to ensure that the shear stress does not depend on the part of velocity gradient arising out of rotation, we have to make π_{ij} depend on the symmetric tensor Λ_{ij} rather than the simple gradient $\partial v_i / \partial x_j$.

For a Newtonian fluid, the shear stress depends linearly on the velocity gradient. The most general second-rank tensor linearly depending on symmetric combinations of velocity gradients is of the form

$$\pi_{ij} = a\left(\frac{\partial v_i}{\partial x_j} + \frac{\partial v_j}{\partial x_j}\right) + b\delta_{ij}\nabla \cdot \mathbf{v}. \tag{5.5}$$

Note that $\nabla \cdot \mathbf{v}$ is the trace of the tensor Λ_{ij}. To proceed further, we have to consider how P_{ij}, as given in (5.1), has to be split between $p\delta_{ij}$ and π_{ij}. The most obvious procedure is to define p as the isotropic part of P_{ij}, i.e.

$$p = \tfrac{1}{3}P_{ii}. \tag{5.6}$$

Although it may not be a priori obvious that p defined in this way is identical to the thermodynamic variable pressure appearing in the energy equation, it is found experimentally to be very nearly true. It then follows from (5.1) and (5.6) that π_{ij} has to be a traceless tensor. This happens if b in (5.5) is taken as $-(2/3)a$ so that we write

$$\pi_{ij} = -\mu\left(\frac{\partial v_i}{\partial x_j} + \frac{\partial v_j}{\partial x_i} - \frac{2}{3}\delta_{ij}\nabla \cdot \mathbf{v}\right), \tag{5.7}$$

where μ is the *coefficient of viscosity*. We note that this form of π_{ij} arrived at from macroscopic considerations is identical with the expression (3.39) derived by kinetic theory. It should now be clear that μ introduced in §3.4 indeed coincides with the usual macroscopic coefficient of viscosity. Moreover, comparison with the kinetic theory results gives us the confidence that p introduced through (5.6) corresponds to the thermodynamic pressure at least in the case of a dilute gas.

5.2 Navier–Stokes equation

On substituting (5.1) in (4.8) with π_{ij} given by (5.7), we obtain

$$\rho \frac{dv_i}{dt} = \rho F_i - \frac{\partial p}{\partial x_i} + \frac{\partial}{\partial x_j} \left[\mu \left(\frac{\partial v_i}{\partial x_j} + \frac{\partial v_j}{\partial x_i} - \frac{2}{3}\delta_{ij}\nabla \cdot \mathbf{v} \right) \right]. \qquad (5.8)$$

This equation is known traditionally as the *Navier–Stokes equation*, although several persons in the early nineteenth century made contributions towards establishing this equation (Navier 1822; Poisson 1829; Saint-Venant 1843; Stokes 1845).

We shall mostly be concerned with situations in which the spatial variation of u is not important. Hence u can be taken outside the spatial derivative so that

$$\rho \frac{d\mathbf{v}}{dt} = \rho \mathbf{F} - \nabla p + \mu \left[\nabla^2 \mathbf{v} + \frac{1}{3}\nabla(\nabla \cdot \mathbf{v}) \right]. \qquad (5.9)$$

The term involving $\nabla(\nabla \cdot \mathbf{v})$ has significance only in the case of flows with variable compression. For example, when we are studying the viscous dissipation of acoustic waves, this term has to be taken into account. For most fluid flow problems of interest to us, however, it can neglected. Hence, by Navier–Stokes equation, we shall usually mean the simpler version of the equation

$$\frac{\partial \mathbf{v}}{\partial t} + (\mathbf{v} \cdot \nabla)\mathbf{v} = \mathbf{F} - \frac{1}{\rho}\nabla p + \nu\nabla^2\mathbf{v}, \qquad (5.10)$$

where

$$\nu = \mu/\rho \qquad (5.11)$$

is called the *kinematic viscosity*. We note that (5.10) is the same as (3.52).

The version (5.10) of the Navier–Stokes equation differs from the Euler equation (4.11) only by virtue of the additional term $\nu\nabla^2\mathbf{v}$. The mathematical characters of the two equations, however, are vastly different due to the fact that the Navier–Stokes contains a higher spatial derivative than the Euler equation. Hence the Navier–Stokes equation requires more boundary conditions for solutions in finite regions. While solving ideal fluid problems with solid boundaries, one usually takes the normal component of velocity to be zero on the solid surface. We have seen in §4.7 that this suffices to find the solution uniquely and additional boundary conditions cannot be implemented. On the other hand, while solving viscous fluid problems, one imposes the extra boundary condition on a solid surface that the tangential component of velocity is also zero making $\mathbf{v} = 0$ there. Observations from everyday life convince us that this is actually the correct boundary condition for fluid flows along solid boundaries. For example, we notice that the blades of a fan collect dust. Even though the fan may be in regular use to produce winds, the air velocity at the surface of the blades remains zero so that the dust can collect there. If you ever try to sweep away fine dust particles from a solid surface by blowing from your mouth, you will realize how difficult it is. No matter how hard you blow, the air velocity exactly at the surface remains zero. If we do not consider viscosity and use the Euler equation, then the realistic boundary condition $\mathbf{v} = 0$ on a solid surface will usually overdetermine the problem and will not allow solutions.

If we take a curl of the Navier–Stokes equation (5.10) in the same way we took the curl of the Euler equation in §4.2, then instead of (4.20) we get

$$\frac{\partial \omega}{\partial t} = \nabla \times (\mathbf{v} \times \omega) + \nu \nabla^2 \omega \qquad (5.12)$$

for incompressible or barotropic fluids. Because of the additional term $\nu \nabla^2 \omega$, Kelvin's vorticity theorem no longer holds for viscous fluids, making the production and decay of vorticity possible. Hence we overcome the unphysical behaviour of ideal fluids pointed out at the end of §4.6. We have seen in §4.2 that the energy equation is redundant for an incompressible ideal fluid. It is straightforward to show that the same considerations hold for incompressible viscous fluids as well. The dynamical theory of incompressible viscous fluids is provided by (5.12) in combination with (4.18) and (4.19).

Since we shall often have occasions to consider hydrodynamic problems in spherical or cylindrical geometries, we write down the Navier–Stokes equation in spherical and cylindrical coordinates in Appendix C. While transforming the Navier–Stokes equation to another coordinate system, particular care should be taken about the term $(\mathbf{v} \cdot \nabla)\mathbf{v}$. This term implies the operator $\mathbf{v} \cdot \nabla$ operating on the vector field $\mathbf{v}$. When this operator $\mathbf{v} \cdot \nabla$ operates on a scalar field like the density ρ, the resultant can be thought of as a scalar product between $\mathbf{v}$ and $\nabla \rho$, i.e. $(\mathbf{v} \cdot \nabla)\rho = \mathbf{v} \cdot (\nabla \rho)$. However, when $\mathbf{v} \cdot \nabla$ operates on a vector field like $\mathbf{v}$, it is no longer possible to think of it as a scalar product between $\mathbf{v}$ and some other quantity. The easiest way of expressing $(\mathbf{v} \cdot \nabla)\mathbf{v}$ in other coordinates is to take recourse to the vector identity

$$(\mathbf{v} \cdot \nabla)\mathbf{v} = \tfrac{1}{2}\nabla(\mathbf{v} \cdot \mathbf{v}) - \mathbf{v} \times (\nabla \times \mathbf{v}) \qquad (5.13)$$

following from (A.6). If we know how to write down all the terms on the R.H.S. in some particular coordinate system, then we find the expression for $(\mathbf{v} \cdot \nabla)\mathbf{v}$. We have also to be careful in expressing $\nabla^2 \mathbf{v}$ in the other coordinate systems and remember that the unit vectors in curvilinear coordinates have many non-vanishing spatial derivatives so that

$$\nabla^2 \mathbf{v} \neq \hat{\mathbf{e}}_r \nabla^2 v_r + \hat{\mathbf{e}}_\theta \nabla^2 v_\theta + \hat{\mathbf{e}}_\phi \nabla^2 v_\phi$$

in spherical coordinates. Perhaps the easiest way of finding $\nabla^2 \mathbf{v}$ in curvilinear coordinates is to use the vector identity (A.12), which gives

$$\nabla^2 \mathbf{v} = \nabla(\nabla \cdot \mathbf{v}) - \nabla \times (\nabla \times \mathbf{v}). \tag{5.14}$$

We leave it as an exercise for the reader to transform the Navier–Stokes equation into spherical and cylindrical coordinates with the help of the vector identities (5.13) and (5.14) (Exercise 5.1). A detailed discussion of the subject can be found in Appendix 2 of Batchelor (1967).

5.3 Flow through a circular pipe

As an illustration of a viscous flow, let us consider the steady flow of an incompressible viscous fluid through a horizontal pipe of circular cross-section. Let the flow be produced by applying a pressure difference Δp between the ends of the pipe of length l. Let us use cylindrical coordinates with the z axis along the axis of the pipe. We expect the velocity to have only the z component which we denote as $v(r)$, where r is the radial distance from the axis of the pipe. For an *incompressible* fluid moving along a tube of uniform cross-section, $v(r)$ should be independent of z. From the z component of the Navier–Stokes equation in cylindrical coordinates, as given in (C.7), we find

$$-\frac{\Delta p}{l} = \mu \frac{1}{r} \frac{d}{dr} \left(r \frac{dv}{dr} \right), \tag{5.15}$$

where we have replaced the partial derivative symbol ∂ by the ordinary derivative symbol d. One boundary condition is that the velocity has to be zero on the surface of the pipe. If a is the internal radius of the pipe, then

$$v = 0 \quad \text{at} \quad r = a. \tag{5.16}$$

The other boundary condition comes from the fact that we expect the profile of $v(r)$ to be smooth on the axis of the pipe so that

$$\frac{dv}{dr} = 0 \quad \text{at} \quad r = 0. \tag{5.17}$$

Integrating (5.15) with the boundary conditions (5.16) and (5.17), we obtain

$$v(r) = \frac{\Delta p}{4\mu l}(a^2 - r^2). \tag{5.18}$$

This velocity profile is parabolic, as sketched in Figure 5.2.

The mass flux rate through the pipe is given by

$$Q = \int_0^a \rho v(r).\, 2\pi r\, dr.$$

Substituting for $v(r)$ from (5.18), we find

$$Q = \frac{\pi \Delta p}{8\nu l} a^4. \tag{5.19}$$

This result is known as *Poiseuille's formula* (Hagen 1839; Poiseuille 1840) and provides a simple method for the determination of the viscosity of liquids. The viscous liquid is made to flow through a horizontal pipe of length l and radius a. By measuring the pressure difference Δp between the ends of the pipe and the corresponding mass flux Q, one can easily find ν from (5.19).

The velocity profile (5.18) has been confirmed experimentally for *not very fast* flow velocities. If, however, the flow velocity is made very fast (for example, by increasing the pressure difference Δp), then it was experimentally found by Reynolds (1883) that the flow becomes irregular in space and time so that (5.18) ceases to hold. Therefore, if one uses (5.19) to measure the viscosity of a liquid, it is necessary to ensure that the flow speed is not too fast so that the flow remains regular. One may raise the question what is meant quantitatively by the phrases 'too fast' or 'not too fast' in the present context. To answer that question, we first have to introduce the very important concept of *Reynolds number*.

Figure 5.2 The velocity profile of a viscous fluid flowing through a circular pipe.

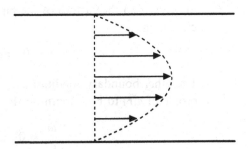

5.4 Scaling and Reynolds number

Suppose we are designing a new aircraft and we want to figure out the air flow pattern around it when it is flown. Is it possible to do laboratory experiments with a miniature model of the aircraft and find out the air flow pattern? It turns out that this is possible to some extent if the aircraft is supposed to move sufficiently subsonically so that the effects of compressibility can be neglected and the flow can be regarded as incompressible.

Let us consider fluid flows around *geometrically similar* objects of different sizes. When applying the fluid dynamical equations to each of these objects, it is useful to take the typical length L of that object as the unit of length. If the typical fluid velocity V is taken as the unit of velocity, then L/V must be taken as the unit of time. If $\mathbf{x}'$, $\mathbf{v}'$, t' and ω' be the values of length, velocity, time and vorticity measured in these scaled units, then they are related to their unscaled values as follows

$$\mathbf{x} = \mathbf{x}'L, \quad \mathbf{v} = \mathbf{v}'V, \quad t = t'\frac{L}{V}, \quad \omega = \omega'\frac{V}{L}. \qquad (5.20)$$

We have already noted that (5.12) is the basic equation for incompressible flows. Substituting from (5.20) into (5.12), we get the equation for the scaled variables

$$\frac{\partial \omega'}{\partial t'} = \nabla' \times (\mathbf{v}' \times \omega') + \frac{1}{\mathscr{R}}\nabla'^2\omega', \qquad (5.21)$$

where $\nabla' = L\nabla$ implies spatial differentiation with respect to the scaled length and

$$\mathscr{R} = \frac{LV}{\nu} \qquad (5.22)$$

is a dimensionless number known as the *Reynolds number*.

What have we gained by introducing the scaled variables? Suppose the Reynolds number $\mathscr{R}$ happens to be the same for two different fluid flows around two *geometrically similar* objects of different sizes. We note that the equation in the scaled variables (5.21) turns out to be identical for these two different cases. The boundary conditions in the scaled variables must also be identical. Hence, for the two different cases with the same Reynolds number $\mathscr{R}$, the flow patterns should satisfy the same equation with the same boundary conditions and hence they must be the same in the scaled units. This is called the *law of similarity* (Reynolds 1883). In order to find out the air flow pattern around our newly designed aircraft, we then have to do experiments with fluid flows around the miniature model, making sure that the

Reynolds number $\mathscr{R}$ of the experimental setup is the same as the Reynolds number in the actual situation in which we are interested.

The Reynolds number is a very powerful tool for unifying and classifying different types of geometrically similar fluid flow patterns. Suppose we want to study flows of different viscous fluids around spheres of different sizes. If we have found out the flow patterns around spheres for all the different Reynolds numbers, then we have got everything we need to know. The flow patterns for different Reynolds numbers can, for example, be obtained with a fixed sphere and a fixed fluid by varying the overall flow velocity around the sphere. The results of such a study will enable us to predict what will happen for other fluids and spheres of other sizes. For flow through a pipe, the Reynolds number can be defined by taking L in (5.22) to be equal to the radius of the pipe and V to be equal to the mean flow velocity through the pipe averaged over a cross-section. The flow through the pipe is found to be laminar when the Reynolds number is less than of the order of 3000 or so. For larger Reynolds numbers, the solution given by (5.18) is unstable and the flow can become turbulent. The subjects of hydrodynamic instability and turbulence will be taken up in later chapters.

We close this section by noting another significance of the Reynolds number. Let us look at the two terms on the R.H.S. of (5.12). If ω is a typical value of the vorticity, then the first term $\nabla \times (\mathbf{v} \times \boldsymbol{\omega})$ has a magnitude of the order $\omega V / L$, whereas the second term is of the order $\nu \omega / L^2$. The ratio of the two terms is the Reynolds number. For low Reynolds number situations, the second term therefore dominates over the first term. We have the naive expectation that the first term will be dominant in the high Reynolds number situations and the viscosity term will be negligible. We shall see in the next section that reality does not always conform to this expectation.

5.5 Viscous flow past solid bodies. Boundary layers

We discussed in §4.7 how the incompressible flows of ideal fluids past solid bodies can be studied. We now wish to consider how similar problems can be studied if the fluid is viscous instead of ideal. In the limit $\mathscr{R} \ll 1$, (5.21) reduces to the much simpler equation

$$\nabla^2 \omega' = 0. \tag{5.23}$$

It was found by Stokes (1851) that this equation admits of an analytical solution for the case of the flow past a sphere. We shall not present the solution here, since the analysis is slightly involved. The interested reader will find the solution presented in the standard textbooks on fluid mechanics (Landau and Lifshitz 1987, §20; Batchelor 1967, §4.9). One of the most important consequences of Stokes's analysis is that a sphere of radius a moving through a viscous fluid with velocity U experiences a drag force of magnitude

$$F_D = 6\pi\mu a U \tag{5.24}$$

in the direction opposite to its motion. Thus we get around d'Alembert's paradox for ideal fluids discussed in §4.7. The expression (5.24) is known as *Stokes's law*.

In the other limit of $\mathscr{R} \gg 1$, it naively appears from (5.12) or (5.21) that the viscosity term will be negligible and consequently we expect the flow pattern to be the same as what we get for ideal fluids. It is, however, found experimentally that the real situation is much more complicated. Figure 5.3 shows flow past a cylinder for different Reynolds numbers. When $\mathscr{R}$ is increased beyond about 20 or so, two vortices seem to appear behind. On increasing $\mathscr{R}$ further to about 60 or 70, the vortices on the two sides seem to be shed alternately so that a string of vortices arises behind the cylinder. This string of vortices is often known as the *Kármán vortex street*, after von Kármán (1911) who studied the properties of such a string of vortices. When $\mathscr{R}$ is finally made larger than 10^4, we find a *turbulent wake* behind the cylinder. It is found experimentally that the drag on the cylinder with a turbulent wake behind it is much larger than what would follow from Stokes's law (5.24). This is shown in Figure 5.4. The drag on a sphere can be written as

$$F_D = \tfrac{1}{2}C_D\pi a^2\rho U^2, \tag{5.25}$$

Figure 5.3 Flow of a viscous fluid past a cylinder for various Reynolds numbers. Reproduced from Batchelor (1967). Originally taken from Homann (1936).

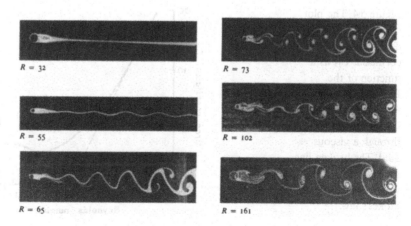

where C_D is a dimensionless coefficient which can be determined experimentally. Figure 5.4 shows a plot of C_D against the Reynolds number $\mathscr{R}$, taken to be aU/ν in the present case. If the drag on the sphere was always given by Stokes's law (5.24), then we would equate the right-hand sides of (5.24) and (5.25) to obtain

$$C_D = \frac{12\mu}{a\rho U} = \frac{12}{\mathscr{R}}.$$

Thus, according to Stokes's law, the coefficient C_D should fall inversely with $\mathscr{R}$ as indicated by the broken line in Figure 5.4. It is clear from the figure that the actual drag on a sphere at high Reynolds numbers is much larger than what is predicted by Stokes's law.

Let us now try to understand why the flows look so different from the simple solution for ideal fluids obtained in §4.7. We know that the ideal fluid problem in §4.7 was solved with the boundary condition that the normal component of velocity is zero on the solid boundary. When the flow past the solid body is sufficiently large, the ideal fluid solutions would imply a large tangential velocity next to the solid surface. On the other hand, we know that a viscous fluid must have $\mathbf{v} = 0$ at a solid boundary. Hence, even when the Reynolds number is very high, the flow pattern of a viscous fluid could not be identical with that of an ideal fluid down to the boundary of the solid. Next to the boundary of the solid, there should be a thin *boundary layer* within which the velocity of the fluid must change from zero at the surface to large values a short distance away from the surface. Since $\nabla^2\mathbf{v}$ must be very large in this layer, we cannot neglect the term $\mu\nabla^2\mathbf{v}$ in the Navier–Stokes equation within this boundary layer. Hence, if there are solid surfaces inside a fluid, then viscosity is always

Figure 5.4 The plot of the experimentally determined drag coefficient C_D as a function of the Reynolds number $\mathscr{R} = aU/\nu$ for a sphere moving through a viscous fluid. The broken line indicates what one would get from Stokes's law.

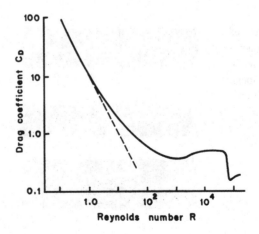

important in layers next to the solid even when the Reynolds number is large—a fundamental point which seems to have been realized first by Prandtl (1905). Prandtl's pathbreaking work cleared up much of the mystery as to why the elegant potential flow solutions obtained by nineteenth century scientists often failed to match experiments. The quotation from Prandtl and Tietjens (1934a) at the beginning of Part I describes the unsatisfactory state of hydrodynamics before Prandtl's work, which finally bridged the gap between theoretical and practical hydrodynamics.

One can give simple arguments that the boundary layer grows in thickness as we go downstream. Figure 5.5 shows a flow past a plane surface. Under steady conditions, the x component of the Navier–Stokes equation gives

$$v_x \frac{\partial v_x}{\partial x} + v_y \frac{\partial v_x}{\partial y} = -\frac{1}{\rho} \frac{\partial p}{\partial x} + v \left(\frac{\partial^2 v_x}{\partial x^2} + \frac{\partial^2 v_x}{\partial y^2} \right). \qquad (5.26)$$

One expects the terms $v_x(\partial v_x/\partial x)$ and $v(\partial^2 v_x/\partial y^2)$ to be larger than the other terms inside the boundary layer so that these two terms must balance each other there in order to satisfy (5.26). If δ be the thickness of the boundary layer at a distance L downstream from the edge of the solid, then we may approximate $v_x(\partial v_x/\partial x)$ by V^2/L and $v(\partial^2 v_x/\partial y^2)$ by vV/δ^2 so that

$$\frac{V^2}{L} \approx \frac{vV}{\delta^2},$$

from which

$$\delta \approx \sqrt{\frac{vL}{V}}, \qquad (5.27)$$

i.e. the boundary layer grows as the square root of the distance downstream.

It is no wonder that the growing boundary layer around a solid of finite extent (like a cylinder or sphere) leaves a wake downstream within which viscosity is important. Since viscosity is important in the boundary layer and the wake, Kelvin's vorticity theorem no longer

Figure 5.5 Viscous flow past a plane surface.

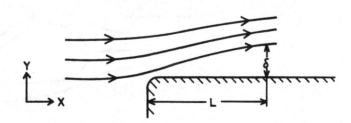

holds in these places and vortices can be produced downstream even though no vortices were present in the upstream flow. When the Reynolds number is made very large for a flow past a sphere or a cylinder, strong velocity shears develop at the sides of the sphere or the cylinder. This makes the flow unstable, giving rise to turbulence which is then carried downstream within the wake. The enhancement of the drag compared to what would follow from (5.24) is also not difficult to understand. The turbulent wake has lots of small-scale fluid motions which eventually decay away downstream. The energy dissipated in this process must have come originally from the kinetic energy of the moving solid. Such a sink of energy gives rise to the additional drag on the moving solid.

5.6 Aerodynamic lift

It is something amazing that objects like aeroplanes made of metals can fly in the air. We now discuss the theory as to how air exerts a vertical lift force on the wings of an aeroplane. For the sake of simplicity, let us consider an infinitely long *aerofoil* of uniform cross-section, as shown in Figure 5.6. The air flow around this aerofoil can be treated as two-dimensional. Suppose the aerofoil was initially at rest and then starts moving horizontally in the forward direction. Certainly the circulation $\oint \mathbf{v} \cdot d\mathbf{l}$ around the aerofoil was initially zero before the aerofoil started moving. If the air around happened to be an ideal fluid with zero viscosity, then Kelvin's theorem would force the circulation $\oint \mathbf{v} \cdot d\mathbf{l}$ around the aerofoil to remain zero even after

Figure 5.6 The streamlines around an aerofoil. (a) The circulation around the aerofoil is zero. (b) The normal situation in the presence of a small viscosity.

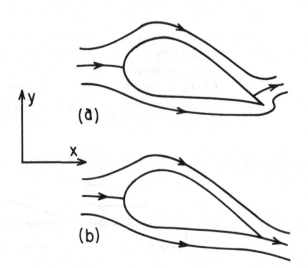

it starts moving. In the frame of the aerofoil, the air flow would then be a steady potential flow (like that we discussed in §4.7) with zero circulation around the aerofoil. Because of the asymmetric shape of the aerofoil, the streamlines have to be as shown in Figure 5.6(a) in order to keep the circulation zero. Obviously $\nabla^2 \mathbf{v}$ would be very large at the rearmost point, and if the air around has even a small viscosity, we expect that the large effect of viscosity at that point will make the streamlines relax to what is shown in Figure 5.6(b). Prandtl and Tietjens (1934b) give a set of rather striking photographs (Plates 17–22) illustrating how the streamlines around an aerofoil change from the configuration of Figure 5.6(a) to that of Figure 5.6(b).

From the frame of the aerofoil, we can now regard it as a long object carrying vorticity

$$\Gamma = \oint \mathbf{v} \cdot d\mathbf{l} = \int \omega \cdot d\mathbf{S} \tag{5.28}$$

placed perpendicularly in a velocity field $\mathbf{v}_\infty$ (i.e. the velocity away from the aerofoil). Since $(\mathbf{v} \cdot \nabla)\mathbf{v}$ gives a term $\omega \times \mathbf{v}$ as seen in (4.16), we can write the Navier–Stokes equation (5.10) as

$$\rho \frac{\partial \mathbf{v}}{\partial t} = \rho \mathbf{v} \times \omega + \text{other terms.}$$

It is clear that there is a hydrodynamic force given by $\rho \mathbf{v} \times \omega$. This can be compared with the Lorentz force $\mathbf{j} \times \mathbf{B}/c$ on a current in a magnetic field. The analogy can be carried further by noting that $\mathbf{j} \propto \nabla \times \mathbf{B}$ in the MHD approximation discussed in §13.6, just as $\omega = \nabla \times \mathbf{v}$. Just as a conductor carrying current I placed perpendicularly in a magnetic field B experiences a transverse force IB/c per unit length, similarly we expect a long aerofoil carrying vorticity Γ placed perpendicularly in a velocity field $\mathbf{v}_\infty$ to experience a transverse force $\rho v_\infty \Gamma$ per unit length, which is vertically upward in Figure 5.6(b). In fact, the reader has been asked to show in Exercise 4.7 that a cylinder with a circulation around it experiences just such a transverse force when moving through a fluid. Since the aerofoil is geometrically more complicated than a right circular cylinder and the equation of motion has other terms besides $\rho \mathbf{v} \times \omega$, it is necessary to do a full analysis to find the net force on the aerofoil. We shall do it using complex variables. It was pointed out in §4.8 that complex variables can be used in two-dimensional potential flow problems. The beautiful and elegant derivation given below should illustrate the power of the complex variables approach.

Viscosity must be important in a thin boundary layer adjacent to the aerofoil. Outside this boundary layer, the effect of viscosity is negligible and it should be possible to approximate the flow outside the boundary layer as ideal flow without vorticity (i.e. potential flow as discussed in §4.7). We have seen in §4.8 that a two-dimensional potential flow can be represented by a complex velocity

$$w = v_x - iv_y, \tag{5.29}$$

where we have taken the x axis in the horizontal direction and the y axis in the vertical direction (as indicated in Figure 5.6). This complex velocity is an analytic function outside the aerofoil and the adjacent boundary layer so that it must be possible to expand it in a Laurent series (see, for example, Arfken 1985, §6.5; Copson 1935, §4.52). Taking the origin to be inside the aerofoil and remembering that the velocity at infinity has to match v_∞, the appropriate Laurent expansion is

$$w = v_\infty + \frac{a_{-1}}{z} + \frac{a_{-2}}{z^2} + \cdots. \tag{5.30}$$

Let us now consider the contour integral

$$\oint w \, dz = \oint (v_x - iv_y)(dx + idy) = \oint [v_x \, dx + v_y \, dy + i(v_x \, dy - v_y \, dx)]$$

along the surface of the aerofoil just outside the thin boundary layer. Since the path of integration is clearly along a streamline, equation (4.62) for streamlines holds so that

$$\oint w \, dz = \oint (v_x \, dx + v_y \, dy) = \oint \mathbf{v} \cdot d\mathbf{l} = \Gamma \tag{5.31}$$

by (5.28). Substituting for w from (5.30), one can easily evaluate the contour integral by Cauchy's residue theorem (see Arfken 1985, §7.2; Copson 1935, §6.1) and obtain the value $2\pi i a_{-1}$. Therefore, from (5.31),

$$\Gamma = 2\pi i a_{-1}. \tag{5.32}$$

The net force on the aerofoil can be obtained by integrating the force exerted by the air pressure p along the boundary, i.e.

$$\mathbf{F} = \oint p \hat{\mathbf{n}} \, dl, \tag{5.33}$$

where $\hat{\mathbf{n}}$ is the inward directed normal and the path of integration can again be taken just outside the boundary layer where the expansion (5.30) is still valid. We now apply Bernoulli's principle (4.39) neglecting compressibility (which is justified if the velocity is small compared to sound speed) and noting that the variation of height h is small over a horizontal aerofoil. Then

$$p = -\tfrac{1}{2}\rho v^2 + \text{constant}$$

on a streamline, and substituting it in (5.33), we get

$$\mathbf{F} = -\tfrac{1}{2} \oint \rho v^2 \hat{\mathbf{n}} \, dl, \tag{5.34}$$

since a constant integrated over a closed contour vanishes. Let us now introduce the complex force

$$Z = F_y + iF_x. \tag{5.35}$$

If θ be the angle between the line element $d\mathbf{l}$ of the contour and the x axis, then the x and y components of the inward normal $\hat{\mathbf{n}}$ are $-\sin\theta$ and $\cos\theta$. On substituting these in (5.34), we can get the x and y components of $\mathbf{F}$ so that the complex force given by (5.35) becomes

$$Z = -\frac{1}{2} \oint \rho v^2 \cos\theta \, dl + \frac{i}{2} \oint \rho v^2 \sin\theta \, dl$$

$$= -\tfrac{1}{2}\rho \oint v^2 (dx - i \, dy), \tag{5.36}$$

since $dx = dl \cos\theta$ and $dy = dl \sin\theta$. We now add under the integral sign the expression

$$2i(v_x \, dy - v_y \, dx)(v_x - iv_y),$$

which is equal to zero by the streamline equation (4.62). A small algebra then gives

$$Z = -\frac{\rho}{2} \oint w^2 \, dz. \tag{5.37}$$

On substituting from (5.30) for w, it is easily seen that the coefficient of z^{-1} in the expansion of w^2 is $2v_\infty a_{-1}$. On applying Cauchy's residue theorem to (5.37), we then get

$$Z = -\tfrac{1}{2}\rho[2\pi i(2v_\infty a_{-1})]. \tag{5.38}$$

From (5.32) and (5.38), we finally have

$$Z = -\rho \Gamma v_\infty.$$

According to (5.35) the real part of Z is F_y so that

$$F_y = -\rho \Gamma v_\infty, \tag{5.39}$$

since it must be clear from (5.29) and (5.30) that the velocity v_∞ at infinity, which is in the x direction, is real. This expression (5.39) for the vertical lift force is known as the *Kutta–Joukowski lift formula* (Kutta 1911; Joukowski 1910).

It is to be noted in Figure 5.6 that the velocity at infinity is in the positive x direction so that v_∞ is positive in this case. Figure 5.6(b) makes it clear that $\Gamma = \oint \mathbf{v} \cdot d\mathbf{l}$ is negative if the line integral is taken in the anti-clockwise direction following the usual convention. Then F_y given by (5.39) must be positive. In other words, the lift force is in the vertically upward direction and acts to counter-balance the weight of the aircraft.

5.7 Accretion disks in astrophysics

Because of the large dimensions of astrophysical systems, the Reynolds number defined by (5.22) usually turns out to be very large in most astrophysical problems. If the Reynolds number is large in a system, we have seen in §5.5 that viscosity plays an important role only in layers next to solid boundaries. If the astrophysical system does not have a solid boundary, which is often the case, then viscosity may well be unimportant everywhere. There is, however, one major astrophysical topic—the theory of accretion disks—in which viscosity holds the centre stage and to which we now turn.

We believe that normal stars produce their energy by nuclear fusion. Apart from nuclear energy, there is one other possible source of energy in the astrophysical Universe. A system may lose its gravitational potential energy which is converted into other forms and radiated away. In fact, Helmholtz (1854) and Kelvin (1861) had proposed earlier that all stars produce energy by slowly shrinking in size and radiating away the excess gravitational potential energy. This theory could not account for the very long lifetimes of stars and, after the advent of nuclear physics, nuclear fusion was suggested as the source of stellar energy (von Weizsäcker 1937; Bethe and Critchfield 1938). There are, however, other situations in astrophysics where the gravitational potential energy may act as the main source of energy output.

Suppose we drop a mass m from a height h in a gravitational field g. The gravitational potential energy mgh is first converted into kinetic energy, and then, on hitting the ground, this energy is transformed into other forms such as heat and sound. Ordinarily, in this process, a very small fraction of the rest mass energy mc^2 is released. If, however, the mass m is dropped from infinity to a star of mass M and radius a, then the gravitational energy lost is

$$\frac{GM}{a}m = \frac{GM}{c^2 a}mc^2.$$

For a typical neutron star of mass $1M_\odot$ and radius 10 km, the factor $GM/c^2 a$ turns out to be about 0.15. Hence the loss of gravitational energy may be a very appreciable fraction of rest mass energy, making such an infall of matter into a deep gravitational well a tremendously efficient process for energy release.

If an object attracts matter distributed isotropically around it, then we have the situation of *spherical accretion* to be discussed in §6.8. More often, however, the accreting material may not be distributed isotropically around the source of attraction and, more importantly, may have some angular momentum associated with it. In such a situation, we have what is called an *accretion disk* in which the accreting matter forms a disk around the central attracting object and a particular parcel of matter gradually spirals inward, thereby losing gravitational potential energy. There is strong circumstantial evidence that this is the process by which energy is released in many astronomical systems. The first astronomical observations in the X-ray band (Giacconi *et al.* 1962) established the existence of compact X-ray sources. After the launch in 1970 of the satellite *Uhuru*, devoted completely to X-ray astronomy, it was possible to identify several compact X-ray sources with binary stellar systems. The most plausible model for such systems is that the binary contains a neutron star (or black hole) and another very large gaseous star, as sketched in Figure 5.7. The gas from the large star flows into the compact star in the form of an accretion disk, and the gravitational energy lost is radiated in the form of X-rays (Prendergast and Burbidge 1968). There are galaxies with what are called *active nuclei* which emit huge amounts of energy from very small volumes, the quasars being the most extreme examples of such active galactic nuclei. The most promising model for the active galactic nucleus is also an accretion disk around a supermassive black hole at the centre of the galaxy (Lynden-Bell 1969). Because of the importance of accretion disks in many astrophysical problems, the theoretical study of accretion disks has become almost an industry from the early 1970s. If the disk is assumed thin, then certain mathematical simplifications follow. A famous paper by Shakura and

Figure 5.7 Sketch of an accretion disk in a binary stellar system.

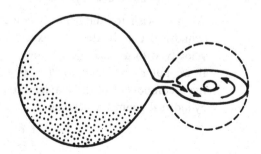

Sunyaev (1973) provided the paradigm for research on *thin accretion disks*.

Suppose some matter is going around a mass M in a nearly circular orbit of radius r. Balancing the gravitational force against the centrifugal force, we find the angular velocity to be

$$\Omega = \left(\frac{GM}{r^3}\right)^{1/2}. \tag{5.40}$$

The angular velocities of planets around the Sun indeed vary as $r^{-3/2}$, leading to Kepler's third law of planetary motion. Hence a circular motion satisfying (5.40) is often called *Keplerian motion* in astronomical jargon. The planets seem to have been moving in this way ever since the formation of the solar system. But now consider a gaseous disk going around a central mass such that the angular velocity within the disk varies in accordance with (5.40). Such a variation of angular velocity would imply the existence of velocity shear within the disk. Due to the action of viscosity, we then expect angular momentum to be transferred from the faster-moving inner regions of the disk to the slower-moving outer regions. As the material in an inner layer loses angular momentum, it moves inward in a spiral path. Hence it is viscosity which determines the rate of radial inflow of matter and therefore the rate at which the gravitational potential energy is converted into other forms. If the gas had no viscosity, then the material in the disk would keep on going in circular orbits and there would be no release of gravitational energy after the formation of the disk.

For several years, the study of accretion disks remained a theoreticians' game. With the help of high-resolution radio observations, however, it has recently been possible to confirm that centres of some galaxies indeed have thin disks with matter going around in nearly Keplerian orbits (Miyoshi *et al.* 1995).

5.7.1 The basic disk dynamics

We now wish to study the dynamics of a thin accretion disk, using cylindrical coordinates. We expect v_θ to be the main component of velocity with a small radial flow v_r caused by the effect of viscosity. If we put $v_z = 0$ and $\partial/\partial\theta = 0$, then the continuity equation and the θ component of the Navier–Stokes equation (as given in (C.6)) in cylindrical geometry become

$$\frac{\partial\rho}{\partial t} + \frac{1}{r}\frac{\partial}{\partial r}(r\rho v_r) = 0, \tag{5.41}$$

$$\rho \left(\frac{\partial v_\theta}{\partial t} + v_r \frac{\partial v_\theta}{\partial r} + \frac{v_r v_\theta}{r} \right) = \mu \left(\frac{\partial^2 v_\theta}{\partial r^2} + \frac{\partial^2 v_\theta}{\partial z^2} + \frac{1}{r} \frac{\partial v_\theta}{\partial r} - \frac{v_\theta}{r^2} \right). \quad (5.42)$$

It may be noted that we are mainly dealing with a version of the Navier–Stokes equation obtained from the general version (5.8) by assuming μ to be constant. The R.H.S. of (5.42) as given here is also valid only when $\mu = \rho\nu$ is constant, which is not the case in an accretion disk. So we shall have to figure out the appropriate expression for the viscous stress from first principles. Integrating both (5.41) and (5.42) over z and writing $\int \rho \, dz = \Sigma$ for the surface density, (5.41) and (5.42) become

$$\frac{\partial \Sigma}{\partial t} + \frac{1}{r} \frac{\partial}{\partial r}(r\Sigma v_r) = 0, \quad (5.43)$$

$$\Sigma \left(\frac{\partial v_\theta}{\partial t} + v_r \frac{\partial v_\theta}{\partial r} + \frac{v_r v_\theta}{r} \right) = \text{terms involving viscosity}, \quad (5.44)$$

where we have neglected the variation of v_r and v_θ with z (although ρ varies with z, we do not expect the components of velocity to vary much). We now add (5.43) multiplied by rv_θ and (5.44) multiplied by r. This gives

$$\frac{\partial}{\partial t}(\Sigma r^2 \Omega) + \frac{1}{r} \frac{\partial}{\partial r}(\Sigma r^3 \Omega v_r) = \mathcal{G}, \quad (5.45)$$

where $\Omega = v_\theta / r$ is the angular velocity and $\mathcal{G}$ is the term involving viscosity, which we now want to find out. Note that $\Sigma r^2 \Omega \cdot 2\pi r \, dr$ is the angular momentum associated with an annular ring from between r and $r + dr$. It is obvious that (5.45) is the equation giving the evolution of the angular momentum, the second term on the L.H.S. being nothing but the divergence of the angular momentum flux $\Sigma r^2 \Omega v_r \hat{\mathbf{e}}_r$ due to the radial flow. If we multiply (5.45) by $2\pi r \, dr$, then we get an equation telling us how the angular momentum of the annulus changes. Obviously

$$\mathcal{G} \cdot 2\pi r \, dr = G(r + dr) - G(r),$$

where $G(r)$ is the viscous torque at radius r. We then have

$$\mathcal{G} = \frac{1}{2\pi r} \frac{dG}{dr}. \quad (5.46)$$

Our job now is to find the viscous torque G.

It may seem that dv_θ / dr would give the velocity shear within the disk. But

$$\frac{dv_\theta}{dr} = \frac{d}{dr}(r\Omega) = \Omega + r \frac{d\Omega}{dr},$$

has a part Ω associated with pure rotation so that only the other part gives the velocity shear. The viscous stress is then $\mu r (d\Omega/dr)$. Multiplying this by r gives the viscous torque per unit area so that

$$G(r) = \int r\, d\theta \int dz\, \mu r^2 \frac{d\Omega}{dr} = 2\pi \nu \Sigma r^3 \frac{d\Omega}{dr}, \qquad (5.47)$$

where we have used $\mu = \rho\nu$. From (5.45), (5.46) and (5.47), we finally have

$$\frac{\partial}{\partial t}(\Sigma r^2 \Omega) + \frac{1}{r}\frac{\partial}{\partial r}(\Sigma r^3 \Omega v_r) = \frac{1}{r}\frac{\partial}{\partial r}\left(\nu\Sigma r^3 \frac{d\Omega}{dr}\right). \qquad (5.48)$$

The dynamics of the accretion disk is governed by the two basic equations (5.43) and (5.48).

Let us consider an accretion disk in which the matter is moving in nearly Keplerian orbits so that (5.40) holds. We can combine (5.40), (5.43) and (5.48) to eliminate v_r, which gives after some algebra

$$\frac{\partial \Sigma}{\partial t} = \frac{3}{r}\frac{\partial}{\partial r}\left[r^{1/2}\frac{\partial}{\partial r}(\nu\Sigma r^{1/2})\right]. \qquad (5.49)$$

The time evolution of an accretion disk can be found from this equation. If ν in (5.49) is constant, then it is possible to solve (5.49) analytically by the method of separation of variables. We shall not discuss the general solution here. Let us consider an initial configuration in the form of a ring of matter at a radius $r = r_0$ given by

$$\Sigma(r, t = 0) = \frac{m}{2\pi r_0}\delta(r - r_0) \qquad (5.50)$$

such that the total mass of the ring is m. The appropriate solution of (5.49), which yields (5.50) in the limit $t \to 0$, is

$$\Sigma(x, \tau) = \frac{m}{\pi r_0^2 \tau x^{1/4}}\exp\left(-\frac{1 + x^2}{\tau}\right) I_{1/4}\left(\frac{2x}{\tau}\right), \qquad (5.51)$$

where $x = r/r_0$ and $\tau = 12\nu t/r_0^2$ are dimensionless variables and $I_{1/4}$ is a modified Bessel function (see, for example, Arfken 1985, §11.5). Figure 5.8 shows the evolution of the ring with time according to (5.51). The effect of viscosity is to spread out the ring into a disk. As time progresses, most of the mass moves inward to smaller radii. But a small amount of mass moves outward. Since there is no net torque, the total angular momentum of the disk has to remain conserved. The angular momentum per unit mass for a Keplerian orbit at radius r is Ωr^2, which goes as $r^{1/2}$. This becomes infinite as $r \to \infty$, which means that a very small amount of mass can move out to infinity carrying a large amount of angular momentum. So most of the matter spirals in losing angular momentum, whereas a negligible amount of matter carries away that angular momentum outward. Some of these properties of a viscous disk were understood well before the study of accretion disks became popular. Several scientists working on the formation of the solar system from a primordial nebula studied the basic physics of viscous disks (Jeffreys 1924; von Weizsäcker 1948; Lüst 1952).

5.7.2 Steady disk

We now consider the possibility of a steady disk, which means that we have to look for time-independent solutions of (5.43) and (5.48). Before proceeding to solve these equations, let us look at the other two components of the Navier–Stokes equation. We assume that the mass of the disk is negligible compared to the mass M of the central gravitating object. Then the r and z components of the gravitational field at a point (r, z) with $z \ll r$ are approximately $-GM/r^2$ and $-GMz/r^3$. Incorporating this gravitational field, the r and z components of the Navier–Stokes equation, as given by (C.5) and (C.7), in the steady state become

$$v_r \frac{\partial v_r}{\partial r} - \frac{v_\theta^2}{r} = -\frac{1}{\rho} \frac{\partial p}{\partial r} - \frac{GM}{r^2}, \qquad (5.52)$$

$$-\frac{1}{\rho} \frac{\partial p}{\partial z} - \frac{GMz}{r^3} = 0, \qquad (5.53)$$

where we have disregarded the viscosity terms in (5.52), because they are going to be very small compared to the other terms. If h is the

Figure 5.8 The evolution of a viscous disk with time. Adapted from Pringle (1981).

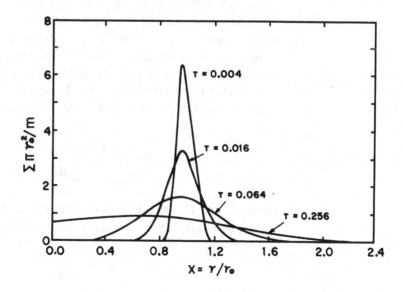

thickness of the disk, then $|\partial p/\partial z|$ can be approximated by p/h so that (5.53) implies

$$\frac{p}{\rho h} \approx \frac{GMh}{r^3},$$

from which

$$\frac{h^2}{r^2} \approx \frac{rp}{GM\rho}. \tag{5.54}$$

We are considering the disk to be thin, i.e. $h \ll r$. It follows from (5.54) that $rp/GM\rho$ has to be very small in order to make a disk thin. Approximating $|\partial p/\partial r|$ by p/r, the ratio of the two terms on the R.H.S. of (5.52) becomes

$$\frac{|(1/\rho)(\partial p/\partial r)|}{GM/r^2} \approx \frac{rp}{GM\rho} \approx \frac{h^2}{r^2}$$

from (5.54). If the disk is thin (i.e. $h \ll r$), it then follows that the pressure term in (5.52) can be neglected compared to the gravity term. In a disk with matter spiraling inward very slowly, the term involving v_r is also expected to be small so that (5.52) can be approximated as

$$\frac{v_\theta^2}{r} = \frac{GM}{r^2}, \tag{5.55}$$

from which (5.40) follows. To sum up, the assumption of nearly Keplerian motion is valid only for thin disks in which the pressure forces are negligible compared to gravity. If pressure forces are important, then the disk becomes thick and (5.52) cannot be approximated by (5.55). The theory of thick accretion disks is a much more difficult subject and will not be considered in this elementary book.

The standard model of a steady thin accretion disk was developed by Shakura and Sunyaev (1973). For a steady thin disk, the r component of the equation of motion can then be replaced by (5.40), whereas we need not bother about the z component. Then we only have to consider (5.43) and (5.48) in addition to (5.40). Putting the time derivative terms equal to zero, the equations (5.43) and (5.48) can easily be integrated to give

$$r\Sigma v_r = C_1, \tag{5.56}$$

$$\Sigma r^3 \Omega v_r - \nu \Sigma r^3 \frac{d\Omega}{dr} = C_2, \tag{5.57}$$

where C_1 and C_2 are constants of integration. For a steady disk, the mass inflow rate is $\dot{m} = -2\pi r \Sigma v_r$. It follows from (5.56) that this is a constant and

$$C_1 = -\frac{\dot{m}}{2\pi}. \tag{5.58}$$

To calculate C_2, we note that the matter at the surface of the gravitating body at $r = r_*$ must be dragged into a rigid rotation so that $d\Omega/dr = 0$ there. Then from (5.56–5.58),

$$C_2 = -\frac{\dot{m}}{2\pi}r_*^2\Omega = -\frac{\dot{m}}{2\pi}(GMr_*)^{1/2} \tag{5.59}$$

on using (5.40). On substituting this in (5.57) and making use of (5.40), (5.56) and (5.58), we finally get

$$\nu\Sigma = \frac{\dot{m}}{3\pi}\left[1 - \left(\frac{r_*}{r}\right)^{1/2}\right]. \tag{5.60}$$

This important equation shows that the mass inflow rate $\dot{m}$ and the viscosity ν depend linearly on each other. If the viscosity is larger, then there will be more mass inflow.

The gravitational potential energy lost by the matter spiralling inward is presumably converted into heat by the viscous dissipation. Hence we can find out the energy emitted from the disk from the rate of viscous dissipation. The viscous dissipation per unit volume within the accretion disk is $\mu r^2 (d\Omega/dr)^2$ (see Exercise 5.5). Integrating this over z gives the energy emitted from unit area of the disk, which is

$$-\frac{dE}{dt} = \int \mu r^2 \left(\frac{d\Omega}{dr}\right)^2 dz = \nu\Sigma r^2 \left(\frac{d\Omega}{dr}\right)^2,$$

where we have used $\mu = \rho\nu$. Substituting from (5.40) and (5.60),

$$-\frac{dE}{dt} = \frac{3GM\dot{m}}{4\pi r^3}\left[1 - \left(\frac{r_*}{r}\right)^{1/2}\right]. \tag{5.61}$$

The total energy emitted by the accretion disk is obtained by integrating this over the entire area of the disk from $r = r_*$ to infinity, i.e.

$$L = \int_{r_*}^{\infty} \left(-\frac{dE}{dt}\right) 2\pi r \, dr = \frac{GM\dot{m}}{2r_*}. \tag{5.62}$$

This is a very elegant result. Since $-GM/r_*$ is the gravitational potential at the surface of the attracting object, $GM\dot{m}/r_*$ is the rate of gravitational energy loss due to the inflow $\dot{m}$. It appears from (5.62) that half of this energy is emitted from the accretion disk. It is easy to show that the other half remains as the kinetic energy of Keplerian motion in the orbit at $r = r_*$ next to the central object. If the infalling material comes to rest at the surface of the central object, then this remaining half of the energy is dissipated in a thin boundary layer next to the object. The fact that half of the energy is emitted from the accretion disk itself is particularly significant for accretion into black holes. Once the material falls inside the event horizon of the black hole, no radiation can come out of it. Although our purely classical treatment has to be modified by general relativity, (5.61) suggests that an appreciable amount of lost gravitational energy would be radiated from the accretion disk around the black hole before the matter disappears behind the event horizon.

The theory sketched above would work beautifully if the material in the accretion disk has a value of viscosity sufficient to cause enough mass inflow as indicated by (5.60). If one uses (3.43) to estimate the viscosity in accretion disks, then it is found that the mass inflow would be completely negligible compared to what we need to explain the luminosities of X-ray binary stars or active galaxies. It was suggested by Shakura and Sunyaev (1973) that the effective viscosity is enhanced by the turbulence in the accretion disk, causing a sufficiently high mass inflow. The enhancement of transport phenomena by turbulence will be discussed in §8.4.

Exercises

5.1   Obtain the Navier–Stokes equation in the cylindrical and spherical coordinates making use of (5.13) and (5.14). Compare your results with the equations as given in Appendix C.

5.2   Consider a viscous fluid between two infinite coaxial cylinders with radii r_a and r_b. If the two cylinders are rotated with angular velocities Ω_a and Ω_b respectively, determine the velocity distribution between the cylinders.

5.3   Suppose there is a unidirectional flow $v_x(y, t = 0)$ inside an infinite viscous fluid at time $t = 0$. Show that the flow remains unidirectional and evolves in time as

$$v_x(y,t) = \frac{1}{2\sqrt{\pi v t}} \int_{-\infty}^{\infty} v_x(y', t = 0) \exp\left[-\frac{(y-y')^2}{4vt}\right] dy',$$

if there is no pressure gradient.

5.4   A ball of radius a falls inside an infinite viscous fluid in a constant gravitational field. Write down an approximate equation of motion and show that the ball eventually reaches a terminal velocity proportional to a^2, which means that a ball of smaller size falls more slowly.

5.5   Consider an *incompressible* fluid in the absence of a body force, i.e. $\mathbf{F} = 0$. Starting from (4.8) with P_{ij} given by (5.1), show that the rate of change of kinetic energy density of the fluid is

$$\frac{\partial}{\partial t}\left(\frac{1}{2}\rho v^2\right) = -\frac{\partial}{\partial x_j}\left[\rho v_j\left(\frac{1}{2}v^2 + \frac{p}{\rho}\right) + v_i\pi_{ij}\right] + \pi_{ij}\frac{\partial v_i}{\partial x_j}.$$

Argue from this that the rate of energy dissipation per unit volume is given by

$$-\pi_{ij}\frac{\partial v_i}{\partial x_j} = \frac{1}{2}\mu\left(\frac{\partial v_i}{\partial x_j} + \frac{\partial v_j}{\partial x_i}\right)^2$$

on using (5.7).

6 Gas dynamics

6.1 Thermodynamic properties of a perfect gas

Most of the problems studied in Chapters 4 and 5 did not involve considerations of compressibility, the only exception being §4.4 where the static equilibrium of compressible fluids was considered. We now wish to study the dynamics of compressible fluids. We saw in §4.7 that the irrotational flow of an incompressible fluid around an object gives rise to the Laplace equation, which is an *elliptic partial differential equation*. The similar problem of high-speed flow of a compressible fluid around an object can give rise to a *hyperbolic partial differential equation*, provided the flow speed is larger than the sound speed. In other words, the mathematical character of the equations governing high-speed compressible flows can be quite different from that of the equations governing incompressible flows (although we begin from the same hydrodynamic equations!), and consequently the solutions can also be of profoundly different nature. We give here only a brief introduction to *gas dynamics*, which is the branch of hydrodynamics dealing with compressible flows. Even the subject of supersonic flows past solid objects just mentioned above, which leads to a two-dimensional problem, is not treated in this elementary introduction. See Landau and Lifshitz (1987, Chapter XII) or Liepmann and Roshko (1957, Chapter 8) for a discussion of this subject. We restrict ourselves to a discussion of one-dimensional gas dynamics problems only.

The mathematical analysis of compressible fluids becomes more manageable if we assume the fluid to behave as a perfect gas. Hence we shall develop the concepts of gas dynamics for a perfect gas. It is straightforward—though mathematically somewhat cumbersome—to generalize these concepts for fluids not obeying perfect gas equations.

104

We begin by listing the basic thermodynamic properties of a perfect gas.

While deriving the macroscopic fluid equations from microscopic physics in Chapter 3, we obtained some relations pertaining to a perfect gas. We combine (3.2) and (3.20) to write

$$p = R\rho T, \tag{6.1}$$

where

$$R = \frac{\kappa_B}{m}. \tag{6.2}$$

This is the general equation of state for a perfect gas. The expression for the internal energy $\epsilon = (3/2)RT$ as given by (3.22), however, is an expression specific to a monatomic gas, since we assumed the energy to be entirely in the form of the kinetic energy of translation (which is true only for monatomic gases). For molecules possessing degrees of freedom other than translatory motion (such as rotation or vibration), the numerical factor is different from 3/2. We write the general expression for the internal energy of a perfect gas as

$$\epsilon = c_V T, \tag{6.3}$$

where c_V is the specific heat per unit mass. The significance of (6.3) is that we are assuming the internal energy to be a function of only one thermodynamic variable, T. It is often customary to use the quantity γ related to c_V through

$$c_V = \frac{R}{\gamma - 1} \tag{6.4}$$

rather than c_V. One can easily show that γ is the ratio of the specific heat at constant pressure to that at constant volume. For a monatomic gas, γ has to be equal to 5/3 to give the correct factor 3/2 in the expression of internal energy.

All the thermodynamic properties of a perfect gas can be obtained from (6.1) and (6.3). Let us introduce the entropy per unit mass s using the usual thermodynamic definition of entropy

$$T ds = d\epsilon + p \, d\left(\frac{1}{\rho}\right). \tag{6.5}$$

One could, of course, introduce entropy in this way for any fluid (not necessarily compressible), although thermodynamics becomes redundant for incompressible fluids as we have seen in §4.2. From (6.1), (6.3), (6.4) and (6.5), it can be easily shown that

$$s = c_V \ln\left(\frac{p}{\rho^\gamma}\right) + s_0, \tag{6.6}$$

where s_0 is a constant of integration. For an adiabatic process, dQ in (4.12) is zero, which is equivalent to

$$\frac{ds}{dt} = 0. \tag{6.7}$$

For the adiabatic evolution of a perfect gas, it follows from (6.6) that

$$\frac{d}{dt}\left(\frac{p}{\rho^\gamma}\right) = 0. \tag{6.8}$$

The enthalpy per unit mass w defined by (4.26) is easily found for a perfect gas to be

$$w = \frac{\gamma}{\gamma - 1} RT. \tag{6.9}$$

We finally note that $\int dp/\rho$ appearing in Bernoulli's principle (4.38) can be replaced by w for adiabatic flows, because

$$dw = T\,ds + \frac{dp}{\rho}.$$

Hence, for the adiabatic flow of a perfect gas, we have

$$\frac{1}{2}v^2 + \frac{\gamma}{\gamma - 1}RT + \Phi = \text{constant} \tag{6.10}$$

along a streamline.

In order to focus our attention on compressibility, we shall neglect the transport processes in most of the calculations in this chapter, i.e. we shall take $\mu = 0$, $K = 0$.

6.2 Acoustic waves

As mentioned in §4.2, compressibility in a fluid plays an important role if there are motions inside with speeds comparable to the sound speed. We now develop the theory of acoustic waves and derive an expression for the sound speed.

Let us assume a homogeneous perfect gas of density ρ_0 and pressure p_0 in the absence of any external force (i.e. we do not consider gravity). Suppose the pressure is perturbed to $p_0 + p_1(\mathbf{x}, t)$, the corresponding density perturbation being $\rho_0 + \rho_1(\mathbf{x}, t)$. We now want to find out how these perturbations will evolve. These perturbations give rise to a velocity field which we denote by $\mathbf{v}_1(\mathbf{x}, t)$ so that the subscript 1 always refers to perturbed quantities. We expect the pressure perturbation to be related to the density perturbation, i.e.

$$p_1 = c_s^2 \rho_1, \tag{6.11}$$

where

$$c_s = \sqrt{\frac{dp}{d\rho}}. \tag{6.12}$$

If the perturbations evolve in times short compared to the time for heat conduction, we may assume adiabatic conditions. In such a situation, it follows from (6.8) and (6.12) that

$$c_s = \sqrt{\frac{\gamma p_0}{\rho_0}} \tag{6.13}$$

for adiabatic perturbations.

The perturbed quantities substituted in the equation of continuity gives

$$\frac{\partial \rho_1}{\partial t} + \nabla \cdot [(\rho_0 + \rho_1)\mathbf{v}_1] = 0. \tag{6.14}$$

The term *perturbation* usually implies that the perturbed quantities are small compared to the unperturbed quantities (i.e. $\rho_1 \ll \rho_0$ and $p_1 \ll p_0$) and the quadratic terms of these quantities can be neglected. The technique of keeping only the linear terms in perturbed quantities and neglecting the higher terms is called the *linearization of perturbation equations* or the *linear perturbation technique*. We shall be using this technique again and again in the next chapter. To linearize the perturbation equation (6.14), we neglect the term involving $\rho_1 \mathbf{v}_1$ so that we are left with

$$\frac{\partial \rho_1}{\partial t} + \rho_0 \nabla \cdot \mathbf{v}_1 = 0. \tag{6.15}$$

We now apply the linear perturbation technique to the Euler equation. As mentioned in §6.1, we neglect all the transport coefficients to concentrate on the effects on compressibility. Hence we use the Euler equation rather than the Navier–Stokes equation. The effects of viscosity and heat conduction, however, have to be incorporated when we want to study the dissipation of acoustic waves. On substituting the perturbed quantities in the Euler equation (4.11) with $\mathbf{F} = 0$,

$$(\rho_0 + \rho_1)\left[\frac{\partial \mathbf{v}_1}{\partial t} + (\mathbf{v}_1 \cdot \nabla)\mathbf{v}_1\right] = -\nabla p_1. \tag{6.16}$$

Linearizing this equation and eliminating p_1 with the help of (6.11), we get

$$\rho_0 \frac{\partial \mathbf{v}_1}{\partial t} = -c_s^2 \nabla \rho_1. \tag{6.17}$$

It is now easy to combine (6.15) and (6.17) to obtain

$$\left(\frac{\partial^2}{\partial t^2} - c_s^2 \nabla^2\right) \rho_1 = 0, \tag{6.18}$$

which is the equation for acoustic waves, and c_s as given by (6.12) is the sound speed. The expression for sound speed was first obtained by Newton (1689), who assumed the perturbations associated with the sound wave to be isothermal. If T is taken to be constant, it follows from (6.1) and (6.12) that $c_s = \sqrt{p_0/\rho_0}$. For air at atmospheric pressure and 0 °C, the sound speed from this expression turns out to be 280 m s^{-1}, much lower than what is found experimentally. This discrepancy between theory and experiment remained a puzzle, until Laplace (1816) realized that one should take the perturbations to be adiabatic and the sound speed is given by (6.13). Taking $\gamma = 1.40$ appropriate for air, the sound speed at standard temperature and pressure as given by (6.13) is 332 m s^{-1}, which agrees well with the experimental value.

It may be noted that (6.12) gives the general expression for the sound speed in any fluid, since we have not used any property of the perfect gas in deriving the wave equation (6.18) (perfect gas relations are used only in evaluating $dp/d\rho$). For liquids like water which are hard to compress, a large excess pressure dp would produce only a small density increase $d\rho$. We then expect from (6.12) that the sound speed in such liquids should be much larger than what it is for more easily compressible fluids. The sound speed in water is indeed about 1430 m s^{-1}, much larger than the sound speed in air. For a 'truly incompressible' fluid, the sound speed is infinite.

Since the perturbation analysis is linear, the principle of superposition holds and any arbitrary perturbation can be decomposed into Fourier components. For a typical Fourier component,

$$\rho_1 = \rho_{1,0} \exp[i(\mathbf{k} \cdot \mathbf{x} - \omega t)], \tag{6.19}$$

substitution into (6.18) gives

$$\omega^2 = c_s^2 k^2. \tag{6.20}$$

Such a relation between ω and $\mathbf{k}$ is known as the *dispersion relation*. We shall see later that dispersion relations can be much more complicated than the simple relation (6.20) for acoustic waves in a homogeneous medium. We assume that the reader is familiar with the concepts of phase velocity and group velocity, which are given by

$$v_p = \frac{\omega}{k}, \qquad \mathbf{v}_g = \nabla_{\mathbf{k}}\omega, \tag{6.21}$$

where ∇_k is the gradient in the k space. The concept of group velocity was first introduced by Stokes (1876) in an examination paper and then developed further by Rayleigh (1881). It is easy to see that both the phase and group velocities of acoustic waves are equal to the sound speed c_s. We have the simple result that acoustic waves of all frequencies travel with the same speed in a homogeneous medium. Such waves, for which the speed does not depend on the frequency, are called *non-dispersive*. Electromagnetic waves in vacuum constitute another example of non-dispersive waves. If, instead of a uniform medium, we consider acoustic waves in a gravitationally stratified atmosphere, then the dispersion relation is more complicated and the waves turn out to be dispersive. The problem of acoustic waves in a stratified isothermal atmosphere, which was first studied by Lamb (1908), is given as Exercise 6.2.

The wave-vector k gives the direction of propagation of the acoustic wave. It follows from (6.17) that v_1 is always in the same direction as k, which implies that acoustic waves are purely *longitudinal* (i.e. waves with displacements in the direction of propagation). One finds $\nabla \cdot v$ to have the same spatio-temporal variation $\exp[i(k \cdot x - \omega t)]$ as all the other perturbed quantities, implying that the acoustic wave consists of alternate compressions and rarefactions.

6.3 Emission of acoustic waves

Any process causing a pressure fluctuation in a gas is expected to generate acoustic waves. Very often movements of foreign bodies placed in a gas happen to be the source of acoustic waves. We may cite the production of sound by a tuning fork or a vibrating membrane as examples. The motion of the gas itself may as well act as a source of sound in certain circumstances. The noise produced by a jet is an example. For many practical purposes, it is important to study the nature of acoustic radiation and the energy output in various processes emitting acoustic waves. We refer the reader to §1.4–10 of Lighthill (1978) for a clear and masterly introduction to the subject. To give a brief glimpse of the subject, we merely sketch the theory of how fluid flows can produce acoustic waves. This theory was developed in a classic paper by Lighthill (1952).

While developing the theory of acoustic waves in §6.2, we neglected the quadratic terms in velocity, since it was assumed that the only velocity in the problem, the velocity field associated with the acoustic waves, is small. If, however, there are fluid velocities which act as sources of acoustic waves, then we have to keep the quadratic terms in them and carry on the analysis in a somewhat different fashion. We have seen in §4.3 that the equation of motion can be put in the form

$$\frac{\partial}{\partial t}(\rho v_i) + \frac{\partial T_{ij}}{\partial x_j} = 0, \tag{6.22}$$

where

$$T_{ij} = p_1 \delta_{ij} + \rho v_i v_j. \tag{6.23}$$

On comparing with (4.24), it will be seen that we have replaced $p = p_0 + p_1$ by p_1 alone, since a constant p_0, as we are assuming here, does not contribute when we substitute T_{ij} in (6.22). We get the equation (6.17) of linear theory if we put $\rho = \rho_0$ in the time derivative term in (6.22) and approximate T_{ij} by $p_1 \delta_{ij} = c_s^2 \rho_1 \delta_{ij}$ neglecting the quadratic term in velocity. Now let us write

$$T_{ij} = c_s^2 \rho_1 \delta_{ij} + Q_{ij}, \tag{6.24}$$

where

$$Q_{ij} = \rho v_i v_j + (p_1 - c_s^2 \rho_1) \delta_{ij}. \tag{6.25}$$

is the departure of T_{ij} from the linear theory. On substituting (6.24) in (6.22), we get

$$\frac{\partial}{\partial t}(\rho v_i) + c_s^2 \frac{\partial \rho_1}{\partial x_i} = -\frac{\partial Q_{ij}}{\partial x_j}. \tag{6.26}$$

On differentiating this equation with respect to x_i and using the equation of continuity

$$\frac{\partial \rho_1}{\partial t} + \frac{\partial}{\partial x_i}(\rho v_i) = 0,$$

we get

$$\left(\frac{\partial^2}{\partial t^2} - c_s^2 \nabla^2\right)\rho_1 = \frac{\partial^2 Q_{ij}}{\partial x_i \partial x_j}. \tag{6.27}$$

Note that it has not been necessary to approximate ρv_i by $\rho_0 v_i$.

The equation (6.27) is the celebrated inhomogeneous wave equation which is discussed and solved in any graduate textbook on electrodynamics (see, for example, Jackson 1975, §6.4–6.6; Panofsky and Phillips 1962, §14.1–2). The solution of (6.27) is

$$\rho_1(\mathbf{x}, t) = \frac{1}{c_s^2} \int \frac{\partial^2 Q_{ij}(\mathbf{x}', t - |\mathbf{x} - \mathbf{x}'|/c_s)/\partial x_i' \partial x_j'}{4\pi|\mathbf{x} - \mathbf{x}'|} dV', \tag{6.28}$$

where $\mathbf{x}$ is the position coordinate of the field point and $\mathbf{x}'$ is the position coordinate of the source over which the integration has to be done. Suppose we have irregular fluid flows in a local region so that

Q_{ij} as given by (6.25) is non-zero only within a finite volume. Then it follows from (6.28) that ρ_1 falls as r^{-1} in regions far away from the source. We expect (6.15) to hold in the faraway regions and v_1 also should fall in the same manner for a wavelike disturbance. As in the electromagnetic theory, such an r^{-1} dependence is a typical signature of the wave field and ensures that the same energy flux passes through successively larger spheres centred around the source region.

The generation of acoustic waves by fluid motions is often known as the *Lighthill mechanism* (Lighthill 1952). For reasons we shall not explain here, the acoustic radiation generated in this fashion is of quadrupolar nature. We again refer the reader to §1.10 of Lighthill (1978) for a discussion of the multipole theory of acoustic radiation. An object changing volume inside a fluid produces monopole radiation, whereas the oscillation of an object of fixed volume gives rise to dipole radiation.

6.4 Steepening into shock waves. The method of characteristics

The analysis of §6.2 is valid only for acoustic waves with small amplitudes. When the wave amplitude is large, we are no longer justified in summarily dismissing the quadratic terms of perturbations as we have done in carrying out the linear perturbation analysis. While passing on from (6.16) to (6.17), we have neglected the nonlinear term $(v_1.\nabla)v_1$. This particular nonlinear term can have far-reaching consequences if the amplitude is not small.

When the wave amplitude is not small, it is no longer meaningful to break up all fluid variables in the unperturbed and perturbed parts as we did in §6.2. Let us go back to the Euler equation, but consider it in one dimension to make life simpler. For a purely longitudinal one-dimensional wave, we can take the x direction to be both the direction of wave propagation and the direction of fluid velocities so that the Euler equation becomes

$$\frac{\partial v}{\partial t} + v\frac{\partial v}{\partial x} = -\frac{1}{\rho}\frac{\partial p}{\partial x}. \tag{6.29}$$

We are interested in finding the effect of the term $v(\partial v/\partial x)$. To solve the full problem, we have to combine (6.29) with the other hydrodynamic equations to obtain the variations of p and ρ. How the term $v(\partial v/\partial v)$ affects the evolution of a wave can, however, be found by solving the simpler equation

$$\frac{\partial v}{\partial t} + v\frac{\partial v}{\partial x} = 0 \tag{6.30}$$

rather than the full equation (6.29). We wish to solve (6.30) to find the values of $v(x, t)$ on the x–t plane. Let us consider the curves

$$\frac{dx}{dt} = v \tag{6.31}$$

in the x–t plane. According to the properties of partial differentiation, the total time derivative of v along a curve in the x–t plane is given by

$$\frac{dv}{dt} = \frac{\partial v}{\partial t} + \frac{\partial v}{\partial x}\frac{dx}{dt}.$$

Hence (6.30) implies that

$$\frac{dv}{dt} = 0 \tag{6.32}$$

along any curve on the x–t plane satisfying (6.31).

The point to note is that we began with the *partial* differential equation (6.30). Now we have ended up with an *ordinary* differential equation (6.32) valid along some curves in the x–t plane. The curves given by (6.31) are called the *characteristic curves* of the equation (6.30). What we have presented here is a rather trivial example of a very powerful method known as the *method of characteristics* for handling a class of partial differential equations. We shall not present a general discussion of this method in this elementary textbook. It is a vast subject of great aesthetic beauty, which cannot be treated adequately in one brief section. The foundations of this subject were laid down in a profound paper by Riemann (1859). Almost the whole of the classic book by Courant and Friedrichs (1948) is devoted to solving gas dynamic problems by the method of characteristics. Let us just mention that it is a method by which one transforms partial differential equations into ordinary differential equations along certain curves in the solution space known as the characteristic curves. The partial differential equations for which such a transformation is possible are called *hyperbolic* partial differential equations. Needless to add, this method does not work for other types of partial differential equations. Once a set of hyperbolic partial differential equations has been transformed into some ordinary differential equations along the characteristic curves, we can easily integrate these ordinary differential equations along those curves. This, in a nutshell, is the basic idea behind the method of characteristics.

The integration of (6.32) gives us the result that v has to be constant on a characteristic curve, which means that the characteristic curves given by (6.31) must be straight lines in the x–t plane. When the method of characteristics is applied to solve hyperbolic partial differential equations, one often finds some quantities to be constant along characteristic curves. They are called *Riemann invariants*. For our simple problem, v is the Riemann invariant. It should also be noted that the characteristic curves in the present case correspond to the Lagrangian trajectories of fluid elements (this is not generally true for other hyperbolic partial differential equations). Hence we have the result that v associated with a fluid element does not change as the element moves. This result would at once follow from (6.30). The application of the method of characteristics happens to be an avoidable detour in the present case. But it is a powerful tool in dealing with more complex problems which cannot be handled otherwise. We have taken this detour to introduce the reader to the basic ideas of the method of characteristics which we are not able to treat fully.

Figure 6.1(a) shows a profile of $v(x)$ which we want to consider. The fluid element at p moves with a velocity $v(p)$ which is larger than the velocity $v(q)$ with which the fluid element at q moves. If, after some time, the fluid element at p moves to p' and the fluid element at q to

Figure 6.1 Three successive profiles of $v(x)$ evolving according to (6.30).

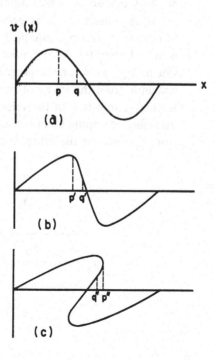

q', we expect the new positions to be closer than the old positions. Since $v(p') = v(p)$ and $v(q') = v(q)$, the new profile looks as sketched in Figure 6.1(b). In other words, the front side of the wave appears steeper and steeper. Eventually the fluid element at p is expected to overtake the fluid element at q so that the velocity profile would look as in Figure 6.1(c). This is a result which would mathematically follow from (6.30). If, however, $v(x)$ is the velocity at the position x, then it is not possible for $v(x)$ to be multi-valued and Figure 6.1(c) is clearly unphysical. Although the simplified equation (6.30) made it clear that the term $v(\partial v/\partial x)$ has the effect of steepening the waves, we have to use the full fluid equations rather than (6.30) in order to understand the dynamics of the steepened wave front which is known as a *shock*. This is done in the next section.

6.5 The structure of shock waves

Even if a wave initially has a reasonably smooth profile as in Figure 6.1(a), we have seen that it is possible for it to develop into a shock wave. A shock wave is a region of small thickness over which different fluid dynamical variables change rapidly. In order to find out the relations between the fluid dynamical variables on two sides of the shock wave, it is useful to regard the shock wave as a mathematical discontinuity across which different variables jump suddenly.

Let us consider a shock propagating into an undisturbed medium of density ρ_1 and pressure p_1. If we go to a frame in which the shock is at rest, then the density, pressure and velocity are ρ_1, p_1, v_1 on one side and ρ_2, p_2, v_2 on the other side, as shown in Figure 6.2. If we are watching from the frame of the undisturbed medium, then the shock is moving into it with the velocity $-v_1$. Under steady conditions, the mass flux, the momentum flux and the energy flux should be conserved from one side to the other. Using the expressions of the fluxes from

Figure 6.2 A shock wave seen from its frame of reference.

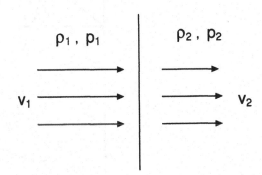

ρ_1, p_1 ρ_2, p_2

v_1 v_2

§4.3, we must have

$$\rho_1 v_1 = \rho_2 v_2, \tag{6.33}$$

$$p_1 + \rho_1 v_1^2 = p_2 + \rho_2 v_2^2, \tag{6.34}$$

$$\frac{1}{2}v_1^2 + \frac{\gamma p_1}{(\gamma - 1)\rho_1} = \frac{1}{2}v_2^2 + \frac{\gamma p_2}{(\gamma - 1)\rho_2}, \tag{6.35}$$

where we have used (6.9) to substitute for enthalpy in the energy flux equation (6.35).

We now have three equations involving six variables. If we eliminate p_2 and v_2 with the help of two equations, then, after a few steps of algebra (left as Exercise 6.3 for the reader), we end up with one equation

$$\frac{\rho_2}{\rho_1} = \frac{(\gamma + 1)\mathcal{M}^2}{2 + (\gamma - 1)\mathcal{M}^2}, \tag{6.36}$$

where

$$\mathcal{M} = \frac{v_1}{\sqrt{\gamma p_1/\rho_1}} = \frac{v_1}{c_{s,1}}. \tag{6.37}$$

From the expression of sound speed (6.13), it is clear that $c_{s,1}$ is the sound speed in the medium 1. The dimensionless quantity $\mathcal{M}$ is known as the *Mach number* and is a measure of the speed with which the shock is propagating into medium 1 in units of the sound speed in that medium. On physical grounds, we expect $\mathcal{M}$ to be larger than 1. If a shock wave were to move slower than the sound speed, then the shock front could produce ordinary acoustic waves moving with the sound speed ahead of the shock, thereby weakening the shock and making it disappear quickly. By writing (6.36) as

$$\frac{\rho_2}{\rho_1} = \frac{\gamma + 1}{(2/\mathcal{M}^2) + (\gamma - 1)},$$

we see that ρ_2/ρ_1 is larger than 1 for $\mathcal{M} > 1$ and increases with an increase in $\mathcal{M}$. This means that the medium behind a shock is always compressed and a shock involving a stronger compression has to move faster. When $\mathcal{M}$ is decreased from larger values down to 1, we note that ρ_2/ρ_1 goes to 1, implying that the shock disappears. The equation (6.36) is not valid if $\mathcal{M} < 1$. If we had eliminated p_2 and v_2 from (6.33–6.35), then we would have ended up with the equation

$$\frac{p_2}{p_1} = \frac{2\gamma \mathcal{M}^2 - (\gamma - 1)}{\gamma + 1}. \tag{6.38}$$

Equations like (6.36) and (6.38) relating the fluid variables before and after a shock are known as *Rankine–Hugoniot relations* (Rankine 1870; Hugoniot 1889).

It is useful to be in the frame of the shock wave to derive the jump conditions across the shock. If we now wish to interpret the results with respect to the frame of the undisturbed medium, we can say the following. As the shock advances, more and more material in the undisturbed medium passes through the shock and gets compressed to higher density and pressure. As more material of the undisturbed medium goes behind the shock, the position of the shock advances. The density compression has a maximum limiting value which can be found when we put $\mathcal{M} \to \infty$ in (6.36) giving

$$\frac{\rho_2}{\rho_1} \to \frac{\gamma+1}{\gamma-1}. \tag{6.39}$$

Using the value $\gamma = 5/3$ for a monatomic gas, the maximum density compression turns out to be 4. In other words, if a disturbance tries to compress such a gas by a factor of 4, that will give rise to a shock wave moving at infinite speed. Hence it is not possible to achieve a higher density compression.

Here we have discussed a shock wave in which the incoming flow is normal to the shock front. Often one encounters the situation of an *oblique shock wave*, in which the inflow is at an angle to the shock wave. Work out Exercise 6.4 to learn more about oblique shocks.

Although the jump conditions may most conveniently be derived by assuming the shock to be a mathematical discontinuity, a shock wave in reality always has a finite thickness. We are neglecting viscosity and heat conduction in our study of gas dynamics. The terms involving these transport coefficients have second spatial derivatives and become important whenever a wave becomes sufficiently steep. These transport coefficients ultimately determine the physical thickness of the shock wave by limiting the steepening process. In astrophysics, we often encounter shock waves behind which the gas becomes hot enough to radiate substantially. Hence radiative losses can become more important than other transport processes. Shu (1992, Chapter 19) gives an introduction to the important subject of radiative shocks.

6.6 Spherical blast waves. Supernova explosions

A sudden explosion within a local region in a gas often gives rise to a blast wave spreading from the centre of the explosion. A bomb exploding in air produces a blast wave. Supernova explosions are also sources of such blast waves, of which many examples are known.

Figure 6.3 shows the image of a blast wave produced by the supernova which was observed by Tycho. The blast wave appears nearly spherical. If the initial explosion was of isotropic nature and the ambient medium is uniform, then the blast wave is naturally expected to be spherically symmetric. The front of the blast is obviously a shock wave across which the Rankine–Hugoniot conditions (6.36) and (6.38) must hold. We want to find out how the front moves with time and how quantities like density or velocity vary with radius inside the sphere of the blast wave. It was independently shown by Sedov (1946, 1959) and Taylor (1950a) that this problem can be solved if we assume that the blast wave evolves in a *self-similar* fashion. This means that the evolution has to be of such nature as if some initial configuration expands uniformly. So a subsequent configuration must appear like an enlargement of a previous configuration. It appears that self-similarity is usually not a bad approximation during an important phase of a blast wave's evolution. The Sedov–Taylor self-similar solution can, therefore, be applied to many practical situations. In fact, the work of Taylor was first done as classified research during World War II to find out the effect of the atomic bomb, which was being developed at

Figure 6.3 Radio image of the blast wave in Tycho's supernova. From Tan and Gull (1985). (©Blackwell Science Ltd. Reproduced with permission from *Monthly Notices of the Royal Astronomical Society*.)

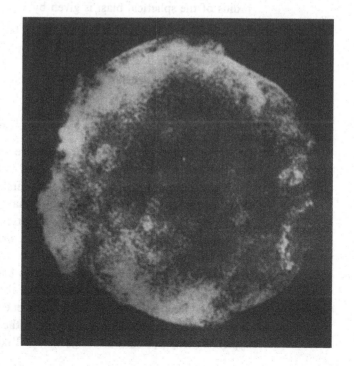

that time. Only a few years after the end of the war, the work was declassified and published openly (Taylor 1950a).

Suppose an energy E is suddenly released in an explosion producing a spherical blast wave, which progresses in an ambient medium of density ρ_1. We shall consider the pressure of the ambient medium to be negligible compared to the pressure inside the blast wave (i.e. $p_1 \approx 0$). Let λ be a scale parameter giving the size of the blast wave at time t after the explosion. Apart from being a monotonically increasing function of time, the evolution of λ may depend on E and ρ_1. Now the only way of combining t, E and ρ_1 to get a dimension of length is

$$\lambda = (Et^2/\rho_1)^{1/5}. \tag{6.40}$$

Let $r(t)$ be the radius of a shell of gas inside the spherical blast. For a self-similar expansion, $r(t)$ has to evolve in the same way as λ. So we can introduce a dimensionless distance parameter

$$\xi = \frac{r}{\lambda} = r\left(\frac{\rho_1}{Et^2}\right)^{1/5} \tag{6.41}$$

such that the value of this parameter for a particular shell of gas does not change with time. Each shell of gas can therefore be labelled by a particular value of ξ. Let ξ_0 correspond to the shock front so that the radius of the spherical blast is given by

$$r_S(t) = \xi_0 \left(\frac{Et^2}{\rho_1}\right)^{1/5}. \tag{6.42}$$

The velocity of expansion is

$$v_S(t) = \frac{dr_S}{dt} = \frac{2}{5}\frac{r_S}{t} = \frac{2}{5}\xi_0 \left(\frac{E}{\rho_1 t^3}\right)^{1/5}. \tag{6.43}$$

The size of the spherical blast therefore increases as $t^{2/5}$ and the velocity of the front goes down as $t^{-3/5}$. A striking confirmation of the $t^{2/5}$ dependence of size comes from the first test atomic explosion in New Mexico in 1945. From a movie of that explosion, Taylor (1950a) estimated the sizes of the blast wave at various times. Figure 6.4 shows the variation of size with time, where the straight line corresponds to the $t^{2/5}$ dependence. This plot provides a convincing proof that self-similarity was a good assumption for this blast wave.

A typical supernova explosion ejects about $1 M_\odot$ with an initial velocity of about 10^4 km s^{-1} so that the energy input can be taken as $E \approx 10^{51}$ erg. The blast wave expands in the interstellar medium with typical density $\rho_1 \approx 2 \times 10^{-24}$ gm cm^{-3}. Taking ξ_0 of order unity, we get from (6.42) and (6.43) that

$$r_S(t) \approx 0.3\, t^{2/5} \quad \text{pc}, \tag{6.44}$$

$$v_S(t) \approx 10^5\, t^{-3/5} \quad \text{km s}^{-1}, \tag{6.45}$$

where t has to be put in years. We note from (6.45) that the velocity becomes about 10^4 km s^{-1} after 100 years, the value of velocity blowing up at very early times. If the initial velocity of ejection was 10^4 km s^{-1}, then it does not make sense to apply (6.45) for $t < 100$ years. Our analysis tacitly assumes that the energy of explosion remains within the blast wave. In reality, however, the energy is continously radiated away from the system. Careful calculations of the cooling rate indicate that the system loses an appreciable amount of energy in 10^5 years. Hence we do not expect (6.44) and (6.45) to hold beyond that time. These equations presumabaly describe the evolution of supernova during only a limited period, which is perhaps the most interesting phase in the evolution of the system.

Let us now calculate the structure of the blast wave. Suppose we are in the frame in which the explosion took place and the shock front

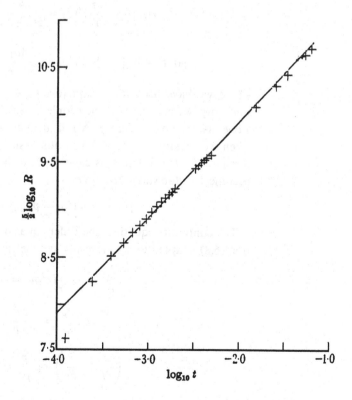

Figure 6.4 Logarithmic plot showing that $r^{5/2}$ was proportional to t for the atomic explosion in New Mexico in 1945. From Taylor (1950a). (©The Royal Society. Reproduced with permission from *Proceedings of the Royal Society*.)

propagates with velocity v_S in that frame. Let ρ_2, v_2 and p_2 be the density, velocity and pressure just inside the shock front. If we were in the frame of the shock, then material would appear to stream in with velocity $-v_S$ and pass out with velocity $-v_S + v_2$. We note that these velocities correspond to v_1 and v_2 as defined in §6.5. Then, in the limit of strong shock ($\mathcal{M} \to \infty$), (6.33) and (6.39) give

$$\rho_2 = \left(\frac{\gamma+1}{\gamma-1}\right)\rho_1, \tag{6.46}$$

$$v_2 = \frac{2}{(\gamma+1)}v_S. \tag{6.47}$$

In the same limit, it follows from (6.37) and (6.38) that

$$p_2 = \frac{2}{(\gamma+1)}\rho_1 v_S^2. \tag{6.48}$$

Let us now introduce the dimensionless variables $\rho'(\xi)$, $v'(\xi)$ and $p'(\xi)$ in the following manner:

$$\rho(r,t) = \rho_2\rho'(\xi) = \rho_1\frac{\gamma+1}{\gamma-1}\rho'(\xi), \tag{6.49}$$

$$v(r,t) = v_2\frac{r}{r_S}v'(\xi) = \frac{4}{5(\gamma+1)}\frac{r}{t}v'(\xi), \tag{6.50}$$

$$p(r,t) = p_2\left(\frac{r}{r_S}\right)^2 p'(\xi) = \frac{8\rho_1}{25(\gamma+1)}\left(\frac{r}{t}\right)^2 p'(\xi), \tag{6.51}$$

where we have used (6.43) and (6.46–6.48). On the ground of self-similarity, we expect the dimensionless variables to be functions of ξ alone, as we have indicated. A full justification, however, comes only when we substitute (6.49–6.51) in our basic equations and everything turns out be consistent. It is clear that all the dimensionless variables become 1 at the shock front, i.e.

$$\rho'(\xi_0) = v'(\xi_0) = p'(\xi_0) = 1. \tag{6.52}$$

The continuity equation, the Euler equation and the adiabatic equation (6.8) adapted to a spherically symmetric adiabatic gas flow are

$$\frac{\partial\rho}{\partial t} + \frac{1}{r^2}\frac{\partial}{\partial r}(r^2\rho v) = 0, \tag{6.53}$$

$$\frac{\partial v}{\partial t} + v\frac{\partial v}{\partial r} = -\frac{1}{\rho}\frac{\partial p}{\partial r}, \tag{6.54}$$

$$\left(\frac{\partial}{\partial t} + v\frac{\partial}{\partial r}\right)\log\frac{p}{\rho^\gamma} = 0. \tag{6.55}$$

We now substitute (6.49–6.51) into (6.53–6.55), noting that

$$\frac{\partial}{\partial t} = -\frac{2}{5}\frac{\xi}{t}\frac{d}{d\xi}, \qquad \frac{\partial}{\partial r} = \frac{\xi}{r}\frac{d}{d\xi}.$$

from (6.41). We get after some straightforward algebra

$$-\xi\frac{d\rho'}{d\xi} + \frac{2}{\gamma+1}\left[3\rho'v' + \xi\frac{d}{d\xi}(\rho'v')\right] = 0, \tag{6.56}$$

$$-v' - \frac{2}{5}\xi\frac{dv'}{d\xi} + \frac{4}{5(\gamma+1)}\left(v'^2 + v'\xi\frac{dv'}{d\xi}\right) = -\frac{2(\gamma-1)}{5(\gamma+1)}\frac{1}{\rho'}\left(2p' + \xi\frac{dp'}{d\xi}\right), \tag{6.57}$$

$$\xi\frac{d}{d\xi}\left(\log\frac{p'}{\rho'^\gamma}\right) = \frac{5(\gamma+1) - 4v'}{2v' - (\gamma+1)}. \tag{6.58}$$

It is to be noted that r and t cancel out from these equations, leaving ξ as the only independent variable. This establishes that self-similar solutions are possible in terms of the dimensionless variables introduced in (6.49–6.51). Our job now is to solve the coupled nonlinear ordinary differential equations (6.56–6.58) with the boundary condition (6.52) to be satisfied at ξ_0. The value of ξ_0 can be obtained from the condition that the total energy of the blast wave (i.e. the sum of kinetic and thermal energies) remains constant, i.e.

$$E = \int_0^{r_s}\left(\frac{\rho v^2}{2} + \frac{p}{\gamma-1}\right)4\pi r^2\,dr,$$

Figure 6.5 The variation of ρ, p and v behind a self-similar blast wave with $\gamma = 1.40$ obtained by solving (6.56–6.58).

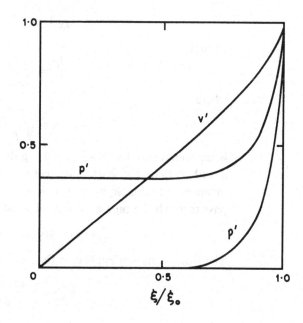

ξ/ξ_0

since the thermal energy density is $c_V \rho T = p/(\gamma - 1)$ by virtue of (6.4). In terms of dimensionless variables, this equation becomes

$$\frac{32\pi}{25(\gamma^2 - 1)} \int_0^{\xi_0} [p' + \rho' v'^2] \xi^4 \, d\xi = 1, \tag{6.59}$$

from which ξ_0 can be found.

One can now find the solution for a given γ numerically, although Sedov (1946) managed to find a general analytical solution. This solution, which involves a rather complicated expression, is given in Landau and Lifshitz (1987, §106). The solution for $\gamma = 1.40$, which is appropriate for air, is shown in Figure 6.5.

6.7 One-dimensional gas flow. Extragalactic jets

We have so far considered disturbances propagating in compressible gases. We now come to the problem of the flow of the compressible gas itself. The simplest flow problem is the one-dimensional flow of a gas through a pipe with varying cross-section. Let us choose the x axis along the axis of the pipe. If the cross-section $A(x)$ of the pipe varies sufficiently slowly, then we can neglect the transverse component of gas velocity and the problem can be regarded as one-dimensional.

Let us consider a *steady, adiabatic* gas flow through such a pipe, i.e. we assume the flow not to vary in time and satisfy the adiabatic relation (6.8), which implies here

$$\frac{d}{dx}\left(\frac{p}{\rho^\gamma}\right) = 0$$

so that

$$\frac{dp}{dx} = c_s^2 \frac{d\rho}{dx}, \tag{6.60}$$

where

$$c_s = \sqrt{\frac{\gamma p}{\rho}} \tag{6.61}$$

is the sound speed and we start using the ordinary derivative notation d rather than the partial derivative notation ∂, as x is the only independent variable in the problem. Since the same mass flux has to pass through the pipe at any x, we must have

$$\rho(x)v(x)A(x) = \text{constant.} \tag{6.62}$$

The Euler equation applied to this problem gives us

$$v\frac{dv}{dx} = -\frac{c_s^2}{\rho}\frac{d\rho}{dx}, \tag{6.63}$$

where we have made use of (6.60). Differentiating (6.62), we obtain

$$\frac{1}{\rho}\frac{d\rho}{dx} + \frac{1}{v}\frac{dv}{dx} + \frac{1}{A}\frac{dA}{dx} = 0. \tag{6.64}$$

Eliminating $(1/\rho)(d\rho/dx)$ with the help of (6.63), we get a very important equation

$$(1 - \mathcal{M}^2)\frac{1}{v}\frac{dv}{dx} = -\frac{1}{A}\frac{dA}{dx} \tag{6.65}$$

containing a great deal of physics. Here

$$\mathcal{M} = \frac{v}{c_s}$$

is the local Mach number at any point along the pipe. When the flow is subsonic (i.e. $\mathcal{M} < 1$), dv/dx and dA/dx in (6.65) have opposite signs. So a narrowing of the pipe will make the flow faster and vice versa. This is what we expect from common sense. The situation, however, becomes rather counter-intuitive if the flow is supersonic (i.e. $\mathcal{M} > 1$), when dv/dx and dA/dx in (6.65) have the same sign. This means that a region of increasing cross-section will accelerate the flow. This goes against our intuition and clearly shows that a supersonic flow can be very different from a subsonic flow. Suppose we wish to arrange things such that a subsonic flow enters from one side of a pipe, accelerates as it flows along the pipe and comes out as a supersonic flow from the other side of the pipe. To achieve this, we must have $\mathcal{M} = 1$ at some place inside the pipe. We see from (6.65) that dA/dx must be zero there. Hence an acceleration from a subsonic flow to a supersonic flow is possible only if the pipe has a shape as shown in Figure 6.6 with a throat where the variation of area is zero. Such a pipe is called a *de Laval nozzle*.

In order for a flow to be accelerated from subsonic to supersonic speeds, the pressure boundary conditions at the two sides have to be

Figure 6.6 A de Laval nozzle.

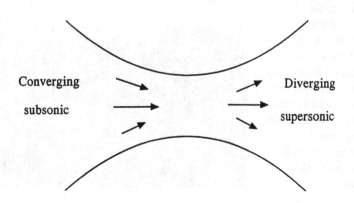

Converging

subsonic

Diverging

supersonic

adjusted suitably. We shall not get into a discussion of the boundary conditions here. The readers are referred to Liepmann and Roshko (1957, Chapter 5) for a full treatment of the problem. We merely point out that (6.65) allows the possibility of a flow which is converging subsonic on the left side of the throat and diverging supersonic on the right side, as shown in Figure 6.6. It follows from (6.65) that such a flow is continuously accelerated as it moves along the nozzle with $\mathcal{M} = 1$ at the throat where the flow makes a transition from subsonic to supersonic. Unless the pressure boundary conditions on the two sides are what are needed for the flow to be smooth and continuous everywhere, shocks may be produced. The appearance of shocks is a quite common feature in supersonic flows. Very often one needs fairly particular boundary conditions for the supersonic flow velocity to be smooth in all space. For other types of boundary conditions, shocks appear in the flow. Figure 5-4 in Liepmann and Roshko (1957) shows the production of shocks in a nozzle flow.

With the help of powerful radio telescopes, astronomers have discovered that many galaxies squirt out vast jets of gas with sizes much larger than the sizes of the parent galaxies. These jets usually emit radio waves. So they can be observed through the radio telescopes, but not through optical telescopes (M87 is the only galaxy having an optical jet). Figure 6.7 shows an image of a jet produced with a radio telescope. Some of the radio-emitting jets are of the order of megaparsecs in size, which makes them by far the largest coherent fluid flow patterns in the Universe. There is evidence that many of these jets are highly supersonic. Hence, by virtue of (6.65), as the jet moves out of the galaxy to regions of lower pressure and the cross-section increases, the gas in the jet is further accelerated. If the gas in the jet first started with a subsonic speed in the interior of the galaxy, then we expect the jet to pass through a de Laval nozzle, which was

Figure 6.7 Radio image of the extragalactic jet Hercules A produced with the radio telescope VLA (Very Large Array). From Dreher and Feigelson (1984). (©Macmillan Magazines Ltd. Reproduced with permission from *Nature*.)

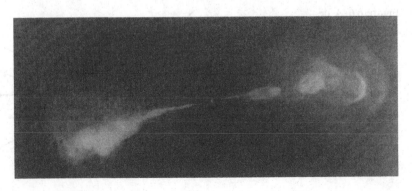

first suggested in a classic paper by Blandford and Rees (1974). They suggested that the light gas, which makes up the jet, is produced by some mechanism near the centre of the galaxy. Then this light gas tries to push its way through the heavier interstellar medium around it and forces two exhausts on the two sides of the galaxy where the thickness of the interstellar medium is least (see Figure 6.8). If the exhausts have structures like the de Laval nozzle, then the jet can be accelerated to supersonic speeds while passing through the exhausts. Although this model played a very influential role in the theoretical understanding of extragalactic jets, it has some limitations. Through VLBI (**V**ery **L**arge **B**aseline **I**nterferometry—a combination of radio telescopes in different continents used together to produce high-resolution images by interferometric techniques) observations, it is found that some jets form within a few parsecs of the centre of the galaxy. On the basis of what we know about the interstellar medium in galaxies, the Blandford–Rees model would predict the nozzle to form at a much larger distance. Since many of the extragalactic jets contain magnetic fields, it is now believed that the magnetic fields probably play an important role in the acceleration and collimation of these jets. More sophisticated models of magnetically driven jets are now being worked out by many astrophysicists. A brief introduction to this subject can be found in §14.9.3.

6.8 Spherical accretion and winds

We saw in §4.4 that the corona of a star can be in static equilibrium only if there is some finite pressure at infinity to stop it from expanding. If the pressure at infinity is less, then there will be an outward flow.

Figure 6.8 The twin-exhaust model of Blandford and Rees (1974).

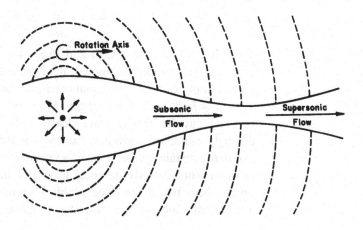

On the other hand, if the pressure at infinity is more than what is needed to maintain static equilibrium, then there should be an inward flow. Parker (1958) was the first person to predict a wind from the Sun and worked out a spherically symmetric model for it. The first spherically symmetric inward accretion flow model was developed by Bondi (1952). Since spherical wind and spherical accretion are mathematically very similar problems, we present them together. It should be borne in mind that the problem of spherical accretion is somewhat of an artificial problem of mainly academic interest. When a gravitating object accretes matter from the space around it, most often the accreting matter has a non-isotropic distribution around the gravitating centre and possesses angular momentum leading to the formation of an accretion disk (discussed in §5.7). Although the solar wind is found to be somewhat non-isotropic (mainly due to the presence of magnetic fields in the solar atmosphere) and the same is expected for the winds from other stars, a spherical wind model probably has more connections with reality than a spherical accretion model.

Let us consider a steady spherical flow such that the velocity v is independent of time and is in the radial direction (either inward or outward). Under steady conditions, the same mass flux has to flow through spherical surfaces at different distances. Hence

$$r^2 \rho v = \text{constant},$$

from which

$$\frac{2}{r} + \frac{1}{\rho}\frac{d\rho}{dr} + \frac{1}{v}\frac{dv}{dr} = 0. \tag{6.66}$$

Assuming the gravitational field to be produced by a central mass M, the Euler equation gives

$$\rho v \frac{dv}{dr} = -\frac{dp}{dr} - \frac{GM}{r^2}\rho. \tag{6.67}$$

We now have to solve (6.66) and (6.67) with an appropriate energy equation. The problem is simplest if we replace the energy equation by the assumption of an isothermal condition, i.e. if the pressure is taken to be $p = R\rho T$ with T a constant. Here we discuss this simplest case to bring out the physics more clearly. Solving the equations for an adiabatic condition is given as Exercise 6.6. It is to be noted that (6.66) and (6.67) remain invariant when we make the transformation $v \to -v$. In other words, steady spherical wind and steady spherical accretion are mathematically the same problem. Once we have a solution for a spherical wind, we can immediately get a solution for a spherical accretion by reversing the velocity at all the points. This symmetry between spherical accretion and wind holds only in the steady state. Time-dependent spherical accretion and wind problems are no longer symmetric.

We write $p = v_c^2 \rho$ in (6.67) where $v_c^2 = RT$ is taken as a constant. Note that v_c is the isothermal sound speed obtained from (6.12). We now eliminate ρ between (6.66) and (6.67), which gives

$$\left(v - \frac{v_c^2}{v}\right) \frac{dv}{dr} = \frac{2v_c^2}{r} - \frac{GM}{r^2}. \tag{6.68}.$$

It is possible for v to be equal to v_c only at the distance

$$r = r_c = \frac{GM}{2v_c^2}$$

so that both sides of (6.68) are zero simultaneously. We can easily integrate (6.68) to obtain

$$\left(\frac{v}{v_c}\right)^2 - \log\left(\frac{v}{v_c}\right)^2 = 4\log\frac{r}{r_c} + \frac{2GM}{rv_c^2} + C, \tag{6.69}$$

where C is the constant of integration. The solutions for different values of C are shown in Figure 6.9. It turns out that the solutions of

Figure 6.9 Different solutions $v(r)$ as given by (6.69).

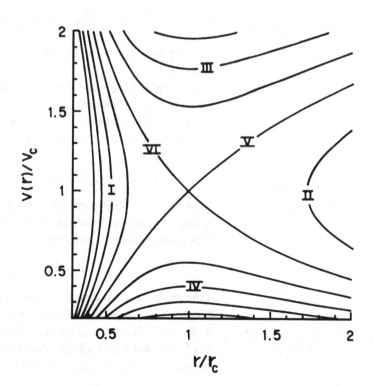

types I and II are double-valued and hence unphysical. The solutions of type III are supersonic everywhere, whereas solutions of type IV are subsonic everywhere (here we are using the words 'subsonic' and 'supersonic' with respect to the isothermal sound speed v_c). Only solutions of types V and VI pass through the critical point $r = r_c$, $v = v_c$, and are subsonic and supersonic in different regions. We easily find from (6.69) that $C = -3$ for these solutions. Which particular solution is realized in a given situation again depends on the boundary conditions. Parker (1958) considered the solar wind to start from subsonic speeds near the solar surface and then to accelerate to high speeds. Hence the solution V is appropriate for Parker's problem. On the other hand, Bondi (1952) studied an inflow starting from small speeds at infinity and becoming faster in the interior. So VI corresponds to the solution for Bondi's problem.

Exercises

6.1   From the adiabatic condition (6.8) for a gas, show that the linearized equation for adiabatic perturbations around a static configuration is

$$\frac{\partial p_1}{\partial t} + \mathbf{v}_1 \cdot \nabla p_0 = c_s^2 \left(\frac{\partial \rho_1}{\partial t} + \mathbf{v}_1 \cdot \nabla \rho_0 \right),$$

where c_s^2 is given by (6.13) and the other symbols have usual meanings.

6.2   First show that the unperturbed pressure p_0 and density ρ_0 in an isothermal atmosphere with constant gravity $-g\hat{\mathbf{e}}_z$ fall off as $\exp(-z/\Lambda)$, where $\Lambda = RT/g$ (R being the gas constant appearing in (6.1)). Consider velocity perturbations of the form

$$\exp(z/2\Lambda) \exp[i(kz - \omega t)]\hat{\mathbf{e}}_z.$$

Show that the kinetic energy density of this moving perturbation remains conserved as it propagates upward and the dispersion relation is given by

$$\omega^2 = k^2 c_s^2 + \frac{g\gamma}{4\Lambda},$$

if the perturbation is assumed adiabatic. You may take $\partial/\partial x = \partial/\partial y = 0$ for all perturbed quantities.

6.3   Begin with the jump conditions (6.33–6.35) across a plane shock and work out all the algebra to derive the Rankine–Hugoniot relations (6.36) and (6.38).

6.4 Let us consider an *oblique shock wave*, i.e. a shock such that the gas approaching from the forward direction is inclined to the shock front at an angle θ. Let the gas in the backward direction leave at an angle χ with respect to the velocity in the forward direction. Apart from the equality of the mass, momentum and energy fluxes on the two sides of the shock, assume that the tangential component of velocity is continuous across the shock. Show that

$$\cot \chi = \tan \theta \left[\frac{(\gamma + 1)\mathcal{M}^2}{2(\mathcal{M}^2 \sin^2 \theta - 1)} - 1 \right].$$

6.5 The jet coming out of a galaxy pushes the surrounding gas away and creates a channel through which the light gas inside the jet flows. The pressure inside the jet has to be equal to the pressure outside. Assuming the flow inside the jet to be one-dimensional and adiabatic, show that the local Mach number at a place of pressure p is given by

$$\mathcal{M}^2 = \frac{2}{\gamma - 1} \left[\left(\frac{p_0}{p} \right)^{(\gamma-1)/\gamma} - 1 \right],$$

where p_0 is the *stagnation pressure*, i.e. the pressure the gas would have at a point if the velocity were zero there. Show that the cross-section A of the jet channel varies with the pressure p as

$$A \propto \left(\frac{p_0}{p} \right)^{(1/2)(1+1/\gamma)} \left[\left(\frac{p_0}{p} \right)^{(\gamma-1)/\gamma} - 1 \right]^{-1/2}.$$

6.6 In §6.8 we obtained the solution for spherical accretion/wind assuming isothermal conditions. Work out the solution if the flow is adiabatic.

7 Linear theory of waves and instabilities

7.1 The philosophy of perturbation analysis

Consider a rock on the top of a hill (Figure 7.1(a)) and a rock at the bottom of a valley (Figure 7.1(b)). If the rock on the top of the hill has its centre of gravity vertically above the point of contact P, then it can easily be shown that all the forces are in balance and hence a static equilibrium is, in principle, possible. We, however, know that it is very unlikely for such an equilibrium configuration to last for long, since a slight disturbance will make the rock roll down the hill. On the other hand, the rock in the valley is in a stable equilibrium configuration. If it is disturbed from its position of rest, then it will come back there again. If the friction between the rock and the surface of the valley can be made sufficiently small, then the disturbed rock may even overshoot while trying to return to its position of rest, thereby giving rise to oscillations.

In the case of the rocks in Figures 7.1(a) and 7.1(b), mere inspection gives us an idea whether the configurations are stable or whether waves

Figure 7.1 Rocks in (a) unstable and (b) stable equilibrium positions.

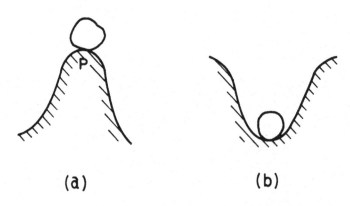

(a)   (b)

can be excited around the configurations. For more complicated situations, one has to carry out a detailed analysis to answer such questions. Consider a fluid configuration satisfying the equations of steady state (i.e. hydrodynamic equations with $\partial/\partial t = 0$ everywhere). We are likely to encounter such a configuration in Nature only if it is stable against arbitrary perturbations. If we find that small perturbations present in the configuration start becoming larger with time, then it will be difficult to realize this configuration in Nature. If the configuration is stable, then there exists the possibility that the perturbations around the configuration may give rise to oscillations or waves. Hence, merely obtaining a steady state solution of the hydrodynamic equations does not guarantee that it can serve as a model for some realistic situation. One has to consider perturbations around the configuration and study their evolution in time to find out if the configuration is stable. If it is stable, then the study of the perturbations will also tell us whether waves are possible around the configuration.

We have already presented an example of a perturbation analysis in §6.2 in our derivation of acoustic waves. There we discussed the technique of linearization of the perturbation equations by neglecting the nonlinear terms in the perturbed quantities. If the perturbations are small, then one is justified in this procedure. The advantage of linearization is that the *principle of superposition*, which holds only for linear equations and hence cannot in general be applied to hydrodynamic equations (because they are nonlinear), becomes applicable for the perturbed quantities. If the growth of perturbations is described by a set of linear equations, then any arbitrary perturbation can be thought of as a superposition of many Fourier components and the evolution of each mode can be studied in isolation by substituting in the perturbation equations. This is the reason why we were justified in trying out a Fourier component of the form (6.19) in our analysis of acoustic waves.

Since the same mathematical techniques are used in the linear theory of waves and the linear theory of instabilities, we plan to discuss them together and present some important examples of waves and instabilities. We shall again and again find that a configuration can be stable or unstable depending on the value of some particular quantity. For example, we may consider a liquid in a vessel being heated from the bottom. We know from everyday experience that convective motions start in the liquid almost suddenly after it has been heated sufficiently (heat a pan of water in your kitchen if you have not watched it closely before!) When the temperature gradient inside the liquid is small, it is in a stable hydrostatic equilibrium and heat is transported from the bottom to the top by conduction. When, however, the temperature gradient becomes larger than a critical value, the liquid becomes unstable to convective motions which then transport the heat. We shall present an analysis of this problem in §7.3. By increasing the temperature gradient, we achieve a transition from a stable to an unstable configuration. In the jargon of applied mathematics, such a transition is often called a *bifurcation*—a situation in which the character of a system is changed by varying some parameter.

We shall now study a few specific systems to find out the conditions under which they are stable or sustain waves. The linear perturbation analysis is *usually* adequate to find out these conditions to a fair degree of reliability. Needless to say, the linear analysis becomes inadequate when the perturbations are large. For example, if a system is unstable, the linear theory just tells us how the perturbations will initially grow if they are very small in the beginning. Eventually the perturbations will grow to large amplitudes and the linear theory will cease to hold. Since some perturbations would be present in any physical system, if the nature of the system is such that small perturbations start growing, then we may finally end up with fluid motions which are irregular in space and time—usually referred to as *turbulence*. Thus there is a connection between hydrodynamic instability and turbulence, but one has to go beyond linear theory to explore this connection. Chapter 8 will present an elementary summary of some of the things (very few!) we understand about turbulence.

In principle, any solution of hydrodynamic equations should be checked for stability before we can be sure that it corresponds to a situation which can be realized in Nature. In practice, however, it is possible to carry out the programme of linear perturbation analysis only for sufficiently simple situations. After the study of several such situations, one develops some intuition for deciding when a configuration may be stable or when it may sustain waves. We hope that this chapter will help the reader in developing a certain amount of intuition.

7.2 Convective instability and internal gravity waves

We begin our investigation of specific waves and instabilities by looking at a problem in which the stability condition can be obtained by fairly simple arguments, without doing a perturbation analysis of the full hydrodynamic equations. This happens to be a problem of great astrophysical importance and also a problem which illustrates beautifully the connection between waves and instabilities.

Suppose we have a perfect gas in hydrostatic equilibrium in a uniform gravitational field. If the z axis is so chosen that the gravity is in the negative z direction, then we expect $\rho(z)$ and $p(z)$ to decrease with z. We assume that a blob of gas has been displaced vertically upward as shown in Figure 7.2. Initially the blob of gas had the same density ρ and the same pressure p as the surroundings. The external gas density and pressure at the new position of the blob are ρ' and p'. We know that pressure imbalances in a gas are rather quickly removed by acoustic waves, but heat exchange between different parts of the gas takes more time. Hence it is not unreasonable to consider the blob to have been displaced *adiabatically* and to have the same pressure p' as the surroundings. Let ρ^* be its density in the new position. If $\rho^* < \rho'$, then the displaced blob will be buoyant and will continue to move upward further away from its initial position, making the system unstable. On the other hand, if $\rho^* > \rho'$, then the blob will try to return to its original position so that the system will be stable. To find the condition for stability, we have to determine whether ρ^* is greater than or less than ρ'.

From the assumption that the blob has been displaced adiabatically,

Figure 7.2 Vertical displacement of a blob of gas in a stratified atmosphere.

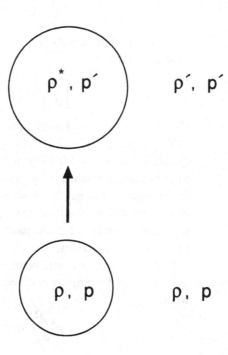

it follows that

$$\rho^* = \rho \left(\frac{p'}{p}\right)^{1/\gamma}. \tag{7.1}$$

If dp/dz is the pressure gradient in the atmosphere, we can substitute

$$p' = p + \frac{dp}{dz}\Delta z$$

and make a binomial expansion keeping terms to the linear order in Δz. This gives

$$\rho^* = \rho + \frac{\rho}{\gamma p}\frac{dp}{dz}\Delta z. \tag{7.2}$$

On the other hand,

$$\rho' = \rho + \frac{d\rho}{dz}\Delta z.$$

Using the fact that $\rho = p/RT$, we get

$$\rho' = \rho + \frac{\rho}{p}\frac{dp}{dz}\Delta z - \frac{\rho}{T}\frac{dT}{dz}\Delta z. \tag{7.3}$$

Here $d\rho/dz$ and dT/dz are the density and temperature gradients in the atmosphere. From (7.2) and (7.3),

$$\rho^* - \rho' = \left[-\left(1 - \frac{1}{\gamma}\right)\frac{\rho}{p}\frac{dp}{dz} + \frac{\rho}{T}\frac{dT}{dz}\right]\Delta z. \tag{7.4}$$

Keeping in mind that dT/dz and dp/dz are both negative, the atmosphere is stable if

$$\left|\frac{dT}{dz}\right| < \left(1 - \frac{1}{\gamma}\right)\frac{T}{p}\left|\frac{dp}{dz}\right|. \tag{7.5}$$

This is the famous *Schwarzschild stability condition* (Schwarzschild 1906). If the temperature gradient of the atmosphere is steeper than the critical value $(1 - 1/\gamma)(p/T)|dp/dz|$, then the atmosphere is unstable to convection, which mainly involves up-and-down displacements as in the case of the gas blob we have considered. While constructing models of stellar structure, the Schwarzschild condition is employed routinely to check if certain parts in the interior of a star are subject to convective instability.

Since the force per unit volume acting inside the displaced blob is $(\rho^* - \rho')(-g)$, the approximate equation of motion of the blob can be written as

$$\rho^* \frac{d^2}{dz^2}\Delta z = -(\rho^* - \rho')g,$$

provided we neglect the fact that any motion of the blob induces motions in the surrounding gas. Substituting from (7.4),

$$\frac{d^2}{dz^2} \Delta z + N^2 \Delta z = 0, \tag{7.6}$$

where

$$N = \sqrt{\frac{g}{T} \left[\frac{dT}{dz} - \left(1 - \frac{1}{\gamma} \right) \frac{T}{p} \frac{dp}{dz} \right]}. \tag{7.7}$$

If the stability condition (7.5) is satisfied, then it is easy to see that N is a real number with the dimension of frequency and is known as the *Väisälä–Brunt frequency* (Väisälä 1925; Brunt 1927). It follows from (7.6) that a blob of gas displaced vertically in a stably stratified atmosphere will have an oscillatory motion. It is, of course, not possible for a blob of gas to have oscillatory motions without disturbing the surrounding atmosphere. In reality, disturbances in stably stratified atmospheres give rise to wave motions known as *internal gravity waves*. The adjective *internal* is used to distinguish such waves from the *surface gravity waves*, which can take place on surfaces of liquids like the air–water interface and will be discussed in §7.4. Our simple analysis merely indicates the possible existence of internal gravity waves. In order to study them properly, one has to carry out a perturbation analysis of the full hydrodynamic equations. In our simplified analysis, we assumed the displaced blob to remain in pressure equilibrium with the surroundings, because acoustic waves quickly establish pressure equilibrium. When full hydrodynamic equations are perturbed, it is no longer necessary to use such an assumption, and one can develop a combined theory of acoustic and internal gravity waves. We refer the reader to the textbook by Lighthill (1978, Chapter 4) for a full treatment of this subject.

7.3 Rayleigh–Bénard convection

Although the condition (7.5) for stability against convection in a gas could be derived fairly easily, the derivation of the similar stability condition in the case of a nearly incompressible liquid like water is more complicated. The theoretical analysis was carried out in a landmark paper by Rayleigh (1916), whereas Bénard (1900) performed careful experimental studies. In the previous section, we have not done a perturbation analysis of the full hydrodynamic equations. Such an analysis following the work of Rayleigh (1916) is presented now for a nearly incompressible liquid. We have to consider first the equilibrium state of a liquid heated from below. Then we shall introduce perturbations to see if the equilibrium state is stable.

Let us choose the z axis in the upward vertical direction, and let $z = 0$, $z = d$ be the bottom and the top of the liquid heated from below. We assume the liquid to be incompressible, which means that its density does not change on the application of pressure. It is, however, necessary to take into account that the density of a liquid decreases on raising the temperature. A liquid heated from below becomes unstable to convection because the hotter liquid at the bottom becomes lighter and wants to come on top of the colder liquid above. The change in internal energy per unit mass for an incompressible liquid is $d\epsilon = c_p dT$, where c_p is the specific heat. Substituting this in (3.53) and assuming the thermal conductivity K to be a constant, we obtain from (3.53) that

$$\frac{\partial T}{\partial t} + \mathbf{v} \cdot \nabla T = \kappa \nabla^2 T, \tag{7.8}$$

where

$$\kappa = \frac{K}{\rho c_p} \tag{7.9}$$

is called the *thermometric conductivity* and $\nabla \cdot \mathbf{v}$ has been neglected because of incompressibility.

If T_b and T_t are the temperatures at the bottom and the top of the liquid, then the equilibrium solution (i.e. the solution with $\partial/\partial t = 0, \mathbf{v} = 0$) of (7.8) is

$$T_0(z) = T_b - \beta z, \tag{7.10}$$

where

$$\beta = \frac{T_b - T_t}{d}.$$

The density corresponding to the temperature distribution (7.10) can be written as

$$\rho_0(z) = \rho_b(1 + \alpha \beta z), \tag{7.11}$$

where α is the coefficient of volume expansion with temperature. The pressure in the equilibrium state satisfies the hydrostatic equation

$$\frac{dp_0}{dz} = -\rho_0(z)g. \tag{7.12}$$

We now introduce perturbations around this equilibrium state. If $T_0 + T_1$ is the perturbed temperature, then the perturbed density has to be $\rho_0 - \rho_b \alpha T_1$. We write the perturbed pressure as $p_0 + p_1$. Since the perturbed state of the liquid does not satisfy the static equilibrium equations, we have to allow for the velocity $\mathbf{v}_1$ arising out of the perturbations. Substituting in the Navier–Stokes equation (5.10), we have

$$(\rho_0 - \rho_b \alpha T_1) \left[\frac{\partial \mathbf{v}_1}{\partial t} + (\mathbf{v}_1 \cdot \nabla)\mathbf{v}_1 \right] = -\nabla(p_0 + p_1) + (\rho_0 - \rho_b \alpha T_1)\mathbf{g} + \mu \nabla^2 \mathbf{v}_1,$$
(7.13)

where $\mathbf{g} = -g\hat{\mathbf{e}}_z$ is the gravitational field. We now linearize the perturbation equation by neglecting the $(\mathbf{v}_1 \cdot \nabla)\mathbf{v}_1$ term. By virtue of (7.12), the terms $-\nabla p_0$ and $\rho_0\mathbf{g}$ cancel each other. We now introduce another approximation known as the *Boussinesq approximation* (Boussinesq 1903). Since the density decrease in the term with $\mathbf{g}$ gives rise to the buoyancy force, we have to include the density decrease due to temperature in that term. On the L.H.S. of (7.13), however, we may neglect the small variation of density and the constant density ρ_b may be taken as the coefficient of $\partial \mathbf{v}_1/\partial t$. With all these simplifications, (7.13) finally becomes

$$\rho_b \frac{\partial \mathbf{v}_1}{\partial t} = -\nabla p_1 - \rho_b \alpha T_1 \mathbf{g} + \mu \nabla^2 \mathbf{v}_1.$$
(7.14)

Substituting the perturbed variables in (7.8) and linearizing, we obtain

$$\frac{\partial T_1}{\partial t} = \beta v_{1z} + \kappa \nabla^2 T_1$$
(7.15)

on making use of (7.10). Equations (7.14) and (7.15) constitute the basic equations describing the evolution of the perturbations. We now obtain an equation more convenient than (7.14) by taking the curl of (7.14) twice and then considering its z component. Remembering that $\nabla \times \nabla \times \mathbf{v} = -\nabla^2 \mathbf{v}$ for incompressible flows,

$$\frac{\partial}{\partial t} \nabla^2 v_{1z} = \alpha g \left(\frac{\partial^2 T_1}{\partial x^2} + \frac{\partial^2 T_1}{\partial y^2} \right) + \nu \nabla^4 v_{1z}.$$
(7.16)

We note that T_1 and v_{1z} are the only two independent variables appearing in the two equations (7.15) and (7.16). Hence these two equations should be sufficient to find out the evolutions of T_1 and v_{1z}.

Since (7.15) and (7.16) are linear, it should be possible to express any arbitrary perturbation as a superposition of Fourier components. We try the following form for a Fourier component

$$v_{1z} = W(z) \exp(\sigma t + ik_x x + ik_y y),$$
(7.17)

$$T_1 = \theta(z) \exp(\sigma t + ik_x x + ik_y y).$$
(7.18)

Because of the symmetry in the x and y directions, we expect the x and y dependences of a Fourier mode to be of the form $\exp(ik_x x + ik_y y)$. The functions $W(z)$ and $\theta(z)$ satisfy equations which can be obtained by substituting (7.17–7.18) into (7.15–7.16). If $\sigma > 0$, then the perturbation grows, whereas $\sigma < 0$ corresponds to a decaying perturbation. Since we are interested in finding the condition for marginal stability, we have to consider the situation $\sigma = 0$. In some stability problems, it is not possible to make both the real and imaginary parts of σ zero. Let us not get into a discussion of such subtleties here, because the problem under study happens to admit of solutions with $\sigma = 0$. Substituting (7.17–7.18) into (7.15–7.16) with $\sigma = 0$, we get the following pair of equations for $W(z)$ and $\theta(z)$:

$$\beta W + \kappa \left(\frac{d^2}{dz^2} - k^2 \right) \theta = 0, \tag{7.19}$$

$$-\alpha g k^2 \theta + \nu \left(\frac{d^2}{dz^2} - k^2 \right)^2 W = 0, \tag{7.20}$$

where $k^2 = k_x^2 + k_y^2$. On eliminating θ between these two equations, we find

$$\nu \kappa \left(\frac{d^2}{dz^2} - k^2 \right)^3 W = -\alpha \beta g k^2 W. \tag{7.21}$$

It is now convenient to substitute $z = z'd$ and $k = k'/d$, where z' is the distance along z axis measured in the unit of the depth d and k' is the horizontal wavenumber in the same unit. Putting these in (7.21),

$$\left(\frac{d^2}{dz'^2} - k'^2 \right)^3 W = -Rk'^2 W, \tag{7.22}$$

where

$$R = \frac{\alpha \beta g d^4}{\kappa \nu} \tag{7.23}$$

is a dimensionless number known as the *Rayleigh number*. The equation (7.22) was first obtained by Jeffreys (1926) who extended the analysis of Rayleigh (1916).

Since (7.22) is a sixth-order differential equation, one needs six boundary conditions to find $W(z)$ by solving (7.22). We refer the reader to Chandrasekhar (1961, pp. 21–22) for a full discussion of the possible boundary conditions under different circumstances. Without getting into that tricky subject, let us just mention that the simplest (not the physically most appropriate!) boundary conditions are satisfied by the solution

$$W(z) = W_0 \sin \pi z'. \tag{7.24}$$

Substituting (7.24) into (7.22), we obtain

$$R = \frac{(\pi^2 + k'^2)^3}{k'^2}. \tag{7.25}$$

Let us now pause for a moment and try to understand where we have got after this lengthy mathematical analysis. Since we have put $\sigma = 0$ in (7.17–7.18), we are basically looking for a Fourier component of the perturbation which is marginally stable. For the Fourier component (7.17) with $W(z)$ given by (7.24), the marginal instability occurs if the Rayleigh number satisfies (7.25). Since the Rayleigh number is seen from (7.23) to be proportional to the temperature gradient, we expect the system to be stable if R as defined by (7.23) is less than the expression $(\pi^2 + k'^2)^3/k'^2$ appearing in (7.25). On the other hand, the Fourier component is unstable if R defined by (7.23) is more than this expression. This can be shown by carrying out a more complete analysis without setting $\sigma = 0$ at the outset as we have done.

Figure 7.3 is a plot of the expression (7.25). The regions above the curve correspond to instability, since a Rayleigh number lying there is higher than the Rayleigh number for marginal instability. If R is sufficiently small, we see from Figure 7.3 that there is no real value of k' satisfying (7.25) and consequently perturbations with all possible horizontal wavenumbers must be stable. If we now increase R from very small values (say by increasing the temperature gradient in the

Figure 7.3 A plot of the Rayleigh number R against the dimensionless wavenumber k' for marginal stability as given by (7.25).

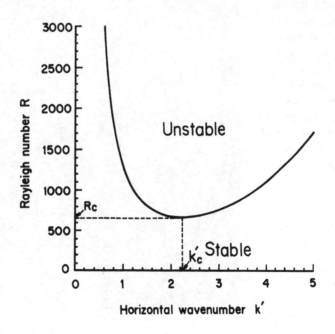

system), then eventually $R = R_c$ touches the lowest point of the curve $(\pi^2 + k'^2)^3/k'^2$. It is easy to show that the minimum of this curve occurs at

$$k'_c = \frac{\pi}{\sqrt{2}} \tag{7.26}$$

and the corresponding value of R is

$$R_c = \frac{27}{4}\pi^4 = 657.5. \tag{7.27}$$

When R is raised to this value R_c, the Fourier component with the wavenumber given by (7.26) becomes unstable. On raising R further, we see from Figure 7.3 that a range of wavenumbers becomes unstable. Hence R_c as given by (7.27) must be the *critical Rayleigh number* for the system. If $R < R_c$, then the system is stable against *all* perturbations. On the other hand, when $R > R_c$, perturbations with certain wavenumbers are unstable, which means that the system is unstable to *arbitrary* perturbations.

Chandrasekhar (1961, §15) presents analyses for two other types of boundary conditions, for which the critical Rayleigh numbers are 1707.7 and 1100.6 in contrast to 667.5 we have obtained. What about the experimental value? Chandrasekhar (1961, §18) gives a summary of experiments also. Most experiments find the critical Rayleigh number

Figure 7.4 Bénard cells in a laboratory experiment. From Koschmieder and Pallas (1974). (©Elsevier Science Ltd. Reproduced with permission from *International Journal of Heat & Mass Transfer*.)

to be around 1700. If the experiment is performed with the Rayleigh number close to its critical value, then we expect only perturbations with one horizontal wavenumber to be unstable. It follows from (7.26) that the horizontal wavelength of the marginally unstable perturbation is of the order of the depth of the liquid layer. This is borne out by experiments. Figure 7.4 shows the pattern of marginal convection in a laboratory experiment. The convection cells appearing at marginal instability are called *Bénard cells* in honour of Bénard (1900) who performed pioneering experiments on convective instability. Finally, for your amusement, we present a photograph of the solar surface in Figure 7.5. This photograph at once reminds one of Bénard cells. What we see are indeed convection cells on the solar surface known as *granules*.

7.4 Perturbations at a two-fluid interface

We have all noticed how the surface of a lake gets ruffled by a gentle wind. To understand such phenomena, we have to study perturbations at an interface between two fluids. Let us consider a plane horizontal interface separating two fluids, with gravity acting downward. If this interface is disturbed, we expect that, under certain circumstances, the interface may oscillate in the form of waves and, under other

Figure 7.5 Granules on the solar surface photographed with the Vacuum Tower Telescope of the Kiepenheuer Institut located in Tenerife. Courtesy: W. Schmidt.

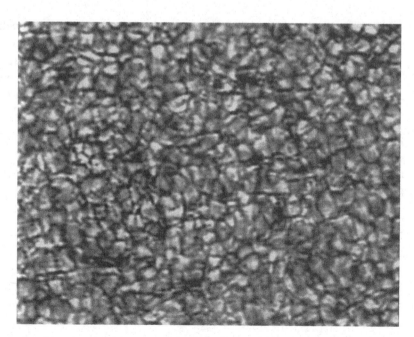

circumstances, the disturbance may grow to give rise to an instability. We shall first present a general mathematical analysis to show how a perturbation at such an interface evolves. Then several special cases of waves and instabilities will follow from the general dispersion relation we derive. To simplify the analysis, we assume the fluids on both sides of the interface to be incompressible and ideal. Hence, if there is no vorticity *inside* one of these fluids at the beginning, Kelvin's vorticity theorem (see §4.6) asserts that the velocities induced inside the fluid as a result of the perturbation remain irrotational. So velocity potentials can be introduced. We neglect the surface tension at the interface between the two fluids. The reader is asked to work out the effect of surface tension in Exercise 7.1.

In §4.5, we derived Bernoulli's principle for steady flows. We now obtain a similar result for a flow which is incompressible and irrotational, but varying in time. Writing $\mathbf{v} = -\nabla\phi$ and substituting in the Euler equation, we get

$$-\nabla\frac{\partial\phi}{\partial t} + \nabla\left(\frac{1}{2}v^2\right) = -\nabla\left(\frac{p}{\rho}\right) - \nabla\Phi, \qquad (7.28)$$

where Φ is the the potential for the body force (gravity in the present case). This equation readily gives rise to the integral

$$-\frac{\partial\phi}{\partial t} + \frac{1}{2}v^2 + \frac{p}{\rho} + \Phi = F(t), \qquad (7.29)$$

where $F(t)$ is constant in space, but can be a function of time. It may be noted that in §4.5 we were not restricting ourselves to irrotational flows. Hence, in order to obtain (4.38), we had to integrate along streamlines to make $d\mathbf{l} \cdot \mathbf{v} \times (\nabla \times \mathbf{v})$ zero. In the present case, $\nabla \times \mathbf{v}$ is zero by the assumption of irrotationality so that it is not necessary to

Figure 7.6 A sketch showing a two-fluid interface being perturbed.

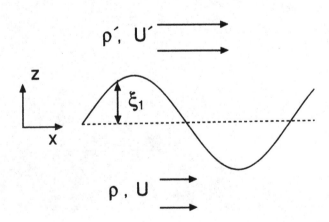

integrate along a streamline. Hence (7.29) should hold between points not lying on a streamline. We shall need (7.29) later in our derivation.

Figure 7.6 shows the configuration we want to study. The horizontal plane $z = 0$ separates the two fluids with densities ρ (fluid below) and ρ' (fluid above). Let us also assume the two fluids to have uniform velocities U and U' in the x direction. It is not difficult to show that this configuration satisfies steady state equations (i.e. hydrodynamic equations with $\partial/\partial t = 0$). Let the interface between the two fluids be perturbed from its undisturbed position $z = 0$ and let us denote the position of the interface by $z = \xi_1(x,t)$. The subscript 1 is to follow our convention that all perturbed quantities have subscript 1. Our aim is to find out whether this perturbation $|\xi_1(x,t)|$ grows with time or decays or oscillates.

The velocity potential in the fluid below can be written as

$$\phi = -Ux + \phi_1, \tag{7.30}$$

where the unperturbed part $-Ux$ would give the uniform velocity U in the x direction and ϕ_1 is the perturbed part satisfying

$$\nabla^2 \phi_1 = 0 \tag{7.31}$$

as a result of the incompressibility condition $\nabla \cdot \mathbf{v} = 0$. Similarly the velocity potential in the fluid above can be written as

$$\phi' = -U'x + \phi_1', \tag{7.32}$$

where ϕ_1' is the perturbed part satisfying

$$\nabla^2 \phi_1' = 0. \tag{7.33}$$

Since the velocity perturbations are caused by the displacements of the interface, we now somehow have to connect the perturbed parts of the velocity potential with the displacements of the interface. To do this, we consider a fluid element of the lower fluid lying infinitesimally close to the interface and find out its vertical velocity. In terms of the velocity potential of the lower fluid, the vertical velocity is given by $-\partial\phi_1/\partial z$. On the other hand, the vertical velocity of the fluid element is also given by the Lagrangian derivative of the displacement $d\xi_1/dt$. Equating these two, we have

$$-\frac{\partial\phi_1}{\partial z} = \frac{\partial\xi_1}{\partial t} + U\frac{\partial\xi_1}{\partial x} \quad \text{at} \quad z = 0, \tag{7.34}$$

where the R.H.S. is just the expansion of the Lagrangian derivative by using (4.2), in which we have kept only the terms linear in the perturbed quantities. Exactly similarly, we may consider a fluid element infinitesimally above the interface and obtain

$$-\frac{\partial \phi_1'}{\partial z} = \frac{\partial \xi_1}{\partial t} + U' \frac{\partial \xi_1}{\partial x} \quad \text{at} \quad z = 0. \tag{7.35}$$

Since we are linearizing the perturbation equations, we can again expand any arbitrary perturbation in Fourier components. Because of the symmetry in the x direction, a Fourier component of the displacement ξ_1 can be written as

$$\xi_1 = A \exp(-i\omega t + ikx). \tag{7.36}$$

The corresponding Fourier components of ϕ_1 and ϕ_1' should have similar x and t dependences. The z dependences of them should be such that the Laplace equations (7.31) and (7.33) are satisfied. Hence they have to be of the form

$$\phi_1 = C \exp(-i\omega t + ikx + kz), \tag{7.37}$$

$$\phi_1' = C' \exp(-i\omega t + ikx - kz), \tag{7.38}$$

where the signs before kz have been chosen in such a fashion that the perturbations vanish as we go far away from the interface. Substituting (7.36–7.38) into (7.34–7.35), we get

$$i(-\omega + kU)A = -kC, \tag{7.39}$$

$$i(-\omega + kU')A = kC'. \tag{7.40}$$

These are two equations relating the amplitudes of perturbed quantities A, C and C'. We need a third equation to solve the problem, since we are dealing with three quantities. This is provided by the condition that the pressure has to be continuous across the interface.

From (7.29), the pressure inside the lower fluid at a point infinitesimally close to the interface can be written as

$$p = -\rho \left(-\frac{\partial \phi_1}{\partial t} + \frac{v^2}{2} + g\xi_1 \right) + \rho F(t), \tag{7.41}$$

where we have written $\Phi = g\xi_1$ for the gravitational potential. Writing a similar expression for the pressure infinitesimally above the interface and equating them,

$$\rho \left(-\frac{\partial \phi_1}{\partial t} + \frac{v^2}{2} + g\xi_1 \right) = \rho' \left(-\frac{\partial \phi_1'}{\partial t} + \frac{v'^2}{2} + g\xi_1 \right) + K \quad \text{at } z = 0. \tag{7.42}$$

Although $K = \rho F(t) - \rho' F'(t)$ can in principle be a function of time, here it has to be a constant due to the boundary condition that the perturbations vanish far away from the interface at all times. We can find K by considering the unperturbed configuration for which $v = U$, $v' = U'$ and ϕ_1, ϕ_1', ξ_1 are all zero. This gives

$$K = \tfrac{1}{2}\rho U^2 - \tfrac{1}{2}\rho' U'^2. \tag{7.43}$$

We note that

$$v^2 = (U\hat{e}_x - \nabla\phi_1)^2 = U^2 - 2U\frac{\partial\phi_1}{\partial x}, \tag{7.44}$$

if we keep only the linear terms in the perturbed quantities. Substituting for v^2 from (7.44) and making a similar substitution for v'^2, we find from (7.42) that

$$\rho\left(-\frac{\partial\phi_1}{\partial t} - U\frac{\partial\phi_1}{\partial x} + g\xi_1\right) = \rho'\left(-\frac{\partial\phi_1'}{\partial t} - U'\frac{\partial\phi_1'}{\partial x} + g\xi_1\right) \quad \text{at} \quad z = 0 \tag{7.45}$$

after making use of (7.43). Substituting from (7.36–7.38) into this equation, we finally get

$$\rho[-i(-\omega + kU)C + gA] = \rho'[-i(-\omega + kU')C' + gA], \tag{7.46}$$

which is the third equation connecting the amplitudes of perturbed quantities.

Combining (7.39), (7.40) and (7.46), we obtain

$$\rho(-\omega + kU)^2 + \rho'(-\omega + kU')^2 = kg(\rho - \rho'). \tag{7.47}$$

We have already mentioned in §6.2 that an equation connecting the frequency and wavenumber of a Fourier component of the perturbation is called a *dispersion relation*. For the two-fluid interface problem, (7.47) then is the dispersion relation. We note that (7.47) is a quadratic equation for ω if k is regarded as given. The solution of this quadratic equation is

$$\frac{\omega}{k} = \frac{\rho U + \rho' U'}{\rho + \rho'} \pm \left[\frac{g}{k}\frac{\rho - \rho'}{\rho + \rho'} - \frac{\rho\rho'(U - U')^2}{(\rho + \rho')^2}\right]^{1/2}. \tag{7.48}$$

We shall now apply this general result to a few special cases.

7.4.1 Surface gravity waves

Let us consider two fluids at rest, with the lighter fluid lying above the heavier fluid. This means $U = U' = 0$ and $\rho > \rho'$. For a perturbation of the interface between these fluids, (7.48) gives

$$\frac{\omega}{k} = \pm\sqrt{\frac{g}{k}\frac{\rho - \rho'}{\rho + \rho'}}. \tag{7.49}$$

It is easily seen that ω is real for a real k. Hence it follows from (7.36) that the disturbance moves on the surface in the form of a wave. Such waves are called *surface gravity waves*. The phase velocity ω/k of these waves depends on the wavenumber k. This means that surface waves of different wavelengths propagate with different velocities, i.e. surface waves are dispersive.

When we consider waves on a water surface, the density of air compared to the density of water can be neglected. Neglecting ρ' compared to ρ in (7.49), we get

$$\omega = \pm\sqrt{gk}. \tag{7.50}$$

A more complete derivation makes the expression for ω somewhat more complicated due to two reasons. Firstly, we have neglected surface tension in our calculations. Secondly, when we use (7.37) for the velocity potential, we tacitly assume the fluid below to have infinite depth. If the depth of water is finite, we have to make sure that the normal component of velocity vanishes at that depth. Modifications for surface tension and for finite depth are given as Exercises 7.1 and 7.2 respectively.

7.4.2 Rayleigh–Taylor instability

Let us still consider the fluids to be at rest so that (7.49) holds. But now assume that we have a heavier fluid above a lighter fluid, i.e. $\rho < \rho'$. Even for such a situation, the hydrostatic equation (4.29) can be satisfied. Hence this is an equilibrium configuration. But we intuitively feel that such an equilibrium will be unstable, because the heavier fluid will try to sink below the lighter fluid. This is what we find from the mathematics also.

When $\rho < \rho'$, we see from (7.49) that ω is imaginary. It is clear from (7.36) that the positive imaginary solution of ω makes the perturbation grow. This is the *Rayleigh–Taylor instability* (Rayleigh 1883; Taylor 1950b). Since uniform acceleration in a mechanical system can be treated mathematically exactly the same way as a gravitational field, the problem of a light fluid accelerating against a heavy fluid is very similar and gives rise to the same type of instability. Whereas Rayleigh's analysis (Rayleigh 1883) was for fluids in a gravitational field, Taylor (1950b) adapted the problem to the situation of accelerating fluids. As we often find a light fluid accelerating against a heavy fluid in the astrophysical Universe, this instability is quite important in astrophysics.

In a supernova explosion, for example, there is an interface between the hot gas ejected in the explosion and the surrounding interstellar medium. Gull (1975) studied the early phases of the evolution of the supernova remnant (i.e. before the system enters the Sedov–Taylor self-similar phase discussed in §6.6) and showed that the ejected gas tends to pile up in a thin shell behind the interface. As the interface is decelerated, we can represent it by an outward gravitational field in the rest frame of the interface. Then we have the dense shell of ejected gas lying 'on top' of the less dense gas outside. It was found by Gull (1975) that this gives rise to the Rayleigh–Taylor instability, leading to the formation of filaments growing in the outward (i.e. the 'downward') direction. Figure 7.7 is a photograph of the Crab Nebula (the remnant of the supernova which was observed from the Earth in 1054) showing an intriguing filamentary structure.

7.4.3 Kelvin–Helmholtz instability

Finally we consider what happens if U and U' are not zero. Let us assume $\rho > \rho'$ so that the system is Rayleigh–Taylor stable. If the expression within the square root in (7.48) is negative, then we note

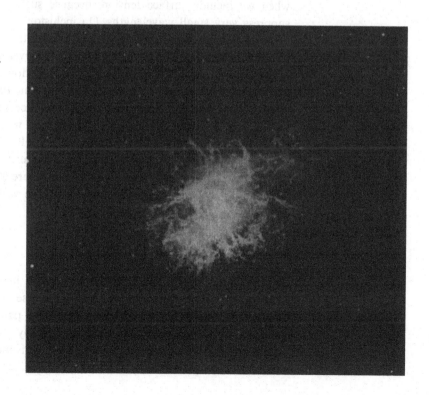

Figure 7.7 An optical photograph of the Crab Nebula taken at Kavalur Observatory, India. Courtesy: Indian Institute of Astrophysics.

that ω has an imaginary part (we are again taking k to be real). We have already seen that a positive imaginary part in ω leads to instability. Hence the instability occurs when the expression within the square root in (7.48) is negative, i.e. when

$$\rho\rho'(U - U')^2 > (\rho^2 - \rho'^2)\frac{g}{k}. \tag{7.51}$$

This instability is known as the *Kelvin–Helmholtz instability* (Helmholtz 1868; Kelvin 1871) and causes an interface between two fluids to wrinkle if the fluids on the two sides are moving with different speeds. The most common example of the Kelvin–Helmholtz instability is provided by the observation that a wind blowing over a water surface causes the water surface to undulate.

Even when $|U - U'|$ is arbitrarily small, we notice from (7.51) that the Fourier components satisfying

$$k > \frac{(\rho^2 - \rho'^2)g}{\rho\rho'(U - U')^2} \tag{7.52}$$

become unstable. Since large k means small wavelengths, we conclude that only perturbations with small wavelengths are unstable if $|U - U'|$ is sufficiently small. This conclusion, however, gets completely modified when we include surface tension, because surface tension tends to suppress very small wavelengths. On inclusion of surface tension, it is found that the Kelvin–Helmholtz instability takes place only when $|U - U'|$ is larger than a critical value. This is given as Exercise 7.1(ii).

It follows from (7.48) that the Kelvin–Helmholtz instability would occur at all wavenumbers for any $|U - U'|$ in the absence of the gravitational field g. An extragalactic jet (as discussed in §6.7) moving out of a galaxy into the intergalactic medium with high speed may be subject to the Kelvin–Helmholtz instability. Radio observations of many extragalactic jets show a variation in intensity and cross-section along the axis of the jet. This can be seen in Figure 6.7 showing the jets in Hercules A. One interpretation is that these variations are results of the Kelvin–Helmholtz instability.

7.5 Jeans instability

We have seen in §7.2 and §7.4 that waves and instabilities are often intimately related, the same system giving rise to either waves or instabilities depending on the value of some parameter. When we derived acoustic waves in §6.2, we did not find any instability associated with them. But we shall demonstrate that there is indeed an instability related to acoustic waves.

When a region of a gas is compressed, the excess pressure there tries to smoothen out the compression, and this gives rise to acoustic waves. The compressed region, however, has enhanced gravitation and this tends to pull more gas into the compressed region. For typical sound waves in the air, the enhancement of gravity in the regions of compression is utterly negligible and we have therefore not considered it in §6.2. For perturbations of gaseous bodies of astronomical size, however, the self-gravity is quite important, and the enhanced gravitation in the region of compression may overpower the expansive tendency of the excess pressure, pulling more material there and triggering an instability. Since Jeans (1902) was the first person to demonstrate the existence of this instability, it is called *Jeans instability* in his honour.

Jeans (1902) considered perturbations in a uniform infinite gas. Apart from the variables we considered in §6.2, we introduce the gravitational potential $\Phi = \Phi_0 + \Phi_1$ broken into unperturbed and perturbed parts. The unperturbed variables should satisfy the hydrostatic equation (4.27) plus the Poisson equation for gravity, i.e.

$$\nabla p_0 = -\rho_0 \nabla \Phi_0, \tag{7.53}$$

$$\nabla^2 \Phi_0 = 4\pi G \rho_0. \tag{7.54}$$

It is trivial to show that a uniform infinite gas does not satisfy these equations. From (7.53), a constant p_0 would imply a constant Φ_0. When a constant Φ_0 is substituted in (7.54), we are driven to the conclusion that the unperturbed density ρ_0 has to be zero everywhere! For a proper stability analysis, one should first find a solution of (7.53–7.54) and then consider perturbations around that solution. Jeans (1902), however, proceeded to perform a perturbation analysis on the uniform infinite gas as if the unperturbed configuration satisfied the equilibrium equations! Hence this approach is often referred as the *Jeans swindle*. We reproduce here the analysis based on the Jeans swindle because of its historical importance and simplicity. It is possible to carry out proper stability analyses for realistic density distributions without recourse to the Jeans swindle. For example, if we consider a slab of gas in static equilibrium under its own gravity, then we can carry out a proper stability analysis. See Spitzer (1978, pp. 283–285) for a discussion of this problem. As it happens, the correct (and much more complicated!) analyses yield results which are qualitatively similar to those we get from the perturbation analysis of the uniform infinite gas with the help of the Jeans swindle.

The perturbation of the equation of continuity still gives (6.15), which we write again

$$\frac{\partial \rho_1}{\partial t} + \rho_0 \nabla \cdot \mathbf{v}_1 = 0. \tag{7.55}$$

Linearizing the Euler equation (with gravity included) and subtracting (7.53) from it, we get an equation similar to (6.17) with an extra term for gravity

$$\rho_0 \frac{\partial \mathbf{v}_1}{\partial t} = -c_s^2 \nabla \rho_1 - \rho_0 \nabla \Phi_1. \tag{7.56}$$

Finally, subtracting (7.54) from the full equation $\nabla^2 \Phi = 4\pi G \rho$, we get

$$\nabla^2 \Phi_1 = 4\pi G \rho_1. \tag{7.57}$$

We now have three equations (7.55–7.57) satisfied by the three perturbation variables ρ_1, $\mathbf{v}_1$ and Φ_1. For a particular Fourier component, let us take all these variables to vary as $\exp[i(\mathbf{k} \cdot \mathbf{x} - \omega t)]$. Then (7.55–7.57) give

$$-\omega \rho_1 + \rho_0 \mathbf{k} \cdot \mathbf{v}_1 = 0,$$

$$-\rho_0 \omega \mathbf{v}_1 = -c_s^2 \mathbf{k} \rho_1 - \rho_0 \mathbf{k} \Phi_1,$$

$$-k^2 \Phi_1 = 4\pi G \rho_1.$$

Combining these three equations, we readily find the dispersion relation

$$\omega^2 = c_s^2 (k^2 - k_J^2), \tag{7.58}$$

where

$$k_J^2 = \frac{4\pi G \rho_0}{c_s^2}. \tag{7.59}$$

When $k < k_J$, we see from (7.58) that ω has to be imaginary, which means that the system is unstable. In other words, if the size of the perturbation is larger than some critical wavelength of the order $\lambda_J = 2\pi/k_J$, then the enhanced self-gravity can overpower the excess pressure so that the perturbation grows. The corresponding critical mass

$$M_J = \frac{4}{3} \pi \lambda_J^3 \rho_0$$

is often referred as the *Jeans mass*. Substituting from (7.59) and using (6.13) for c_s with γ taken as 1 for large-wavelength slowly evolving perturbations (which can regarded as isothermal), we get

$$M_J = \frac{4}{3} \pi^{5/2} \left(\frac{\kappa_B T}{Gm} \right)^{3/2} \frac{1}{\rho_0^{1/2}}, \tag{7.60}$$

where m is the mass of the gas particles. If a perturbation in a uniform gas involves a mass larger than the Jeans mass, then we expect the gas in the perturbed region to keep contracting due to the enhanced gravity. Thus an initially uniform distribution of gas may eventually fragment into pieces due to the Jeans instability.

Jeans instability is the basic reason why the matter in the Universe is not spread uniformly. Stars and galaxies are believed to be the end-products of perturbations which initially started growing due to the Jeans instability. Astronomers think that new stars are still born out of the interstellar matter. Assuming the interstellar matter to have 1 hydrogen atom per cm^3 at temperature 100 K, (7.60) gives a Jeans mass of about 8×10^{38} gm. This is several orders of magnitude larger than the typical mass of a star (about 10^{33} gm). Presumably the interstellar matter first breaks into large chunks with masses corresponding to clusters of stars rather than individual stars. Then somehow these contracting chunks of gas have to break further to produce stars. The presence of angular momentum or magnetic field can make the process quite complicated. We shall say a few words about the role of angular momentum and magnetic field in §14.9.1 and §14.10. See Spitzer (1978, §13.3) for an introduction to the complex subject of star formation.

A related problem of great astrophysical significance is the study of gravitational instability in a self-gravitating rotating disk, like the disk of a galaxy. Under certain circumstances, perturbations of the form of spiral arms in such disks may become unstable. It is possible that the arms of spiral galaxies are caused due to such reasons—an idea propounded by Lin and Shu (1964).

7.6 Stellar oscillations. Helioseismology

The theory of stellar oscillations is one important area of astrophysics in which the linear perturbation analysis happens to be the standard technique. It has been known for a long time that certain stars (such as Cepheid variables) show a periodic variation in luminosity. The most obvious explanation for such stars with periodically varying luminosity is that these are pulsating stars. The simplest type of stellar oscillation is a purely radial oscillation. We shall first discuss such oscillations and then make a few comments on non-radial oscillations. The theory of radial stellar oscillations was developed by Eddington (1918) to explain the Cepheid variability.

Oscillations in a system of finite size are usually of the nature of standing waves, for which only certain discrete frequencies are possi-

ble. A stretched string tied at the two ends or an air column in a pipe
are examples of such systems which support standing waves of definite
frequencies. We expect the radial pulsations of a star also to be of the
nature of many normal modes with definite eigenfrequencies. We shall
not present a proof here. The interested reader may look up Kippenhahn and Weigert (1990, §38) for a proof assuming the oscillations
to be adiabatic. It should be apparent to anybody familiar with the
theory of normal oscillations that the radial perturbation associated
with a certain eigenfrequency must be of the form $\exp(-i\omega_n t)R_n(r)$,
where ω_n is an eigenfrequency of the star. Finding ω_n or $R_n(r)$ for a
realistic star is a complicated problem. The calculations become somewhat simpler if we can assume a relation like $p = K\rho^\gamma$ throughout the
star.

If ω_n has a positive imaginary part, then it is easy to see that the
star is unstable towards radial perturbations. Without presenting a
full stability analysis, let us do a simple calculation to illuminate the
nature of the problem. From the Euler equation (4.11), we can write
down the acceleration of a fluid element at a distance r from the centre
of the star:

$$\frac{dv}{dt} = -\frac{1}{\rho}\frac{dp}{dr} - \frac{GM}{r^2}, \tag{7.61}$$

where v is the radial component of velocity (we are considering purely
radial motions here) and M is the mass inside the radius r. For a
star in equilibrium, the R.H.S. of (7.61) must vanish. Suppose the star
is *uniformly* expanded from an initial equilibrium configuration such
that the position of a fluid element changes from r_0 to $r_0(1 + \delta)$. It
is easily seen that the various quantities at the position of the fluid
element change in the linear order to

$$r = r_0(1 + \delta), \quad \rho = \rho_0(1 - 3\delta), \quad p = p_0(1 - 3\gamma\delta), \tag{7.62}$$

where we have assumed the pressure to vary as ρ^γ. The mass M inside
the radius r does not change. On substituting (7.62) in (7.61) and
keeping terms to the linear order, we get

$$\frac{dv}{dt} = -\frac{1}{\rho_0}\frac{dp_0}{dr_0}(1 + 2\delta - 3\gamma\delta) - \frac{GM}{r_0^2}(1 - 2\delta). \tag{7.63}$$

If the initial configuration was one of equilibrium, then

$$-\frac{1}{\rho_0}\frac{dp_0}{dr_0} - \frac{GM}{r_0^2} = 0.$$

Using this, we finally have from (7.63) that

$$\frac{dv}{dt} = -\frac{GM}{r_0^2} 3\delta \left(\gamma - \frac{4}{3} \right). \tag{7.64}$$

Suppose we consider a uniform expansion of the star so that δ is positive. If the acceleration given by (7.64) is positive, then the expansion rate will be accelerated leading to an instability. Hence the star is stable only if the acceleration is negative. It then follows from (7.64) that the condition for stability is $\gamma > 4/3$.

It should be emphasized that we have *not* carried out a general stability analysis here. Our calculations apply only to uniform expansions or contractions. We have merely obtained the stability condition under such specialized (and very artificial) perturbations. To find the stability condition under arbitrary perturbations, we have to calculate the eigenfrequencies and see if any of them have a positive imaginary part. This more complete (and much more complicated) analysis eventually gives exactly the same stability condition we derived above! For an equation of state $p = K\rho^\gamma$, the star is stable if $\gamma > 4/3$ (see, for example, Kippenhahn and Weigert 1990, §38). A large value of γ implies that pressure increases quickly with density. Suppose something causes a volume contraction in a star. If the central pressure increases sufficiently rapidly with the enhanced density, then the star may be able to 'bounce back' to its original configuration. This explains why a larger γ makes a star more stable.

Because of the spherical geometry, we expect that the perturbation associated with a non-radial normal mode must be of the form

$$\mathbf{v}(t, r, \theta, \phi) = \exp(-i\omega_{nlm}t) \, \xi_{nlm}(r) Y_{lm}(\theta, \phi), \tag{7.65}$$

where $Y_{lm}(\theta, \phi)$ is a spherical harmonic. To calculate the eigenfrequency ω_{nlm}, we have to substitute expressions like (7.65) for various perturbed quantities in the linearized equations. If the amplitude of the oscillation is small, then we expect the perturbation in the gravitational field to be negligible. Cowling (1941) showed that the problem becomes more tractable if the perturbation in the gravitational field is neglected. This is known as the *Cowling approximation.*

With the realization that the Sun is full of non-radial oscillations, interest in the study of such oscillations has grown tremendously in recent decades. The presence of oscillations at the solar surface was first established by Leighton, Noyes and Simon (1962). Initially it was not clear if these oscillations are of the nature of mere local disturbances or if they are true normal modes of the whole Sun. With the availability of better data, it was eventually possible to do a proper Fourier analysis and show that the power in the oscillations is concentrated at discrete frequencies for any given wavenumber (Deubner 1975). This established the global nature of the oscillations. The study of these oscillations is known as *helioseismology*.

Just as the energy levels of an atom give valuable information about the structure of the atom, the eigenfrequencies of the Sun can be used to probe the interior of the Sun. One dramatic development in the recent years is that the distribution of angular velocity in the interior of the Sun has been mapped. If the Sun were non-rotating, then ω_{nlm} would have been independent of m. In other words, the eigenfunctions with the same n and l, but different m, would have the same frequencies. But rotation causes frequencies with different m to be split (Gough 1978). We point out the analogy from atomic physics that the energy levels of the hydrogen atom for different m are degenerate in the absence of a magnetic field. But a magnetic field lifts this degeneracy and splits the levels. In exactly the same way, the rotation of the Sun lifts the degeneracy of eigenfrequencies with different m. The amount of splitting of a mode depends basically on the angular velocity in the region where the mode has the largest amplitude. By studying the splittings of different modes having the largest amplitudes in different regions of the Sun, one can then obtain

Figure 7.8 The contours of constant angular velocity inside the Sun, as obtained by helioseismology. The contours are marked with rotation frequency in nHz. It may be noted that frequencies of 340 nHz and 450 nHz correspond, respectively, to rotation periods of 34.0 days and 25.7 days. See Tomczyk, Schou and Thomson (1996) for details. Courtesy: J. Christensen-Dalsgaard and M. J. Thomson.

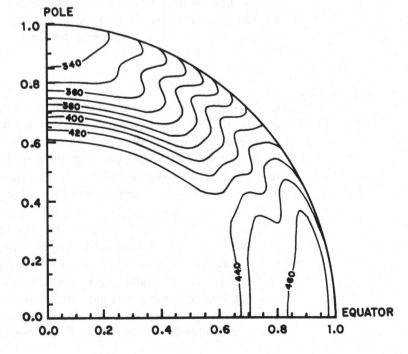

a map of how the angular velocity varies in the interior of the Sun. Figure 7.8 shows a recent map giving the distribution of the angular velocity in the interior of the Sun. From surface observations, it was known for a long time that the Sun does not rotate like a solid body. Helioseismology has given us the very valuable information as to how the angular velocity varies inside the Sun.

7.7 A few general comments. Beyond the linear theory

We have introduced a few important waves and instabilities in this chapter. This should have given the reader a general idea of the subject. But there are many other types of waves and instabilities, which we could not even touch upon. For example, we mentioned in §5.3 that the flow through a pipe becomes unstable when the Reynolds number is sufficiently high. Since the mathematical theory of the viscous shear instability is a fairly complicated subject, we have not attempted to cover it in this elementary book. Interested readers will find an introduction to this subject in the book by Drazin and Reid (1981, Chapter 4).

To understand many waves and instabilities better, it is necessary to go beyond the linear theory and consider the nonlinear terms in the lowest order. The linear theory does not tell us anything about the amplitudes of waves and instabilities. The linearized equations would be satisfied by any value of the amplitude. It often happens that an instability saturates at a certain amplitude due to the effect of the nonlinear terms, instead of growing indefinitely as suggested by the linear theory. This possibility was first discussed by Landau (1944). Malkus and Veronis (1958) carried out the nonlinear analysis of the Rayleigh–Bénard problem for Rayleigh numbers larger than the critical value R_c. It was found that the amplitude of the convective motions increases monotonically with $|R - R_c|$. So, when $|R - R_c|$ is increased further, more heat is tranported by convection.

We pointed out in §6.4 that the nonlinear term causes an acoustic wave to steepen. If we keep the similar nonlinear term in the study of surface gravity waves discussed in §7.4.1, then there also the same kind of steepening takes place. The acoustic waves in a homogeneous medium are particularly simple in the sense that the phase velocity is independent of wavelength. This is not true for surface waves. If we have a surface wave packet made up of different wavelengths, then the linear theory suggests that the wave packet will get dispersed due the different phase velocities of different Fourier components.

Nonlinearity has exactly the opposite effect of steepening the wave packet. It may happen for a particular surface wave packet that the velocity dispersion is exactly balanced by the nonlinearity. Such a wave packet, known as a solitary wave or a *soliton*, may be able to propagate without changing its shape. Scott Russell (1844) is reported to have observed such a surface wave in a channel and followed it on horseback for some time.

We have given it as Exercise 7.2 to show that the dispersion relation for surface waves on a liquid layer of thickness h is

$$\omega = \pm(gk \tanh kh)^{1/2}, \tag{7.66}$$

which reduces to (7.50) in the limit of large h. Let us now consider the opposite limit of h being small compared to the wavelength so that $kh \ll 1$. Then the phase velocity can be written as

$$c_{\mathrm{ph}} = \frac{|\omega|}{k} = (gk^{-1} \tanh kh)^{1/2} \approx c_0 \left(1 - \frac{1}{6}k^2 h^2\right), \tag{7.67}$$

where $c_0 = \sqrt{gh}$. It is easy to see that a disturbance of wavenumber k propagating in the positive x direction with velocity $c_0(1 - \frac{1}{6}k^2 h^2)$ has to satisfy the equation

$$\frac{\partial v}{\partial t} + c_0 \frac{\partial v}{\partial x} + \sigma \frac{\partial^3 v}{\partial x^3} = 0, \tag{7.68}$$

where

$$\sigma = \tfrac{1}{6}c_0 h^2.$$

If we want to include the effect of nonlinearity as well, then we should add a term $v(\partial v/\partial x)$ to (7.68). If we further use the coordinate $X = x - c_0 t$ instead of x, then we end up with the equation

$$\frac{\partial v}{\partial t} + v \frac{\partial v}{\partial X} + \sigma \frac{\partial^3 v}{\partial X^3} = 0. \tag{7.69}$$

This is the celebrated *Korteweg–de Vries equation* (Korteweg and de Vries 1895) which incorporates the effects of both nonlinearity and dispersion, the second term in (7.69) giving the nonlinearity and the third term giving the dispersion. This is the basic equation for studying solitary waves in which the spreading due to dispersion is balanced by the nonlinearity. Problems involving many types of nonlinear, dispersive waves can ultimately be reduced to the Korteweg–de Vries equation. Hence the Korteweg–de Vries equation has come to be regarded as the one of the fundamental nonlinear equations of mathematical physics.

Exercises

7.1   We neglected surface tension in the analysis of §7.4. The effect of surface tension is to make the pressure at a point infinitesimally below the surface higher than the pressure at a point infinitesimally above the surface by an amount $-T(\partial^2\xi_1/\partial x^2)$, where T is the surface tension. So such a pressure difference has to be included in equation (7.42). Show that this leads to the dispersion relation

$$\frac{\omega}{k} = \frac{\rho U + \rho' U'}{\rho + \rho'}$$

$$\pm \left[\frac{g}{k} \left\{ \frac{\rho - \rho'}{\rho + \rho'} + \frac{k^2 T}{g(\rho + \rho')} \right\} - \frac{\rho \rho'(U - U')^2}{(\rho + \rho')^2} \right]^{1/2}$$

in the place of (7.48).

(i) What is the phase velocity of surface waves on the air–water interface? Show that this velocity is minimum for a particular wavelength. Given $T = 74$ dynes cm^{-1}, calculate this wavelength.

(ii) Now consider the Kelvin–Helmholtz instability on the air–water interface. What is the critical value of $|U - U'|$ above which the instability starts in the sense of some wavenumber becoming unstable?

7.2   Consider surface waves at the air–water interface, with the water having a finite depth h. Assuming that the normal component of velocity becomes zero at the bottom and neglecting surface tension, show that the dispersion relation is given by

$$\omega^2 = gk \tanh kh.$$

If the water is shallow (i.e. the depth is small compared to the wavelength), show that the propagation velocity of the surface wave is proportional to $\sqrt{h}$. Use this result to explain the formation of surf on the seashore.

7.3   Show that the Korteweg–de Vries equation (7.69) admits of a solution of the form

$$v = 3v_0 \operatorname{sech}^2 \left[\tfrac{1}{2}(v_0/\sigma)^{1/2}(X - v_0 t) \right].$$

This famous solution corresponds to a solitary wave moving without any change of shape.

8 Turbulence

8.1 The need for a statistical theory

The linear perturbation theory presented in Chapter 7 makes it clear that a fluid configuration can, under certain circumstances, be unstable to perturbations. Once the perturbations grow to sufficiently large amplitudes, the linear theory is no longer applicable. Hence the linear theory is unable to predict what eventually happens to an unstable fluid system.

To understand the effect of an instability on a general dynamical system, let us employ the notion of a phase space introduced in Chapter 1. Figure 8.1 is a schematic representation of the phase space of a dynamical system, within which let P be a point corresponding to an *unstable equilibrium*. If the state of the system is represented exactly by P, then the state does not change by virtue of equilibrium. If, however, there is some perturbation around the equilibrium, then the state of the system is represented by some point in the neighborhood of P, and as the perturbation grows, the point in the phase space moves away from P. Thus, depending on whether the initial state was exactly at P or slightly away, the final state after some time can lie in very different regions of the phase space. Because of the limited accuracy in any measurement in a realistic situation, one can only assert that the initial state of a system lies in some finite region of the phase region. If this region happens to be around a point of instability, then it becomes effectively impossible to predict the future evolution of the system, even though its dynamics may be governed by deterministic equations.

In an unstable fluid system, we expect that the growth of perturbations may lead to a loss of our predictive capability. One indeed encounters situations in which fluid velocities seem to vary randomly in space and time. Such a state of a fluid is called *turbulence*. Although a hydrodynamic instability may be a natural way of producing turbulence, it is not the only way. A fluid stirred with random forces at different points may develop turbulence inside.

Since it is not possible to develop a deterministic theory of turbulence, one wishes to develop a statistical theory based on the average properties of turbulence. As soon as we talk of averages, one question immediately comes up: what kind of average? If the turbulence is homogenous in space, then we can take spatial averages of various fluctuating quantities over some region of space. On the other hand, if the turbulence is invariant in time, one can take a time average. From a conceptual point of view, perhaps an ensemble average is the most general and most satisfactory kind of average. We can think of many hypothetical replicas of the same system having the *same statistical properties* of turbulence, but the actual value of some quantity like velocity at the same point at the same time in the different members of the ensemble may be different. By averaging the values of the same quantity in different members of the ensemble, we get the ensemble average. If a particular system admits of more than one kind of averaging, are theories based on these different averaging procedures equivalent? This is the vexing question of ergodicity. It is not easy to give a rigorous answer to this question. Most statistical theories of physics are based on the tacit assumption that the different possible averaging procedures are probably ultimately equivalent. Whenever we talk of averages in this chapter, we shall normally mean the ensemble average and denote it by an overline. It should be noted that this is very different from the molecular average in kinetic theory introduced through (2.36), which is denoted by an angular bracket $\langle \ldots \rangle$ in this book.

The velocity at a point in the fluid can be broken into two parts

$$\mathbf{v} = \bar{\mathbf{v}} + \mathbf{v}', \tag{8.1}$$

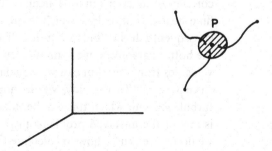

Figure 8.1 The phase space of a system with a point of unstable equilibrium P.

where $\bar{\mathbf{v}}$ is the average and $\mathbf{v}'$ is the fluctuation around it. We often have turbulence around a mean flow. In such a situation, $\bar{\mathbf{v}}$ refers to mean flow and $\mathbf{v}'$ is the turbulent velocity field, of which we want to study the statistical properties. From (8.1), it follows that

$$\bar{\mathbf{v}}' = 0.$$

Hence, in order to investigate the statistical properties of turbulence, we need to consider more complicated averages than $\bar{\mathbf{v}}'$. Taylor (1935) suggested in a classic paper that we need to consider the averaged quantity $\overline{\mathbf{v}'(\mathbf{x}, t) \cdot \mathbf{v}'(\mathbf{x} + \mathbf{r}, t)}$. If $\mathbf{r} = 0$, then this is simply the average value of $v'^2(\mathbf{x}, t)$, which is a measure of the kinetic energy in the turbulence and hence a measure of the strength of turbulence. On the other hand, if $\mathbf{r}$ is very large, then the velocities $\mathbf{v}'(\mathbf{x}, t)$ and $\mathbf{v}'(\mathbf{x} + \mathbf{r}, t)$ will not have any correlation so that

$$\lim_{r \to \infty} \overline{\mathbf{v}'(\mathbf{x}, t) \cdot \mathbf{v}'(\mathbf{x} + \mathbf{r}, t)} = \overline{\mathbf{v}'(\mathbf{x}, t)} \cdot \overline{\mathbf{v}'(\mathbf{x} + \mathbf{r}, t)} = 0. \qquad (8.2)$$

We conclude that $\overline{\mathbf{v}'(\mathbf{x}, t) \cdot \mathbf{v}'(\mathbf{x} + \mathbf{r}, t)}$ has substantial non-zero values only if r lies within a range. This is called the *correlation length* of turbulence. Thus the velocity correlation functions like $\overline{\mathbf{v}'(\mathbf{x}, t) \cdot \mathbf{v}'(\mathbf{x} + \mathbf{r}, t)}$ contain information about both the strength and the correlation length of the turbulence. In other words, such correlation functions are the appropriate quantities with which one may try to build a statistical theory of turbulence. From now on, we shall not indicate the dependence on t explicitly while writing the correlation functions. It will be assumed that the velocities at the different points are considered at the same time t. The only exception will be a derivation in §8.4, where we shall caution the reader appropriately.

The velocity correlation tensor $\overline{v_i'(\mathbf{x})v_j'(\mathbf{x} + \mathbf{r})}$ has nine components. If there are some symmetries in the system, we shall see that the number of independent components is reduced drastically. One can consider more complicated three-point correlation tensors like $\overline{v_i(\mathbf{x}_1)v_j(\mathbf{x}_2)v_k(\mathbf{x}_3)}$. The goal of the statistical theory of turbulence is to determine such correlation functions in a particular situation. For example, consider a fluid heated from below, which is subject to convective instability and consequently has turbulence inside. If the temperatures at the top and the bottom are given, we cannot calculate the instantaneous turbulent velocities from that, but can we calculate the correlation functions like $\overline{v_i'(\mathbf{x})v_j'(\mathbf{x} + \mathbf{r})}$? To this day, we do not possess a statistical theory of turbulence with which this can be done. When we say that turbulence is one of the unsolved problems in physics, we technically mean that we do not yet know how to calculate the various turbulent correlation functions in a particular situation from a fundamental theory. In the remaining sections of this chapter, we shall present the bits and pieces of theory developed in the last few decades. We shall restrict our discussion only to turbulence in incompressible fluids satisfying $\nabla \cdot \mathbf{v} = 0$.

8.2 Kinematics of homogeneous isotropic turbulence

If the turbulence is homogeneous and isotropic, then we can derive some properties of the velocity correlation tensors simply from symmetry. Firstly, the mean flow $\bar{\mathbf{v}}$ has to be zero, since a mean flow in a direction is not consistent with isotropy. It is then no longer necessary to break the velocity in two parts as in (8.1). We drop the prime and use $\mathbf{v}$ for the turbulent velocity.

Homogeneity demands that the correlation function $\overline{v_i(\mathbf{x})v_j(\mathbf{x}+\mathbf{r})}$ has to be independent of $\mathbf{x}$, whereas isotropy dictates that it can only depend on the magnitude of $\mathbf{r}$. Hence we write

$$R_{ij}(r) = \overline{v_i(\mathbf{x})v_j(\mathbf{x}+\mathbf{r})}. \tag{8.3}$$

It it easy to see that

$$\frac{\partial R_{ij}}{\partial r_j} = \overline{v_i(\mathbf{x})\frac{\partial v_j(\mathbf{x}+\mathbf{r})}{\partial r_j}}$$

must be zero due to the incompressibility condition $\nabla.\mathbf{v} = 0$. Then, from symmetry,

$$\frac{\partial R_{ij}}{\partial r_i} = \frac{\partial R_{ij}}{\partial r_j} = 0. \tag{8.4}$$

Figure 8.2
Longitudinal and lateral velocity correlations.

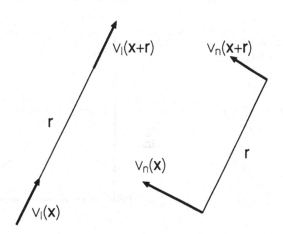

It was realized by von Kármán and Howarth (1938) that the most general form of the tensor function $R_{ij}(r)$ is

$$R_{ij}(r) = A(r)r_i r_j + B(r)\delta_{ij}, \tag{8.5}$$

where $A(r)$ and $B(r)$ are two scalar functions. We now consider the longitudinal and lateral velocity correlation functions as indicated in Figure 8.2. We note that the longitudinal component of $\mathbf{r}$ is $r_l = r$, whereas the lateral component is $r_n = 0$. Hence

$$R_{ll}(r) = A(r)r^2 + B(r) = \tfrac{1}{3}\overline{v^2}\, f(r), \tag{8.6}$$

$$R_{nn}(r) = B(r) = \tfrac{1}{3}\overline{v^2}\, g(r), \tag{8.7}$$

where $f(r)$ and $g(r)$ are the two scalar functions corresponding to the longitudinal and lateral correlations. They have been defined in such a way that

$$f(0) = g(0) = 1 \quad \text{at} \quad r = 0. \tag{8.8}$$

It is to be noted that l and n repeated twice in (8.6–8.7) do not imply summation, but the ll and nn components of the tensor $R_{ij}(r)$. Using (8.6–8.7) to express $A(r)$ and $B(r)$ in terms of $f(r)$ and $g(r)$, we find from (8.5) that

$$R_{ij}(r) = \frac{1}{3}\overline{v^2}\left[\frac{f(r) - g(r)}{r^2}r_i r_j + g(r)\delta_{ij}\right]. \tag{8.9}$$

It readily follows from (8.4) that

$$g(r) = f(r) + \frac{1}{2}r\frac{df}{dr}. \tag{8.10}$$

Hence, if we can somehow determine the one scalar function $f(r)$, then all the components of the correlation function $R_{ij}(r)$ can be obtained

Figure 8.3 A sketch of $f(r)$ in a typical situation.

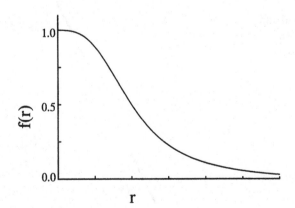

(from (8.9) and (8.10))—an important point first made by von Kármán and Howarth (1938). Since $f(r)$ is the longitudinal correlation function, we expect on physical grounds that, in some typical situation, it will look as sketched in Figure 8.3.

We now consider the Fourier transform of $R_{ij}(r)$:

$$\Phi_{ij}(\mathbf{k}) = \frac{1}{(2\pi)^3} \int R_{ij}(r)e^{i\mathbf{k}\cdot\mathbf{r}}\, d^3r. \tag{8.11}$$

From the spherical symmetry of $R_{ij}(r)$ in the **r**-space, we conclude that $\Phi_{ij}(\mathbf{k})$ has to be spherically symmetric in **k**-space so that we write it as $\Phi_{ij}(k)$. The inverse Fourier transform is

$$R_{ij}(r) = \int \Phi_{ij}(k)e^{-i\mathbf{k}\cdot\mathbf{r}}\, d^3k. \tag{8.12}$$

The incompressibility condition (8.4) implies

$$k_i\Phi_{ij} = k_j\Phi_{ij} = 0. \tag{8.13}$$

Considerations of symmetry again dictate that the most general form of $\Phi_{ij}(k)$ must be

$$\Phi_{ij}(k) = C(k)k_ik_j + D(k)\delta_{ij}, \tag{8.14}$$

where the two scalar functions $C(k)$ and $D(k)$ have to be related by virtue of (8.13) in the following way

$$D(k) = -C(k)k^2. \tag{8.15}$$

Hence we can write down Φ_{ij} in terms of just one scalar function $E(k)$ as follows

$$\Phi_{ij}(k) = \frac{E(k)}{4\pi k^4}(k^2\delta_{ij} - k_ik_j). \tag{8.16}$$

The significance of this expression becomes clear by considering the kinetic energy density of the turbulent fluid

$$\tfrac{1}{2}\overline{v^2} = \tfrac{1}{2}R_{ii}(0) = \tfrac{1}{2}\int \Phi_{ii}(k)\, d^3k \tag{8.17}$$

using (8.12), where a summation over i is implied. Substituting for $\Phi_{ij}(k)$ from (8.16) and noting that d^3k can be replaced by $4\pi k^2\, dk$ for a spherically symmetric integrand, we obtain

$$\tfrac{1}{2}\overline{v^2} = \int_0^\infty E(k)\, dk. \tag{8.18}$$

This expression suggests that we may regard the function $E(k)$ as the *energy spectrum* of turbulence. Just as the blackbody radiation in a region can be thought of as made up of contributions at different wavelengths, similarly the turbulence field can be thought of as being made up of different Fourier components and $E(k)$ has to be regarded as their energy spectrum such that the integration over it gives the total energy in the unit volume.

Before leaving the subject of the kinematics of isotropic turbulence, we note the possibility of adding a term proportional to $\epsilon_{ijl}r_l$ in (8.5) or a term proportional to $\epsilon_{ijl}k_l$ in (8.14). It is easy to see that such a term would preserve isotropy, but would not be invariant under reflection (i.e. parity inversion). We normally expect turbulence to be reflectionally symmetric and hence omit such a term, as we are doing in this chapter. Turbulence in a rotating frame, however, may not be reflectionally symmetric, and it may be necessary to keep that extra term for such turbulence. Work out Exercise 8.2 to learn more about this. We shall see in Chapter 16 that dynamo action in plasmas requires turbulence which is not symmetric under reflection.

We want to emphasize that all the deductions in this section have been based on purely geometrical considerations. Apart from the incompressibility condition, we have never used any physical law. We find that homogeneous isotropic turbulence can be described either by a scalar function $f(r)$ or a scalar function $E(k)$. These two scalar functions are of course related to each other, which we ask the reader to show in Exercise 8.1. This is as far as we can go on the basis of geometrical arguments. In order to draw some inferences about the functions $f(r)$ or $E(k)$, we need a physical theory. We now turn to the discussion of such a theory.

8.3 Kolmogorov's universal equilibrium theory

When developing the statistical theory of a gas kept in isolation, we calculate the equilibrium distribution of energy amongst different molecules by assuming that the most probable distribution corresponds to equilibrium. A turbulent fluid *kept in isolation*, however, cannot have an equilibrium distribution, because turbulence is inherently a dissipative process. The total energy of the molecules in an isolated gas remains constant in time. On the other hand, the total kinetic energy of an isolated turbulent fluid decreases with time due to viscous dissipation. Hence a turbulent fluid can be maintained in a steady state only if energy is continuously fed into the system such that the energy input rate equals the rate of energy dissipation. If the energy is fed by stirring the fluid in such a fashion that the turbulence produced is homogenous and isotropic, then such a system provides perhaps as close an analogy with a gas in thermodynamic equilibrium as we can have in the present context. Kolmogorov (1941a,b) proposed a bold theory to calculate the energy spectrum of such a system and thereby heralded a new era in the theory of turbulence.

The turbulent velocity field can be thought of as being made of many eddies of different sizes. The input energy is usually fed into the system in a way to produce the largest eddies. Kolmogorov (1941a,b) had the great intuition to realize that these large eddies can feed energy to the smaller eddies and these in turn feed the still smaller eddies, resulting in a cascade of energy from the larger eddies to the smaller ones. To have an understanding of exactly how this cascade may take place, consider the following simple application of Kelvin's vorticity theorem (proved in §4.6), which should hold for the larger eddies for which viscous dissipation is not very important. Let P and Q be two fluid elements on a vortex tube as shown in Figure 8.4(a). Any fluid element in a turbulent situation moves randomly. When the two fluid elements P and Q move randomly, the separation between them will increase in most cases. According to Kelvin's theorem, these fluid elements should 'carry' the vorticity they had. Hence the vortex tube will be stretched as shown in Figure 8.4(b). If the fluid is assumed incompressible, then the cross-section of the vortex tube must decrease in size. In other words, the size of the vortex, as estimated from its transverse dimensions, becomes smaller. This simple argument shows how larger vortices or eddies in a turbulent fluid can give rise to smaller ones.

Eddies of a certain size l are expected to have some typical velocity v associated with them. The corresponding Reynolds number lv/v should be large for the larger eddies, as we anticipate the viscosity to be not very important for them. The energy is cascaded from these large eddies to smaller and smaller eddies. We, however, cannot have

Figure 8.4 The stretching of a vortex tube by turbulence.

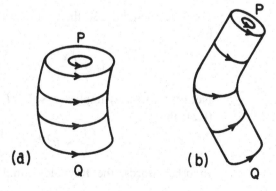

(a) (b)

eddies of indefinitely small size. For sufficiently small eddies of size l_d and velocity v_d, the Reynolds number is of order unity, i.e.

$$l_d v_d \sim \nu, \tag{8.19}$$

and the energy in these eddies is dissipated by viscosity. Hence (8.19) sets a rough limit to the smallest size eddies possible. Now we have the following scenario of the turbulence. Energy must be fed at some rate ϵ per unit mass per unit time at the largest eddies of size L and velocity V, for which the Reynolds number is

$$R \sim \frac{LV}{\nu} \gg 1. \tag{8.20}$$

This energy then cascades to smaller and smaller eddies until it reaches eddies satisfying (8.19), which dissipate this energy ϵ per unit mass per unit time in order to maintain the equilibrium. The intermediate eddies merely transmit this energy ϵ to the smaller eddies. These intermediate eddies are characterized only by their size l and velocity v. Since they are able to transmit the energy at the required rate ϵ, Kolmogorov (1941a,b) postulated that it must be possible to express ϵ in terms of l and v. On dimensional grounds, there is only one way of writing ϵ in terms of l and v:

$$\epsilon \sim \frac{v^3}{l}, \tag{8.21}$$

from which

$$v \sim (\epsilon l)^{1/3}. \tag{8.22}$$

This means that the velocity associated with eddies of a particular size is proportional to the cube root of the size—a result known as Kolmogorov's scaling law. Since (8.21) should be valid for eddies down to the smallest size, we have

$$\epsilon \sim \frac{v_d^3}{l_d}. \tag{8.23}$$

From (8.19) and (8.23), the characteristics of smallest eddies are given by

$$l_d \sim \left(\frac{\nu^3}{\epsilon}\right)^{1/4}, \qquad v_d \sim (\nu\epsilon)^{1/4}. \tag{8.24}$$

Noting that (8.21) implies $\epsilon \sim V^3/L$ and using (8.20), we get from (8.24) that

$$\frac{L}{l_d} \sim R^{3/4}, \qquad \frac{V}{v_d} \sim R^{1/4}. \tag{8.25}$$

In other words, the Reynolds number associated with the largest eddies determines how small the smallest eddies will be compared to them.

We now at last come to the question of the spectrum $E(k)$. Since the largest eddies have the size L corresponding to some wavenumber $k_L \sim 1/L$, we expect that there will be a cutoff at wavenumbers smaller than k_L as indicated in the sketch of the spectrum in Figure 8.5. On the other side also, there will be a cutoff at wavenumbers larger than $k_d \sim 1/l_d$ associated with the smallest eddies. The range of wavenumbers from k_L to k_d is called the *inertial range*. Within this range, the energy ϵ per unit mass per unit time is transferred to smaller eddies (i.e. larger wavenumbers). Hence $E(k)$ in the inertial range is expected to depend only on the two quantities ϵ and k. Dimensional considerations again dictate that this dependence can only be of the form

$$E(k) = C\epsilon^{2/3}k^{-5/3} \tag{8.26}$$

in the inertial range. This dependence of $E(k)$ on the $-5/3$ power of k is the famous $-5/3$ law of Kolmogorov.

We now argue that (8.26) expresses the same thing as (8.21). It is implied by (8.18) that the kinetic energy density v^2 associated with some wavenumber around k is $E(k)\,dk$, which we roughly write as $E(k)k$. Then

$$E(k)k \sim v^2.$$

Substituting for v from (8.22) with l replaced by $1/k$, we get

$$E(k)k \sim \epsilon^{2/3}k^{-2/3},$$

from which (8.26) readily follows.

Many readers may quite justifiably feel somewhat uncomfortable

Figure 8.5 A typical spectrum of turbulence according to Kolmogorov's theory.

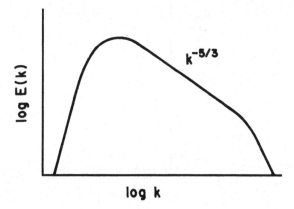

and unsure with what we are doing in this section. What we have presented cannot qualify to be called a 'derivation'. We have merely presented a string of plausible arguments that certain things could be true. But are they really true? Since we do not have a proper theory of turbulence, it is not possible to put Kolmogorov's theory on a completely rigorous theoretical foundation. The only way of checking the validity of the theory is to go to Mother Nature directly and determine whether experiments confirm the theoretical predictions. The −5/3 power law can be verified only by doing experiments on a turbulent fluid with a sufficiently large inertial range over which measurements can be made. It follows from (8.25) that a larger Reynolds number associated with the biggest eddies will give rise to a larger inertial range. In laboratory experiments, it is very difficult to reach high enough Reynolds numbers to produce a sufficiently broad inertial range. One of the first confirmations of Kolmogorov's theory was reported by Grant, Stewart and Moilliet (1962) by conducting experiments in a tidal channel between Vancouver Island and mainland Canada. Figure 8.6 shows the spectrum obtained by them. Since they had too many data points within the region indicated by the square, that region was enlarged into a separate figure in their paper. We, however, are more interested in the inertial range where the straight

Figure 8.6 The spectrum of turbulence obtained by Grant, Stewart and Moilliet (1962). (©Cambridge University Press. Reproduced from *Journal of Fluid Mechanics*.)

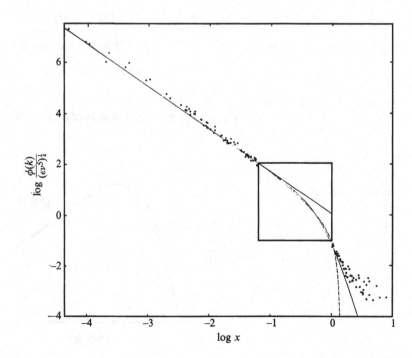

line corresponds to the $k^{-5/3}$ dependence, and we note that data fit the straight line quite well. After this pioneering work of Grant, Stewart and Moilliet (1962), many experiments have established the basic soundness of Kolmogorov's ideas, and the dimensionless constant C in (8.26) appears to be a universal constant with a value close to 1.5 (Monin and Yaglom 1975, p. 485).

8.4 Turbulent diffusion

Although the theory of homogeneous isotropic turbulence may have great mathematical elegance, it is somewhat of an idealized subject and often not very useful in handling practical problems involving turbulence, in which inhomogeneities often play important roles. So we now turn to turbulence in inhomogeneous systems. One of the consequences of inhomogeneity in a fluid system is that it gives rise to transport phenomena. What does turbulence do to these phenomena? Since fluid elements carrying different physical quantities move around randomly in a turbulent situation, we expect turbulence to enhance the transport processes. The theory of *turbulent diffusion* was first formulated by Taylor (1921a). Even people who have never heard of Taylor's theory make use of the notion of turbulent diffusion whenever they stir coffee to mix sugar in it. If one puts sugar in the coffee and does not stir it, then the sugar mixes with the coffee only by molecular diffusion, which takes a very long time. On the other hand, if the coffee is stirred, then we generate turbulence inside it and turbulent diffusion mixes up the sugar much more efficiently than molecular diffusion.

We assume that some markers are introduced in a region in a turbulent fluid at time $t = 0$. If the markers are carried along with the fluid velocity, then the displacement of a marker after time $t = T$ is given by

$$\mathbf{x}(T) = \int_0^T \mathbf{v}_L(t) \, dt, \qquad (8.27)$$

where $\mathbf{v}_L(t)$ is the fluid velocity at the position of the marker at time t. The subscript L is a reminder that we are employing a Lagrangian approach by following a marker in the fluid. It is obvious that the mean displacement averaged over all the markers must be zero. The mean square displacement, however, is not zero and is given by

$$\overline{x^2(T)} = \int_0^T dt \int_0^T d\tau \, \overline{\mathbf{v}_L(t) \cdot \mathbf{v}_L(\tau)}. \qquad (8.28)$$

Here $\overline{\mathbf{v}_L(t) \cdot \mathbf{v}_L(\tau)}$ is again a velocity correlation function, which is, however, quite different from the correlation functions we have considered previously. It is a measure of the correlation between the velocities at the positions of a marker at times t and τ. If we have a steady state turbulence, then this correlation function depends on the difference of time $\tau - t$ alone. Hence we write

$$\overline{\mathbf{v}_L(t) \cdot \mathbf{v}_L(\tau)} = \overline{v^2}\, R(\tau - t). \tag{8.29}$$

It is clear from this definition that

$$R(0) = 1. \tag{8.30}$$

We also assume the symmetry $R(-\tau) = R(\tau)$. We expect the turbulence to have some correlation time τ_{cor} such that $R(\tau)$ is substantially different from zero only if τ is of the order or smaller than τ_{cor}. Substituting (8.29) in (8.28), we have

$$\overline{x^2(T)} = \int_0^T dt\,\overline{v^2} \int_0^T d\tau\, R(\tau - t). \tag{8.31}$$

If we allow the markers to move for some time much shorter than the turbulence correlation time, i.e if $T \ll \tau_{\mathrm{cor}}$, then we write $R(\tau - t) \approx 1$ during that time such that

$$\overline{x^2(T)} = \overline{v^2}\, T^2. \tag{8.32}$$

We, however, should look at the other limit of $T \gg \tau_{\mathrm{cor}}$ in order to study the statistical effects of turbulence. In that limit, we can change the limits of integration of τ in (8.31) to $-\infty$ and ∞ without much error, since the integrand does not contribute much when $|\tau - t|$ is much larger than the correlation time. Hence

$$\overline{x^2(T)} = \int_0^T dt\,\overline{v^2} \int_{-\infty}^{\infty} d\tau\, R(\tau - t). \tag{8.33}$$

Writing

$$D_T = \tfrac{1}{3}\overline{v^2} \int_0^{\infty} R(\tau)\, d\tau \tag{8.34}$$

and neglecting its spatio-temporal variations, we have from (8.33) that

$$\overline{x^2(T)} = 6 D_T T. \tag{8.35}$$

The linear dependence of the mean square displacement on time is a clear signature of a diffusive process. We now argue that D_T can be interpreted as a diffusion coefficient.

Let $n(\mathbf{x}, t)$ be the density of the markers. If the dispersion of the markers is a diffusive process, then $n(\mathbf{x}, t)$ should satisfy a diffusion equation

$$\frac{\partial n}{\partial t} = D\nabla^2 n, \tag{8.36}$$

where D is the diffusion coefficient. Let the markers be introduced near the origin of the coordinates in a symmetric fashion such the evolution is spherically symmetric. Then the mean square displacement at some time t is given by

$$\overline{x^2(t)} = \frac{\int_0^\infty r^2 n(r,t)\, 4\pi r^2\, dr}{\int_0^\infty n(r,t)\, 4\pi r^2\, dr}. \tag{8.37}$$

It readily follows from (8.36) that

$$\frac{\partial}{\partial t} \int_0^\infty r^2 n\, 4\pi r^2\, dr = D \int_0^\infty \frac{\partial}{\partial r}\left(r^2 \frac{\partial n}{\partial r}\right) 4\pi r^2\, dr. \tag{8.38}$$

Integrating by parts, the R.H.S. of (8.38) is found to be $6D \int_0^\infty n\, 4\pi r^2\, dr$ so that

$$\int_0^\infty r^2 n\, 4\pi r^2\, dr = 6Dt \int_0^\infty n\, 4\pi r^2\, dr.$$

Therefore (8.37) implies

$$\overline{x^2(t)} = 6Dt,$$

which is the same as (8.35). Therefore D_T, as defined in (8.34), can justifiably be interpreted as a diffusion coefficient and is called the coefficient of turbulent diffusion.

When we derived the expressions for molecular transport processes by kinetic theory, we found that the different transport coefficients have different expressions. For example, look at the expressions (3.36) and (3.40) for thermal conductivity and viscosity. The coefficients for different turbulent transport processes, however, are believed to be the same. The cause for this contrast between molecular and turbulent transport is that molecular motions are related to the quantities they transport. Molecules carrying more thermal energy move faster. This is not true in the case of a turbulent fluid. The temperature of a blob of fluid does not have a direct functional relation with the turbulent velocity of the blob. Hence all the diffusion coefficients for the passive transport of any quantity in a turbulent fluid have the approximate value

$$D_T \approx \tfrac{1}{3}\overline{v^2}\tau_{\text{cor}} \tag{8.39}$$

from (8.34), since $\int_0^\infty R(\tau)d\tau \approx \tau_{\text{cor}}$. We may mention that a turbulent transport coefficient can have a value different from that given by (8.39) if the quantity being transported can react back on the turbulence.

8.5 The mean equations

Let us now consider how the turbulent viscosity may affect the Navier–Stokes equation. For an incompressible fluid, we can write (5.10) in the form

$$\frac{\partial}{\partial t}(\rho v_i) = \rho F_i + \frac{\partial}{\partial x_j}\left(-p\delta_{ij} - \rho v_i v_j + \mu\frac{\partial v_i}{\partial x_j}\right). \qquad (8.40)$$

To apply this equation to a turbulent situation, we follow Reynolds (1895) and break up all the quantities into mean and fluctuating parts: $v_i = \overline{v_i} + v_i'$, $p = \overline{p} + p'$, $F_i = \overline{F_i} + F_i'$. Let us now substitute them in (8.40) and then take an average. All the terms linear in fluctuating variables give zero on averaging. We have to be a little bit careful about the only nonlinear term

$$\rho\overline{v_i v_j} = \rho\overline{(\overline{v_i} + v_i')(\overline{v_j} + v_j')} = \rho(\overline{v_i}\,\overline{v_j} + \overline{v_i}\overline{v_j'} + \overline{v_i'}\overline{v_j} + \overline{v_i'v_j'}).$$

Although $\overline{v_i'} = \overline{v_j'} = 0$, the term $\overline{v_i'v_j'}$ is generally non-zero. Hence

$$\rho\overline{v_i v_j} = \rho(\overline{v_i}\,\overline{v_j} + \overline{v_i'v_j'}). \qquad (8.41)$$

On taking the average of (8.40) applied to a turbulent situation, we therefore have

$$\frac{\partial}{\partial t}(\rho\overline{v_i}) = \rho\overline{F_i} + \frac{\partial}{\partial x_j}\left(-\overline{p}\delta_{ij} - \rho\overline{v_i}\,\overline{v_j} - \rho\overline{v_i'v_j'} + \mu\frac{\partial\overline{v_i}}{\partial x_j}\right). \qquad (8.42)$$

This equation was first derived by Reynolds (1895) and is often called the *Reynolds equation*. It is very similar to the original equation (8.40) apart from the additional term $\overline{v_i'v_j'}$. This additional term $\overline{v_i'v_j'}$, however, is the crucial term which describes how turbulence affects the mean quantities and is known as the *Reynolds stress*. If we want to study the evolution of the mean quantities by using (8.42), then we have to evaluate the Reynolds stress.

One possibility of evaluating $\overline{v_i'v_j'}$ is to employ the equation

$$\frac{\partial}{\partial t}(\overline{v_i'v_j'}) = \overline{v_i'\frac{\partial v_j'}{\partial t}} + \overline{\frac{\partial v_i'}{\partial t}v_j'}. \qquad (8.43)$$

On substituting for $\partial v_j'/\partial t$ and $\partial v_i'/\partial t$ (obtained by subtracting (8.42) from (8.40)), it is straightforward to see that this equation will involve the three-correlation functions $\overline{v_i'v_j'v_k'}$. If one tries to derive an evolution equation for the three-correlation function, that will in turn involve four-correlation functions. This goes on indefinitely, posing what is called the *closure problem*. Much work was done in the 1960s and 1970s on various approximation schemes to handle the closure problem. This is a highly technical subject and its achievements are of questionable value as far as practical applications are concerned. Since one effect of turbulence on the mean quantities is to introduce turbulent transport processes, one crude closure scheme is to assume that the Reynolds stress has a form resembling (5.7) with the ordinary viscosity replaced by turbulent viscosity. Many astrophysics, geophysics and engineering problems are often tackled by using (8.42) with the Reynolds stress given by

$$\overline{v_i' v_j'} = -D_T \left(\frac{\partial \overline{v_i}}{\partial x_j} + \frac{\partial \overline{v_j}}{\partial x_i} \right). \qquad (8.44)$$

This approach *usually* gives results in fair agreement with experiments or observations, thus giving us confidence that this approach is not completely unreasonable. But one should keep in mind that this very crude closure scheme can occasionally lead to erroneous conclusions in complicated situations. Chapters 4–5 of Tennekes and Lumley (1972) discuss many examples how the mean equations can be applied to study practical problems.

8.6 Turbulence in astrophysics

Turbulence appears quite ubiquitous in the astrophysical Universe. If we ever have a full theory of turbulence in future, then that will make a tremendous impact on many areas in astrophysics.

We saw in §7.2 that if the temperature gradient of an atmosphere goes beyond a critical value, then the atmosphere becomes unstable to convection. Inside most main-sequence stars, there is some region where the temperature gradient is unstable and we have turbulent motions associated with convection. In the case of very massive stars, the central temperatures are very high. Since the nuclear energy generation rate is a rather sensitive function of temperature at those temperatures, the energy production rate increases very sharply as we move towards the centre. This tends to produce a steep temperature gradient in the central region, and the massive stars, as a result, have convectively unstable cores. In the case of lighter stars, the outer layers are relatively cool (by stellar standards!). Stellar matter at such temperatures happens to be rather opaque to radiation. If energy had to be transported by radiation in these outer layers, then it would have been a rather inefficient process and the temperature gradients would have to be pretty large. Since such temperature gradients would make the system convectively unstable, we expect the lighter stars to have convectively unstable outer envelopes full of turbulent motion. This is the case for the Sun. We presented a photograph of the solar surface in Figure 7.5, which shows clear evidence of convective turbulence just below the visible surface.

Since most stars have some region of turbulent convection, it is essential to handle that region in an adequate fashion in order to develop stellar models, i.e. models which tell us how quantities like density $\rho(r)$ or temperature $T(r)$ vary with radius inside the star. Biermann (1948) and Vitense (1953) adapted some ideas of Prandtl (1925) to the astrophysical situation and developed the *mixing length theory*, which has become the standard tool of modelling stellar convection zones. Any textbook on stellar structure provides an account of the mixing length theory (see, for example, Schwarzschild 1958, §7; Kippenhahn and Weigert 1990, Chapter 7). We shall not discuss this theory here. The basic idea is the following. A hot blob of gas is assumed to rise through a typical distance l before it mixes with the surroundings and gives up its excess heat to the regions around. Similarly a cold blob of gas also descends through a distance l before it is mixed up with the surroundings. Taking the mixing length l to be a fraction of the pressure scale height, one can carry out calculations to derive models of stellar structure. When the convective energy transport is very efficient, the temperature gradient can be shown to be rather close to the critical value given in (7.5). In other words, we can get reasonably good stellar models if we use the equation

$$\frac{dT}{dr} = \left(1 - \frac{1}{\gamma}\right) \frac{T}{p} \frac{dp}{dr}. \tag{8.45}$$

So the stellar structure models are *not* sensitively dependent on the details of turbulence. This is something very lucky. Stellar structure happens to be one of the most solid and well-understood areas of modern astrophysics, even though our understanding of turbulence in the interiors of stars is very primitive.

Understanding stellar structure, however, is not the only concern in stellar astronomy. Turbulence may play an important role in transporting and distributing angular momentum in different parts of a rotating star. We have pointed out in §7.6 that the angular velocity varies inside the Sun in a complicated fashion. It is believed that the Reynolds stress associated with turbulence in the solar convection is what causes this distribution of angular momentum, although we are very far from having a proper understanding of the subject. We shall further see in Chapter 16 that turbulence is the cause of the dynamo effect, which produces magnetic fields in astronomical objects. In contrast to the stellar structure theory which is insensitive to the details of turbulence, the theory of stellar rotation and the dynamo theory depend crucially on some subtle aspects of turbulence. In these subjects, one usually proceeds with some suitable mean equations like the ones discussed in the previous section. Not only are we unable to ascertain the values of some important terms in the equations, sometimes we are even unable to conclude if they are positive or negative (see §16.6)! The theory of rotating stars and the theory of magnetic stars are still at very unsatisfactory stages of development. Unless there is a breakthrough in our understanding of convective turbulence, these subjects will remain of the nature of parameter-fitting exercises.

Stars are not the only astrophysical systems within which we find turbulence. The interstellar medium between the stars also seems to be usually in a turbulent state. Turbulence in the interstellar medium gives rise to the dynamo action in the galaxy, thereby producing the galactic magnetic field. One of the very strange effects of the galactic magnetic field is that a few particles get accelerated to very high energies. We shall discuss this curious problem of cosmic ray acceleration in §10.5. The kinetic energy density of turbulence in the interstellar medium, the energy density of the interstellar magnetic field and the energy density of the cosmic ray particles per unit volume appear to have fairly close values. In the interstellar space around the solar system, all these different energy densities have values of the order of 10^{-12} erg cm^{-3}. In other words, an equipartition of energy takes place in the interstellar medium amongst the turbulence, the magnetic field and the cosmic rays. We are still far from having a proper theoretical understanding of interstellar turbulence with its different manifestations.

Exercises

8.1   Define $R(r)$ as

$$R(r) = \tfrac{1}{2} R_{ii}(r),$$

where $R_{ij}(r)$ is the correlation function introduced in (8.3) and show that

$$R(r) = \frac{1}{2}\overline{v^2}\left[f(r) + \frac{1}{3}r\frac{df}{dr} \right],$$

where $f(r)$ is the scalar function introduced in (8.6).

Prove that the energy spectrum $E(k)$ has the following

reciprocal relations with $R(r)$:

$$E(k) = \frac{2}{\pi} \int_0^\infty R(r)kr \, \sin(kr) \, dr,$$

$$R(r) = \int_0^\infty E(k)\frac{\sin(kr)}{kr} \, dk.$$

8.2   If the turbulence is isotropic, but not reflectionally symmetric, then one can introduce an additional term

$$-\frac{iF(k)}{8\pi k^4}\epsilon_{ijk}k_k$$

in the expression of $\Phi_{ij}(k)$ in (8.16). Define helicity as the dot product of the velocity **v** and the vorticity ω. Prove that the average helicity

$$\overline{\mathbf{v}.\omega} = \int_0^\infty F(k) \, dk$$

so that $F(k)$ can be interpreted as the helicity spectrum. Justify that the average helicity and, therefore, $F(k)$ will be zero if the turbulence is reflectionally symmetric.

8.3   The convective granules near the solar surface have the typical size of about 10^3 km and the typical velocity of about 1 km s^{-1}. Make an estimate of the coefficient of turbulent diffusion. Sunspots are structures on the solar surface with sizes of about 10^4 km. If the sunspots decay due to turbulent diffusion, estimate the lifetime of a sunspot.

9 Rotation and hydrodynamics

9.1 Introduction

Most astrophysical bodies have some angular momentum associated with them. Hence, in many astrophysical problems, we have to consider fluid flows in the presence of rotation.

Suppose we take water in a bucket and start rotating the bucket around its axis. Initially only the water near the walls of the bucket starts rotating with the bucket, whereas the water in the central part still remains at rest. In other words, the angular velocity varies within the water. Such a variation of angular velocity within a rotating fluid is referred to as *differential rotation*. If we wait for some time, then viscosity eventually stops the relative motions amongst different layers of water, and the whole water spins with the same angular velocity as that of the bucket. When a fluid mass rotates with the angular velocity constant within it, we call it *solid-body rotation*, since a rigid solid body rotates with constant angular velocity inside.

In the astrophysical Universe, we find many objects rotating not like a solid body. This may be due to two reasons. Firstly, viscous forces may not have enough time to establish a solid-body rotation. Secondly, there may be some physical mechanism which maintains the differential rotation. To understand more about this, let us consider how the centrifugal force in a rotating body of fluid is balanced. Let us think of a steady, axisymmetric rotation such that we can put $\partial/\partial t = 0$, $\partial/\partial\theta = 0$ and $v_r = 0$ in the r component of the Navier–Stokes equation in cylindrical coordinates, as given in equation (C.5). This leads to

$$-\frac{v_\theta^2}{r} = g_r - \frac{1}{\rho}\frac{\partial p}{\partial r}, \tag{9.1}$$

where g_r is the r component of the gravitational field. In some astrophysical systems such as accretion disks or the disk of a spiral galaxy, the pressure gradient force may be quite unimportant. In such situations, the centrifugal force is balanced by gravity so that the angular velocity given by $v_\theta/r = \sqrt{|g_r|/r}$ generally varies with radius, making differential rotation inevitable in such systems. The effect of viscosity in such systems is to induce a slow radial inflow of matter rather than to produce solid-body rotation, as we have seen in our discussion of accretion disks (§5.7). The situation is completely different inside a slowly rotating star, where the gravitational field and the pressure gradient nearly balance each other. But the pressure can adjust itself in such a fashion that there remains an unbalanced part of pressure gradient, which counteracts the centrifugal force. Since there is no obvious mechanism within such stars to sustain differential rotation, we may expect such stars eventually to rotate like solid bodies due to the effect of viscosity. Usually, however, the effect of molecular viscosity inside stars is completely negligible. Viscosity appears in the Navier–Stokes equation in the term $\nu\nabla^2 \mathbf{v}$, which is insignificant in systems with large length scales. If, however, there is convection inside a star, we expect the turbulent diffusion of various quantities like the angular momentum as pointed out in §8.4. The effect of turbulent diffusion may *not* be like a scalar viscosity in all respects. It may sometimes be appropriate to introduce an anisotropic viscosity to model the turbulent diffusion. The theory of rotating stars with anisotropic viscosity was formulated by Kippenhahn (1963). The steady state with anisotropic viscosity is usually a state of differential rotation. If the viscosity is made more isotropic, then the star rotates more like a solid body. The Sun happens to be the only star of which the differential rotation is observationally well studied, as pointed out in §7.6. The angular velocity between the pole and the equator of the Sun differs by more than 10%.

It should be noted that all types of differential rotation are not stable, an important result obtained by Rayleigh (1917). Let us consider a fluid of uniform density rotating differentially around an axis of symmetry. To find the codition for its stability, we suppose that a fluid ring at a distance r_0 from the axis moving with velocity v_0 is interchanged with a fluid ring at a greater distance r_1 (i.e. $r_1 > r_0$) moving with velocity v_1. The system is stable if the displaced fluid rings tend to return to their initial positions, whereas it is unstable if the displaced rings move further away. Assuming the conservation of angular momentum, we conclude that the fluid ring displaced from r_0 to r_1 acquires a velocity $(r_0/r_1)v_0$. The ring previously at r_1 had a centripetal acceleration v_1^2/r_1, which must have been provided by the various forces there such as the part of the pressure gradient left after balancing gravity. The ring brought to r_1 now requires a centripetal acceleration $r_0^2 v_0^2/r_1^3$ to remain in its new position. If this is less than v_1^2/r_1, then we expect the forces present there to push the ring inward towards its initial position. The condition for stability is then

$$\frac{r_0^2 v_0^2}{r_1^3} < \frac{v_1^2}{r_1},$$

so that

$$(r_0^2 \Omega_0)^2 < (r_1^2 \Omega_1)^2,$$

where $\Omega_0 = v_0 r_0$ and $\Omega_1 = v_1 r_1$ are the angular velocities at r_0 and r_1. This stability condition can be written more conveniently as

$$\frac{d}{dr}[(r^2\Omega)^2] > 0. \tag{9.2}$$

This is known as *Rayleigh's criterion* (Rayleigh 1917).

When an astronomical object rotates like a solid body, it is often convenient to study hydrodynamic phenomena from a frame of reference rotating with the same angular velocity. For the Earth, one can introduce a rotating frame of reference without any difficulty. Even for stars like the Sun which do not rotate exactly like a solid body, it is useful in many problems to introduce a frame rotating with an average angular velocity. The next section will provide an elementary introduction to hydrodynamics in rotating frames. Then §9.3 will be devoted to a topic which has a long and distinguished history of research starting with Newton (1689). It is the problem of finding the shape of a self-gravitating rotating body of fluid, which we anticipate to be flattened near the poles. We shall restrict ourselves mainly to incompressible and ideal fluids in §9.2 and §9.3. We pointed out in §4.2 that we can avoid thermodynamic considerations when a fluid is incompressible. This simplification turns out to be particularly convenient in rotating frames, since the thermodynamics of a moving gas in a rotating frame is a problem of formidable difficulty. The subject of compressible rotating fluids is outside the scope of this elementary textbook, although it is a subject of considerable importance in astrophysics and a few comments will be made in §9.4.

9.2 Hydrodynamics in a rotating frame of reference

Suppose we are in a frame of reference rotating with constant angular velocity Ω. It is well known that the usual acceleration has to be replaced in the rotating frame by

$$\frac{d\mathbf{v}}{dt} \rightarrow \frac{d\mathbf{v}}{dt} + 2\mathbf{\Omega} \times \mathbf{v} + \mathbf{\Omega} \times (\mathbf{\Omega} \times \mathbf{r}), \tag{9.3}$$

where $\mathbf{r}$ is the position vector with respect to some origin on the axis of rotation (see, for example, Goldstein 1980, §4–10). Making this replacement in the Navier–Stokes equation (5.10), we obtain the equation of motion in the rotating frame

$$\frac{\partial \mathbf{v}}{\partial t} + (\mathbf{v} \cdot \nabla)\mathbf{v} = -\frac{1}{\rho}\nabla p + \mathbf{F} + \nu\nabla^2\mathbf{v} - 2\mathbf{\Omega} \times \mathbf{v} - \mathbf{\Omega} \times (\mathbf{\Omega} \times \mathbf{r}), \tag{9.4}$$

where $-2\mathbf{\Omega} \times \mathbf{v}$ is the Coriolis force and $-\mathbf{\Omega} \times (\mathbf{\Omega} \times \mathbf{r})$ is the centrifugal force, which can also be written as $\frac{1}{2}\nabla(|\mathbf{\Omega} \times \mathbf{r}|^2)$. If the body force $\mathbf{F}$ is of gravitational origin, we can write it as $-\nabla\Phi$, where Φ is the gravitational potential. Then (9.4) becomes

$$\frac{\partial \mathbf{v}}{\partial t} + (\mathbf{v} \cdot \nabla)\mathbf{v} = -\frac{\nabla p}{\rho} - \nabla\left(\Phi - \frac{1}{2}|\mathbf{\Omega} \times \mathbf{r}|^2\right) + \nu\nabla^2\mathbf{v} - 2\mathbf{\Omega} \times \mathbf{v}. \tag{9.5}$$

This is the basic equation of motion for fluids in a rotating frame of reference. It is clear that the effect of the centrifugal force is to introduce a potential force that modifies gravity. Equation (9.5) readily suggests that we can introduce an effective gravitational potential

$$\Phi_{\text{eff}} = \Phi - \frac{1}{2}|\mathbf{\Omega} \times \mathbf{r}|^2. \tag{9.6}$$

The effect of the Coriolis force, which comes into action only when there are motions relative to the rotating frame, is more complicated.

When considering fluid flows through pipes in the laboratory, we do not usually have to be concerned that the Earth is a rotating frame of reference. On the other hand, large ocean currents and monsoon winds are very much influenced by the Earth's rotation. We can get an idea from (9.5) when rotational effects become important. It is easy to see that the centrifugal force has to be taken into account when it modifies the effective gravitational potential significantly (which is not the case for the Earth's rotation). To figure out how important the Coriolis force is, we should compare it with the other term $(\mathbf{v} \cdot \nabla)\mathbf{v}$ in (9.5) which, like the Coriolis force, is non-zero only when there are motions with respect to the rotating frame. If V and L are the typical velocity and length scales, then $(\mathbf{v} \cdot \nabla)\mathbf{v}$ is of order V^2/L and the Coriolis force $-2\mathbf{\Omega} \times \mathbf{v}$ is of order ΩV. The ratio of the two

$$\epsilon = \frac{V^2/L}{\Omega V} = \frac{V}{\Omega L} \tag{9.7}$$

is a dimensionless number known as the *Rossby number* (named after C. G. Rossby, a pioneer in geophysical fluid dynamics). The Coriolis force is then important if the Rossby number is of order unity or less. This is the case for large-scale motions in the atmosphere or the oceans, but not for most fluid phenomena in the laboratory. The study of large-scale motions in the atmosphere and the oceans is the subject of *geophysical fluid dynamics*, where one considers the dynamics of a thin spherical shell of fluid with small Rossby number.

9.2.1 The geostrophic approximation

When considering large-scale atmospheric or oceanic circulations, one can usually simplify the equations in the following way. The fluid flows associated with such large-scale circulations are nearly horizontal because of the the thinness of the atmosphere or the ocean compared to the overall dimensions of the Earth. If the flows are changing slowly and have low Rossby numbers, then the terms on the L.H.S. of (9.5) are very small compared to the leading terms on the R.H.S.. Amongst the terms on the R.H.S. also, the centrifugal and viscosity terms are small and can be neglected. Then (9.5) reduces to

$$-\frac{\nabla p}{\rho} - g\hat{\mathbf{e}}_r - 2\Omega \times \mathbf{v} = 0, \tag{9.8}$$

where we have written $-\nabla\Phi = -g\hat{\mathbf{e}}_r$ for gravity.

We now consider the vertical and the horizontal components of this equation. The magnitude of the Coriolis force is usually small compared to the gravity. Hence the Coriolis force does not play an important role in the force balance in the vertical direction, in which the gravity, therefore, has to be balanced by the pressure gradient, i.e.

$$\frac{1}{\rho}\frac{\partial p}{\partial r} = -g. \tag{9.9}$$

Since gravity is absent in the horizontal direction, the weaker Coriolis force gets the chance to play a dominant role in the horizontal force balance. According to (9.8), it has to be balanced by the horizontal component of pressure gradient, i.e.

$$\nabla_h p = -2\rho(\Omega \times \mathbf{v})_h, \tag{9.10}$$

where the subscript h implies the horizontal component. We expect the horizontal pressure gradient to be much smaller in magnitude compared to the vertical pressure gradient. The large vertical gradient balances the large force of gravity, whereas the smaller horizontal gradient balances the much weaker Coriolis force.

Equations (9.9) and (9.10) for the vertical and the horizontal directions constitute what is known as the *geostrophic approximation*. It is used extensively in geophysical fluid dynamics, and its consequences are discussed in standard textbooks on the subject such as Pedlosky (1987). We merely point out one intriguing consequence. It should be obvious from (9.10) that the horizontal flow velocity $\mathbf{v}$ has to be *perpendicular* to $\nabla_h p$ if its cross product with Ω is to balance $\nabla_h p$. This is contrary to what we expect in common situations. If pressure drops in a region, we normally expect the surrounding fluid to flow into that region. In other words, we expect the velocity $\mathbf{v}$ to be in the direction of $-\nabla p$. In geophysical circulations, however, we find $\mathbf{v}$ to be perpendicular to $-\nabla_h p$! Hence, if there is a depression in a region of the atmosphere, then the air in the surrounding regions, instead of rushing towards the depression, skirts around it.

9.2.2 Vorticity in a rotating frame

In the rest of this section and the next section, we want to consider ideal, incompressible fluids. We therefore drop the viscosity term in (9.5) and write $\nabla p/\rho$ as $\nabla(p/\rho)$. Making use of the vector identity (4.16), we can write (9.5) in the form

$$\frac{\partial \mathbf{v}}{\partial t} - \mathbf{v} \times \omega = -\nabla \left(\frac{p}{\rho} + \frac{1}{2}v^2 + \Phi - \frac{1}{2}|\Omega \times \mathbf{r}|^2 \right) - 2\Omega \times \mathbf{v}. \quad (9.11)$$

On taking the curl of this equation, we get

$$\frac{\partial \omega}{\partial t} = \nabla \times (\mathbf{v} \times \omega) + \nabla \times (\mathbf{v} \times 2\Omega).$$

Since Ω is constant in time, this equation can be put in the form

$$\frac{\partial}{\partial t}(\omega + 2\Omega) = \nabla \times [\mathbf{v} \times (\omega + 2\Omega)]. \quad (9.12)$$

Figure 9.1 The spreading of a volume of fluid into a thinner layer in a rotating frame of reference.

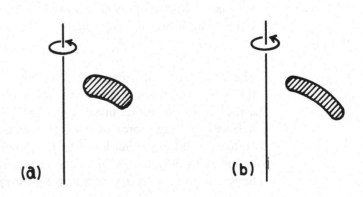

(a)   **(b)**

This equation has the same form as (4.42), and we at once conclude

$$\frac{d}{dt} \int_S (\omega + 2\Omega).d\mathbf{S} = 0. \qquad (9.13)$$

This is the generalization of Kelvin's vorticity theorem in rotating frames and is often known as *Bjerknes's theorem* (Bjerknes 1937).

This theorem gives an idea how vortices are generated quite naturally in rotating frames. Figure 9.1(a) shows a volume of fluid which is at rest in a rotating frame. Suppose the fluid is squeezed to spread into a thinner layer as indicated in Figure 9.1(b). The flux $\int_S \Omega \cdot d\mathbf{S}$ associated with the fluid certainly becomes larger after the fluid has been spread into the thinner layer. Even though there was no vorticity initially, the fluid after spreading must develop a vorticity ω opposite to Ω so that the integral $\int_S (\omega + 2\Omega) \cdot d\mathbf{S}$ remains conserved. This gives an idea how cyclonic storms may be produced in the Earth's atmosphere.

9.2.3 Taylor–Proudman theorem

Often we have fairly steady fluid flows in a rotating frame of reference. It follows from (9.12) that for such flows

$$\nabla \times [\mathbf{v} \times (\omega + 2\Omega)] = 0.$$

Suppose the flows are slow such that any vorticity associated with the flow is small compared to Ω. Then we must have

$$\nabla \times (\mathbf{v} \times \Omega) = 0, \qquad (9.14)$$

from which

$$(\Omega \cdot \nabla)\mathbf{v} = 0. \qquad (9.15)$$

This means that $\mathbf{v}$ does not change in the direction of Ω. In other words, slow steady flows in rotating frames tend to be invariant parallel to the rotation axis. This result is known as the *Taylor–Proudman theorem* (Proudman 1916; Taylor 1921b).

Suppose we have water in a cylindrical vessel rotating about its axis with a solid object fixed to the bottom of the vessel. In the steady state, the water rotates like a solid body. Suddenly we increase the rotation speed of the vessel slightly. Water will take some time to pick up the increased rotation (due to the small viscosity of water). During that time, we have a slow rotation of the body of water with respect to the vessel and we expect the Taylor–Proudman theorem to hold. Figure 9.2 shows a photograph of the system in which there is a suspension of reflecting powder in the water so that a light shining on it makes the flow pattern visible. The water at the bottom naturally skirts around the solid object. Since the Taylor–Proudman theorem suggests that the flow pattern should be invariant parallel to the rotation axis, we find the striking phenomenon of water even in the higher layers moving in a way as if it is skirting around an invisible obstacle.

9.3 Self-gravitating rotating masses

Let us consider a self-gravitating fluid mass of constant density ρ. If there is no rotation in the system, we know from symmetry that the equilibrium configuration of the system must be spherical. Now suppose we put in some angular momentum. We expect that the rotation would cause some flattening near the poles. The calculations become somewhat manageable if the fluid undergoes solid-body rotation and we can go to a frame with angular velocity $\boldsymbol{\Omega}$ in which the fluid is everywhere at rest. Then, from (9.11),

$$\nabla\left(\frac{p}{\rho} + \Phi - \frac{1}{2}|\boldsymbol{\Omega} \times \mathbf{r}|^2\right) = 0,$$

which implies

$$\frac{p}{\rho} + \Phi - \frac{1}{2}|\boldsymbol{\Omega} \times \mathbf{r}|^2 = \text{constant}. \tag{9.16}$$

On the outer surface of the fluid mass, the pressure can be taken to be zero. Choosing the z axis along the axis of rotation, we then have

Figure 9.2 A photograph showing a Taylor–Proudman column in a rotating fluid. Reproduced from Greenspan (1968). (©Cambridge University Press.)

from (9.16) that

$$\Phi - \tfrac{1}{2}\Omega^2(x^2 + y^2) = \text{constant} \tag{9.17}$$

on the outer surface.

We now want to establish that the fluid takes up the shape of an ellipsoid. To proceed further, we have to obtain the gravitational potential Φ due to an ellipsoid. It is one of the triumphs of potential theory that this problem can be solved exactly for an ellipsoid of uniform density. Let us consider an ellipsoid with uniform density ρ inside the boundary surface

$$\frac{x^2}{a^2} + \frac{y^2}{b^2} + \frac{z^2}{c^2} = 1. \tag{9.18}$$

The gravitational potential at any point inside this ellipsoid is given by

$$\Phi = \pi G \rho(\alpha_0 x^2 + \beta_0 y^2 + \gamma_0 z^2 - \chi_0), \tag{9.19}$$

where

$$\alpha_0 = abc \int_0^\infty \frac{d\lambda}{(a^2 + \lambda)\Delta}, \quad \beta_0 = abc \int_0^\infty \frac{d\lambda}{(b^2 + \lambda)\Delta},$$

$$\gamma_0 = abc \int_0^\infty \frac{d\lambda}{(c^2 + \lambda)\Delta}, \tag{9.20}$$

and

$$\chi_0 = abc \int_0^\infty \frac{d\lambda}{\Delta}, \tag{9.21}$$

Δ being given by

$$\Delta = [(a^2 + \lambda)(b^2 + \lambda)(c^2 + \lambda)]^{1/2}. \tag{9.22}$$

Here we merely quote this famous result without proof. The reader is referred to Chandrasekhar (1969, Chapter 3) or Binney and Tremaine (1987, §2.3) for further details.

If the rotating mass of fluid takes up the shape of an ellipsoid, then the expression (9.19) for the gravitational potential Φ at the bounding surface has to satisfy (9.17). In other words, we must have

$$\left(\alpha_0 - \frac{\Omega^2}{2\pi G \rho}\right) x^2 + \left(\beta_0 - \frac{\Omega^2}{2\pi G \rho}\right) y^2 + \gamma_0 z^2 = \text{constant} \tag{9.23}$$

on the surface given by (9.18). In order for (9.18) and (9.23) to hold simultaneously, the coefficients of x^2, y^2 and z^2 in these two equations must be proportion This implies

$$\left(\alpha_0 - \frac{\Omega^2}{2\pi G \rho}\right) a^2 = \left(\beta_0 - \frac{\Omega^2}{2\pi G \rho}\right) b^2 = \gamma_0 c^2. \tag{9.24}$$

If we can find solutions of this equation, then we would have established that rotating fluids take up ellipsoidal configurations.

9.3.1 Maclaurin spheroids

We intuitively expect the rotating fluid to take up a symmetric and flattened configuration around the rotation axis. This means that the axes of the ellipsoid given by (9.18) must satisfy

$$a = b > c. \tag{9.25}$$

An ellipsoid with such axes is called a spheroid, of which the eccentricity e is defined by

$$e^2 = 1 - \frac{c^2}{a^2}. \tag{9.26}$$

We now wish to solve (9.24) for such a spheroid. Since Maclaurin (1740) first demonstrated the existence of such spheroidal solutions, they are called *Maclaurin spheroids* in his honour.

When $a = b$, the integrals in (9.20) can be worked out in closed forms. They are

$$\alpha_0 = \beta_0 = \frac{(1 - e^2)^{1/2}}{e^3} \sin^{-1} e - \frac{1 - e^2}{e^2}, \tag{9.27}$$

$$\gamma_0 = \frac{2}{e^2} \left[1 - (1 - e^2)^{1/2} \frac{\sin^{-1} e}{e^2} \right]. \tag{9.28}$$

Figure 9.3 The relation between $\Omega^2/\pi G\rho$ and eccentricity e of Maclaurin spheroids. Adapted from Chandrasekhar (1969).

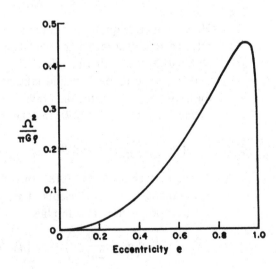

On substituting (9.27) and (9.28) into (9.24), we obtain

$$\frac{\Omega^2}{\pi G \rho} = \frac{2(1 - e^2)^{1/2}}{e^3}(3 - 2e^2)\sin^{-1}e - \frac{6}{e^2}(1 - e^2). \qquad (9.29)$$

If the fluid mass is made to rotate with angular velocity Ω, then it takes up a spheroidal configuration with eccentricity e, the relation between Ω and e being given by (9.29). The expression is somewhat complicated. To understand its nature, a plot of $\Omega^2/\pi G \rho$ is shown against e in Figure 9.3. As expected, a spherical configuration ($e = 0$) corresponds to no rotation ($\Omega = 0$). Initially we see that e increases with increasing Ω so that the spheroid becomes more flattened with more rotation. It may at first seem surprising that eventually $\Omega^2/\pi G \rho$ reaches a maximum value of 0.449 at $e = 0.930$ and then decreases with further increase of e. In other words, less rotation makes the system more flattened. To understand what is happening, we have to look at the angular momentum. Since the moment of inertia of a spheroid of mass M and equatorial radius a is $(2/5)Ma^2$, the angular momentum is

$$L = \tfrac{2}{5}Ma^2\Omega, \qquad (9.30)$$

where M is given by

$$M = \tfrac{4}{3}\pi a^2 c\rho = \tfrac{4}{3}\pi a^3(1 - e^2)^{1/2}\rho. \qquad (9.31)$$

The angular momentum can be made dimensionless by dividing (9.30)

Figure 9.4 The relation between dimensionless angular momentum $\overline{L}$ and eccentricity e of Maclaurin spheroids. Adapted from Chandrasekhar (1969).

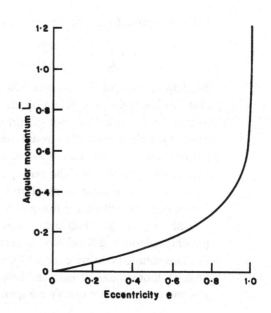

by $[GM^3(a^2c)^{1/3}]^{1/2}$, which has the dimension of angular momentum and is a constant for a given mass of rotating fluid. Then the dimensionless angular momentum is

$$\bar{L} = \frac{L}{[GM^3(a^2c)^{1/3}]^{1/2}} = \frac{\sqrt{3}}{5}\left(\frac{a}{c}\right)^{2/3}\left(\frac{\Omega^2}{\pi G\rho}\right)^{1/2}. \qquad (9.32)$$

On substituting for a/c from (9.26) and for $\Omega^2/\pi G\rho$ from (9.29), we can express $\bar{L}$ as a function of e alone. This functional dependence is plotted in Figure 9.4. We see that angular momentum and eccentricity are monotonically increasing with each other. So more angular momentum always makes the system more flattened. Beyond a certain stage, however, additional angular momentum makes the system so flattened and increases the moment of inertia at such a fast rate that the angular velocity goes down in spite of the increasing angular momentum (which is the product of moment of inertia and angular velocity). This explains the decrease of rotation with eccentricity when the system has become sufficiently flattened, as seen in Figure 9.3.

9.3.2 Jacobi ellipsoids

Jacobi (1834) established the surprising fact that (9.24) can admit of ellipsoidal solutions with three unequal axes under certain circumstances. It easily follows from (9.24) that

$$(\alpha_0 - \beta_0)a^2b^2 + \gamma_0 c^2(a^2 - b^2) = 0. \qquad (9.33)$$

On substituting for α_0, β_0 and γ_0 from (9.20) into (9.33), we get

$$(a^2 - b^2)\int_0^\infty \left[\frac{a^2b^2}{(a^2 + \lambda)(b^2 + \lambda)} - \frac{c^2}{c^2 + \lambda}\right]\frac{d\lambda}{\Delta} = 0. \qquad (9.34)$$

To demonstrate the existence of *Jacobi ellipsoids*, we have to show that this integral equation can be satisfied for real and unequal values of a, b and c. We shall not go into the details of that calculation here. We only mention the result that, if the dimensionless angular momentum $\bar{L}$ is less than 0.304 (corresponding to $e < 0.813$), then Maclaurin spheroids are the only possible solutions of the problem. Only when the angular momentum is increased beyond 0.304, does a Jacobi ellipsoid with three unequal axes become a possibility.

When $\bar{L}$ is larger than the critical value 0.304, we then have two possible solutions: a Maclaurin spheroid and a Jacobi ellipsoid. Which of these solutions do we expect to be realized in Nature? For a given angular momentum, it can be shown that the Jacobi ellipsoid has *less* rotational kinetic energy compared to the Maclaurin spheroid. If there is some dissipation mechanism like viscosity, then the Maclaurin spheroid may become unstable and ultimately relax to a Jacobi ellipsoid after some dissipation of energy. This is a subject of immense complexity and richness, and has been investigated by some of the greatest mathematicians and physicists of the nineteenth century. We refer the reader to Chandrasekhar (1969) for a full account of the subject. The first chapter of the book traces the fascinating history of the subject as well. The intriguing aspect of Jacobi's result is that we get solutions without axisymmetry even though we had originally posed a problem with axisymmetry. In other words, the solution spontaneously breaks the symmetry which was present in the formulation of the problem. Nowadays we know many other examples of spontaneous symmetry-breaking in physics.

Are there some objects in the astronomical Universe which are examples of Jacobi ellipsoids? There is evidence that some elliptical galaxies are triaxial in shape (see Mihalas and Binney 1981, pp. 329–333). One might have been tempted to regard these galaxies as stellar dynamic analogues of Jacobi ellipsoids. Elliptical galaxies, however, are known to have very little angular momentum. It appears that the triaxial nature of these galaxies is due rather to the anisotropic velocity distribution of the stars than to rotation.

9.4 Rotation in the world of stars

The discussion in the previous section should make it clear that any problem involving rotation along with self-gravity is extremely complex. We have given an outline of the simplest situation where we consider the solid-body rotation of a mass of incompressible fluid. Most stars are made of compressible gases with differential rotation and other types of internal motion so that the whole star may not be at rest in any frame of reference. The differential rotation in the interior of the Sun was shown in Figure 7.8. The theory of rotating stars is a notoriously difficult subject. Although most stars are rotating, the majority of the stellar structure textbooks proceed as if there is no such thing as rotation in the world of stars. One exception is the excellent book by Kippenhahn and Weigert (1990), in which Part IX provides a comprehensive introduction to the subject of stellar rotation. Although this subject is beyond the scope of this book, we make a few brief remarks.

If we assume that a compressible star can be completely at rest in a rotating frame of reference, then (9.5) gives us

$$\frac{\nabla p}{\rho} = -\nabla \Phi_{\text{eff}}, \tag{9.35}$$

where Φ_{eff} is given by (9.6). On taking the curl of this equation, we get

$$\nabla p \times \nabla \rho = 0.$$

This implies that the contours of constant p and constant ρ must coincide. It is then clear from (9.35) that these contours should also be the contours of constant Φ_{eff}, which are expected to be of spheroidal shape. It follows from the perfect gas law that these must be contours of constant temperature as well. Because of the spheroidal shapes of these contours, the temperature gradient near the pole of a particular contour should be larger than that near the equator. Hence the radiative energy flux coming through the poles of a star must be larger than that coming through the equator. Thus purely hydrodynamic considerations make some demands on the energy flux in different parts of the star. Now the energy flux is ultimately generated by nuclear reactions in the interior, which have no connections with hydrodynamics. Are the demands on energy flux from hydrodynamic considerations consistent with the nuclear generation rates? A simple calculation by von Zeipel (1924) showed that these hydrodynamic and nuclear requirements are mutually incompatible! This is known as the *von Zeipel paradox*. We refer to Chapter 42 of Kippenhahn and Weigert (1990) for a derivation of this paradox and a discussion of its consequences. How do we get around this paradox? If there is an internal flow in the meridional plane of the star, then there would be extra terms in (9.35), and Eddington (1925) showed that one can get around the von Zeipel paradox in this fashion. Such a meridional flow driven by the requirements of force balance is known as the *Eddington–Sweet circulation* (Eddington 1925; Sweet 1950).

In the case of the Sun and most main-sequence stars, the rotation is not too strong, in the sense that the centrifugal force is a small fraction of gravity. Hence stellar models neglecting rotation are expected to give reasonably good results. But rotation becomes increasingly important in collapsed stars. As we decrease the radius a of a star, the moment of inertia decreases as a^2. If the angular momentum is conserved, then the angular velocity Ω goes as a^{-2} and the centrifugal force $\Omega^2 a$ goes as a^{-3}. Hence the centrifugal force increases faster with decreasing radius than gravity, which increases only as a^{-2}. Even if the centrifugal force is initially negligible compared to gravity in a star, it may become important after the star has collapsed sufficiently.

Suppose we collapse a star like the Sun with a radius of about 10^{11} cm such that its final radius becomes about 10^6 cm, the typical radius of neutron stars. An initial rotation period of about a month (as the Sun has) would then decrease by a factor of 10^{10} to become less than 10^{-3} s. The centrifugal force for such a rotation is so large that it would overcome the gravity and make the stellar material fly away, leading to a disruption of the star.

We have seen in §9.3.1 that there is a maximum limit of the rotation rate for a mass of incompressible fluid. It is

$$\Omega^2 < 0.449 \, \pi G \rho. \tag{9.36}$$

The fluid mass would fall apart if it is made to rotate faster. The corresponding condition for the period $T = 2\pi/\Omega$ comes out to be

$$T > \frac{2.05 \times 10^4}{\sqrt{\rho}} \text{ s}, \tag{9.37}$$

if ρ is in gm cm^{-3}. Taking the density of white dwarfs to be about 10^8 gm cm^{-3}, the lower limit of period turns out be about 2 s. On the other hand, the rotation period of neutron stars with density 10^{14} gm cm^{-3} has the limit of about 2×10^{-3} s. When Hewish *et al.* (1968) discovered the first pulsar with a period of 1.377 s, almost immediately Gold (1968) suggested that pulsars must be rotating neutron stars. One can rule out the possibility that the period of a pulsar is an oscillation period rather than a rotation period, since the observed periods do not match the fundamental oscillation periods of either white dwarfs or neutron stars. The most plausible explanation is that the period of a pulsar is a rotation period. It is not possible for a white dwarf to rotate as fast as many of the pulsars. We need something with higher density, and the neutron star seemed the natural candidate. This was one of the first convincing identifications of some observed astronomical objects as neutron stars, which were theoretically proposed long time ago (Baade and Zwicky 1934). The fastest rotating pulsar known today is the so-called millisecond pulsar with a period of 1.558×10^{-3} s discovered by Backer *et al.* (1982). It seems that this period is lower than the rough limit of 2×10^{-3} s estimated above. It should be kept in mind that the material inside a neutron star is not an incompressible fluid. So an accurate estimate of the limit of rotation period can be made only by using appropriate equations of state and incorporating general relativistic corrections. The point to note is that the period of the millisecond pulsar must be close to the theoretical lower limit and can be used to put constraints on possible equations of state.

Just as rotation is important in the end stages of a star's life, it seems that rotation must be important during the star's birth as well. We have pointed out in §7.5 that stars are born out of interstellar clouds due to the Jeans instability. Now a typical interstellar cloud has some angular momentum. Since the effect of rotation is enhanced on the shrinking of size, we expect that eventually the collapsing cloud should rotate so fast that further gravitational collapse is halted by the centrifugal force. We therefore need some mechanism for removing the angular momentum in order for the star to form. We shall discuss in §14.9.1 the possibility that magnetic stresses may carry away the angular momentum from the inner parts of the collapsing cloud, thereby allowing the gravitational collapse to proceed till stars are formed.

9.5 Rotation in the world of galaxies

If a galaxy is regarded primarily as a collection of stars, then we have to apply the equations of stellar dynamics rather than the equations of hydrodynamics to study galactic rotation. We have seen in (9.1) that the gravity in a hydrodynamic system can be balanced either by the pressure gradient or by the Coriolis force. The same holds for a stellar dynamical system, where the analogue of pressure is the random velocities of stars.

The reader was asked to derive the appropriate stellar dynamical force balance equation in Exercise 3.3. It is

$$\frac{\partial}{\partial r}(n\langle \Pi^2\rangle) + \frac{\partial}{\partial z}(n\langle \Pi Z\rangle) + \frac{n}{r}[\langle \Pi^2\rangle - \langle \Theta^2\rangle] = ng_r, \qquad (9.38)$$

where Π, Θ and Z are the r, θ and z components of the velocities of stars, which have the number density n. Now consider a situation in which all stars move in regular circular orbits without any random motion. Then we have $\Pi = Z = 0$ with

$$\Theta = \Theta_c = \sqrt{r|g_r|} \qquad (9.39)$$

from (9.38). In this case, the gravity is completely balanced by the centrifugal force. On the other hand, if the stars have random velocities, then the gravity can be partially or fully balanced by the 'pressure' forces arising out of these random motions. In such a situation, only a part of the gravitational force remains to be balanced by the centrifugal force, and we expect $\langle \Theta\rangle$ to be less than Θ_c. This can actually be shown by suitably manipulating (9.38) with some reasonable assumptions. See Binney and Tremaine (1987, §4.2) for details.

By analyzing the random motions of stars in the solar neighbourhood, Oort (1928) drew the fundamental conclusion that these stars can be divided into two categories: (i) stars with low random velocities which move in nearly circular orbits; and (ii) stars with high random velocities. For the stars in category (ii), it was found that they have a much less average rotation around the galactic centre compared to the stars in category (i), presumably due to the fact their random motions balance the gravity and they need not have as much rotation as the stars in category (i). Our Galaxy then appears to be a combination of two stellar systems. Even within category (i), there are some groups of stars having more velocity dispersion compared to the others. The groups of stars with more velocity dispersion seem to lag behind the other stars while going around the galactic centre (Mihalas and Binney 1981, §6-4). In contrast to the spiral galaxies like our Galaxy, the elliptical galaxies appear to have much less rotation and the gravity is balanced mainly by the velocity dispersion of stars.

Let us end by making one final comment on the rotation patterns of spiral galaxies. In these galaxies, most of the visible stars in the disk move in nearly circular orbits with Θ given by (9.39). If the gravity drops as r^{-2} in the outer regions of the galaxy, then we expect Θ to fall off as $r^{-1/2}$. In the majority of spiral galaxies, however, such a fall in rotation velocity has not been observed (Mihalas and Binney 1981, §8-4). The most plausible explanation is that these galaxies have large amounts of unseen dark matter in regions beyond the visible disk so that the gravitational field does not fall off as rapidly as we may naively expect.

Exercises

9.1   Suppose a cylindrical vessel containing water is rotated around its axis with a constant angular velocity. Show that the water surface eventually takes up the shape of a parabola.

9.2   Consider fluid displacements in a rotating fluid body such that the Coriolis force is the dominant force on the R.H.S. of (9.5). In the linear theory, show that such displacements lead to circularly polarized oscillations with frequency 2Ω. Such oscillations are known as *inertial oscillations*.

Part 2 Plasmas

It seems very probable that electromagnetic phenomena will prove to be of great importance in cosmic physics. Electromagnetic phenomena are described by classical electrodynamics, which, however, for a deeper understanding must be combined with atomic physics ... No definite reasons are known why it should not be possible to extrapolate the laboratory results in this field to cosmic physics. Certainly, from time to time, various phenomena have been thought to indicate that ordinary electrodynamic laws do not hold for cosmic problems. For example, the difficulty of accounting for the general magnetic fields of celestial bodies has led different authors, most recently Blackett (1947), to assume that the production of a magnetic field by the rotation of a massive body is governed by a new law of nature. If this is true, Maxwell's equations must be supplemented by a term which is of paramount importance in cosmic physics. The arguments in favour of a revision are still very weak. Thus it seems reasonable to maintain the generally accepted view that all common physical laws hold up to lengths of the order of the 'radius of the universe' and times of the order of the 'age of the universe', limits given by the theory of general relativity.

— H. Alfvén (1950)

In many respects the astronomical universe has reached the stage of middle-age, with its violent youth behind it and its final stages of senility still safely in the future ... It is with some surprise, therefore, that examination of the universe on a small-scale shows so much activity. After 10 billion years new stars are still being formed. After 5 billion years, the sun is still popping and boiling, unable to settle down into the decadent middle age that simple theoretical considerations would suggest. Perhaps the most singular property of the general activity is the continual acceleration of a small fraction of atoms to high speeds ...
It appears that the radical element responsible for the continuing thread of

cosmic unrest is the magnetic field. Magnetic fields are familiar in the laboratory, and indeed in the household, where their properties are well known; they are easily controlled, and they serve at our beck and call. In the large dimensions of the astronomical universe, however, the magnetic field takes on a role of its own, quite unlike anything in the laboratory. The magnetic field exists in the universe as an 'organism', feeding on the general energy flow from stars and galaxies. The presence of a weak field causes a small amount of energy to be diverted into generating more magnetic field, and that small diversion is responsible for the restless activity in the solar system, in the galaxy, and in the universe. Over astronomical dimensions the magnetic field takes on qualitative characteristics that are unknown in the terrestrial laboratory. The cosmos becomes the laboratory, then, in which to discover and understand the magnetic field and to apprehend its consequences.

— E. N. Parker (1979)

10 Plasma orbit theory

10.1 Introductory remarks

We now begin the study of plasmas, which are gases in which the constituent particles are charged. In a gas made up of neutral particles, two particles are assumed to interact only when they collide, i.e. are physically very close. Between collisions, the neutral particles move along straight lines. In contrast, the particles in a plasma always interact with each other through long-range electromagnetic interactions and the trajectories of individual particles can be quite complicated. Before investigating how collections of charged particles behave, it is worthwhile developing some ideas about the motions of individual charged particles in electromagnetic fields. This topic is referred as the *plasma orbit theory* and often turns out to be very useful in handling problems involving plasmas. While developing the theory of neutral gases in Chapters 2–3, it was not necessary to pay much attention to motions of individual neutral particles, as these motions are quite simple. After discussing motions of individual plasma particles in this chapter, we shall, however, follow a course of development roughly similar to that which we followed for neutral fluids. In the next chapter, we shall begin developing theoretical techniques for treating plasmas as collections of charged particles and eventually we shall end up with continuum models in Chapters 14–16.

As soon as we start discussing electromagnetic quantities, we face the vexing question of choosing units. Two different systems of units—the Gaussian system and the SI system—are of wide use, and each system has its strong advocates claiming its superiority over the other system. Several of the pioneers in the study of astrophysical plasmas such as Alfvén, Cowling, Chandrasekhar, Spitzer and Parker used Gaussian units. Due to the influence of these pioneers, the Gaussian system still seems to be more widely used than the SI system, especially by theoretical astrophysicists. Hence we have decided to use the Gaussian system, with apologies to those readers who might have been more comfortable with the SI system. All electrical quantities including the current density **j** will be assumed to be in e.s.u., whereas all the magnetic quantities will be assumed to be in e.m.u.

A particle of mass m and charge q moving in an electromagnetic field satisfies the equation of motion

$$m\frac{d\mathbf{u}}{dt} = q\left(\mathbf{E} + \frac{\mathbf{u}}{c} \times \mathbf{B}\right). \tag{10.1}$$

A plasma particle is subjected to the electromagnetic fields produced by the other particles in the plasma. In addition, there may be external magnetic fields imposed on the plasma. As we shall see in the discussion of Debye shielding in the next chapter, the interior of a plasma is usually shielded from external electric fields. In this chapter, we shall concentrate on the motion of a charged particle due to given electromagnetic fields. It may be worthwhile pointing out that (10.1) is incomplete in one sense. It is a well-known result of classical electrodynamics that an accelerated charged particle emits electromagnetic radiation. The emission of radiation involves a loss of energy, which introduces an effective drag in the equation of motion known as the *radiation reaction*. A proper treatment of the radiation reaction is notoriously difficult (Panofsky and Phillips 1962, §21-6; Jackson 1975, §17.2). For most of the applications to be discussed in this book, the radiation reaction happens to be unimportant, and we shall use (10.1) which neglects the radiation reaction.

Let us consider a uniform magnetic field **B** and a particle moving with velocity $\mathbf{u}_\perp$ perpendicular to **B**. The simplest way to study the motion of the particle is to choose the plane of motion as the xy plane, write down the x and y components of the equation of motion

$$m\frac{d\mathbf{u}_\perp}{dt} = \frac{q}{c}\mathbf{u}_\perp \times \mathbf{B}$$

and solve these equations. This is given as Exercise 10.1. It is easy to see that the particle moves in a circle lying in the plane transverse to **B**, the circular frequency being the gyrofrequency or the cyclotron frequency

$$\omega_c = \frac{|q|B}{mc}. \tag{10.2}$$

If the particle has a component of velocity $\mathbf{u}_\parallel$ parallel to **B**, it is easily seen from (10.1) that the magnetic field does not affect this component. Hence this component leads to a uniform translation of the circular trajectory with velocity $\mathbf{u}_{||}$, making the path of the particle helical. Thus we can think of the motion of the particle as resulting from the combination of two motions: (1) a circular motion around a central point we call the *guiding centre*; and (2) a translatory motion of the guiding centre.

Even when a charged particle moves in a non-uniform magnetic field, it is often possible to break up the motion into (1) a circular motion around the guiding centre and (2) the motion of the guiding centre. In order for this to be possible, the non-uniformities of the magnetic field have to be small over the region through which the particle is making the circular motion. In other words, the length scale of the magnetic field inhomogeneities has to be larger than the gyroradius (i.e. the radius associated with the circular motion of the particle), also called the *Larmor radius*—a slightly confusing terminology in view of the fact that there is something called Larmor frequency (see, for example, Goldstein 1980, §5-9) which is different from the cyclotron frequency ω_c associated with the motion (actually the Larmor frequency turns out to be half of the cyclotron frequency, although it arises out of very different considerations!). If the magnetic field does not change much within the gyroradius, then the particle moves through a nearly uniform magnetic field while making a circular round. The non-uniformities, however, can make the guiding centre move in a way different from a simple translatory motion. The aim of the plasma orbit theory is to find equations describing the motion of the guiding centre, which we expect to be simpler to handle than the full equation (10.1) for the charged particle.

Unfortunately, no general equation of motion exists for the guiding centre in an arbitrary electromagnetic field. We have, rather, a set of rules prescribing how the guiding centre moves in a few particular situations. Some of the more important rules will be derived in this chapter. Once these rules have been mastered, one can often figure out the motion of the guiding centre in complicated magnetic configurations by a judicious combination of the basic rules. Figure 10.1 is taken from the classic monograph *Cosmical Electrodynamics* of Alfvén (1950), who laid down the foundations of the guiding centre theory (Alfvén 1940). This figure shows the trajectory of a charged particle in a dipolar magnetic field. The detailed trajectory was calculated by Störmer (1907) by integrating (10.1). Before the advent of electronic computers, such numerical integrations were tedious and time-consuming. Alfvén (1940) used his guiding centre theory to calculate an approximate trajectory of the charged particle. The guiding centre calculations involved considerably less amount of numerical

work and produced trajectories in good agreement with Störmer's
detailed calculations.

When the guiding centre theory was first introduced by Alfvén
(1940), one of its main appeals was that it was a labour-saving trick.
Nowadays, however, electronic computers can solve equations like
(10.1) in a routine manner and provide trajectories of charged particles.
So should one still bother to learn the guiding centre theory? Although
no longer so useful as a short cut to numerical computations, the
guiding centre theory helps us to develop an intuition about the
motions of charged particles and this intuition turns out to be very
useful in solving many practical problems. We all have an intuition
about how masses move in gravitational fields. When we see a ball in
motion, we instantly form an idea where the ball is going to hit the
ground and hence we can play ball games. But imagine a group of
players playing with a charged ball in a non-uniform magnetic field!

Figure 10.1 Motion
of a charged particle
in a dipole field
calculated by
Störmer from the full
equations (dashed
lines) and by Alfvén
using the guiding
centre theory (solid
lines indicate the
trajectory of the
guiding centre).
Projection upon a
plane through the
axis of the dipole
(*above*) and upon the
equatorial plane
(*below*). (©Oxford
University Press.
Reproduced with
permission from
Alfvén's *Cosmical
Electrodynamics*
1950.)

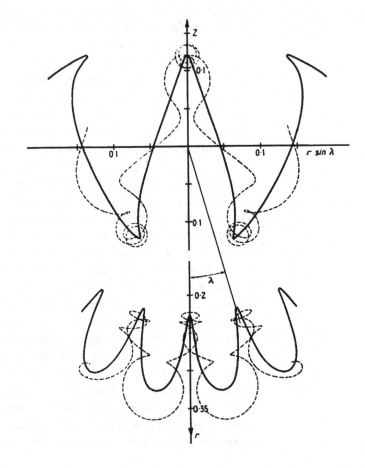

This will be a much harder game, as it will be much more difficult to figure out where the charged ball will end up by watching its present movements. The results of guiding centre theory give us some feeling for such motions.

As we have already pointed out, the guiding centre theory is more like a bag of tricks rather than a general theory. Its results exist in the form of a few rules, some of which will be derived in the next sections.

10.2 The effect of a perpendicular force

We begin our discussion of the motion of the guiding centre by considering a situation where there is a uniform force **F** acting on a charged particle in a direction perpendicular to a uniform magnetic field. The results of this analysis will be applicable to many problems as we shall see below.

The equation of motion is

$$m\frac{d\mathbf{u}}{dt} = \mathbf{F} + \frac{q}{c}\mathbf{u} \times \mathbf{B},\qquad(10.3)$$

where **F** and **B** are perpendicular. In the absence of a magnetic field, the effect of **F** would have been to accelerate the particle in the direction of **F**. It is easy to see that the motion of the guiding centre is going to be quite different. Figure 10.2 sketches the trajectory of a positively charged particle with **B** coming out of the page perpendicularly and **F** in the upward direction. When the particle is at some point P at the bottom of its trajectory, **F** and the Lorentz force $(q/c)\mathbf{u} \times \mathbf{B}$ both act in the upward direction. This enhanced normal acceleration makes the trajectory more sharply bent than it would have been in the absence of **F**. On the other hand, when the particle is at some top point Q, the

Figure 10.2 The trajectory of a positively charged particle subjected to a magnetic field **B** and a perpendicular force **F**.

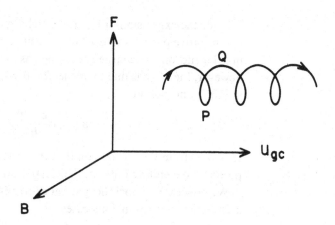

Lorentz force is diluted by $\mathbf{F}$, thereby causing the trajectory to be less sharply bent. As a result, we expect the trajectory to appear as shown in Figure 10.2, clearly indicating that there is a drift $\mathbf{u}_{gc}$ of the guiding centre in a direction perpendicular to both $\mathbf{B}$ and $\mathbf{F}$. We now derive an expression for $\mathbf{u}_{gc}$.

It is easy to see from (10.3) that there is no acceleration parallel to $\mathbf{B}$. So, if there is any initial velocity in that direction, it remains unaffected. Let us therefore focus our attention on the velocity $\mathbf{u}_\perp$ in the plane perpendicular to $\mathbf{B}$. We now break up $\mathbf{u}_\perp$ into two parts

$$\mathbf{u}_\perp = \mathbf{u}_c + \mathbf{u}_{gc}, \tag{10.4}$$

where $\mathbf{u}_c$ is the velocity of circular motion in the magnetic field satisfying

$$m\frac{d\mathbf{u}_c}{dt} = \frac{q}{c}\mathbf{u}_c \times \mathbf{B} \tag{10.5}$$

and $\mathbf{u}_{gc}$ is the drift of the guiding centre. Since $\mathbf{u}_{gc}$ is expected to be uniform, it must satisfy

$$0 = \mathbf{F} + \frac{q}{c}\mathbf{u}_{gc} \times \mathbf{B} \tag{10.6}$$

such that the sum of (10.5) and (10.6) give the transverse component of (10.3). To obtain $\mathbf{u}_{gc}$ from (10.6), we take a cross product of it with $\mathbf{B}$, which gives

$$\mathbf{F} \times \mathbf{B} = \frac{q}{c}\mathbf{B} \times (\mathbf{u}_{gc} \times \mathbf{B}) = \frac{q}{c}B^2\mathbf{u}_{gc}$$

so that

$$\mathbf{u}_{gc} = \frac{c}{q}\frac{\mathbf{F} \times \mathbf{B}}{B^2}. \tag{10.7}$$

Since the expression of $\mathbf{u}_{gc}$ has the charge q in it, we expect positive and negative particles to drift in opposite directions if the force $\mathbf{F}$ does not depend on the charge (as in the case of a gravitational force). If, however, the force is due to an electric field, then we substitute $\mathbf{F} = q\mathbf{E}$ in (10.7) and obtain

$$\mathbf{u}_{gc} = c\frac{\mathbf{E} \times \mathbf{B}}{B^2}. \tag{10.8}$$

It is clear from this expression that the drift is the same for all charged particles, depending only on the electric and magnetic fields.

We now discuss how the general expression (10.7) can be applied to a few other not so obvious cases.

10.2.1 Gradient drift

Consider a uni-directional magnetic field $B(y)\hat{e}_z$ of which the strength is varying in the y direction. Figure 10.3 shows a positively charged particle moving in the xy plane. If $B(y)$ increases with y, then we expect the Lorentz force to be larger in the upper part of the trajectory so that the curvature at the uppermost points is larger (i.e. the trajectory is more sharply bent) than that at the lowermost points. The trajectory, therefore, looks as shown in Figure 10.3, i.e. there is a drift in the negative x direction.

It is clear from the symmetry of the problem that the average force on the particle in the x direction is zero, although there is an average force in the y direction. The instantaneous y component of the Lorentz force is

$$F_y = -\frac{q}{c}u_x B_z(y).$$

If the y coordinate is measured from the position of the guiding centre and the variation of $B_z(y)$ over the trajectory of the particle is small, then we write

$$B_z(y) = B_0 + y\frac{dB_z}{dy}$$

within the region of the particle trajectory where we keep only the first-order term in the Taylor expansion. Hence the instantaneous Lorentz force is

$$F_y = -\frac{q}{c}u_x\left[B_0 + y\frac{dB_z}{dy}\right] \tag{10.9}$$

which we now want to average. For a circular motion (i.e. if we neglect the effect of drift while calculating averages), the average of qu_xB_0 is clearly zero, since $\overline{u_x} = 0$. It is easy to show that the average of u_xy is

Figure 10.3 The trajectory in the xy plane of a positively charged particle subjected to a non-uniform magnetic field $B(y)\hat{e}_z$.

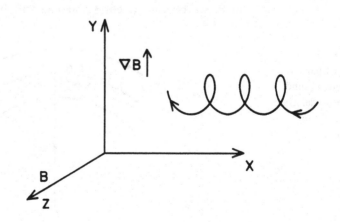

$\pm\frac{1}{2}u_\perp r_L$, where r_L is the gyroradius (or the Larmor radius) and the $\pm$ signs depend on whether the trajectory is clockwise (for positive q) or anti-clockwise (for negative q). Hence the average of (10.9) gives

$$\overline{\mathbf{F}} = \mp\frac{q}{2c}u_\perp r_L \nabla B. \tag{10.10}$$

Substituting this average force in (10.7), we get the expression of the drift

$$\mathbf{u}_{\nabla B} = \pm\frac{1}{2}u_\perp r_L \frac{\mathbf{B}\times\nabla B}{B^2}, \tag{10.11}$$

which is often called the gradient drift, since it is caused by the gradient of the magnetic field.

It may be noted that the gradient drift has opposite signs for positive and negative charges. Hence, if ions and electrons are present in a region of transverse magnetic gradient, the ions and the electrons would drift in opposite directions and give rise to a current.

10.2.2 Curvature drift

Let us consider the magnetic field to be constant in magnitude in a certain region, but to have a constant radius of curvature R_C. Such a magnetic configuration is sketched in Figure 10.4. Since a charged particle gyrates around a field line, there should be an average centrifugal force

$$\mathbf{F}_C = -mu_\parallel^2 \frac{\mathbf{R}_C}{R_C^2} \tag{10.12}$$

acting on it as it moves along the field line. Here $u_\parallel$ is the component of the velocity parallel to $\mathbf{B}$ and $\mathbf{R}_C$ is the radius of curvature vector directed towards curvature centre. Since this force $\mathbf{F}_C$ is perpendicular to $\mathbf{B}$, we obtain the corresponding drift by substituting it for $\mathbf{F}$ in

Figure 10.4 A charged particle gyrating around magnetic field lines with constant curvature.

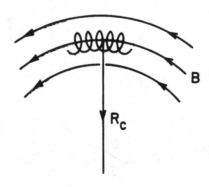

(10.7), i.e.

$$\mathbf{u}_{\mathrm{C}} = -\frac{cmu_{\parallel}^2}{qB^2}\frac{\mathbf{R}_{\mathrm{C}} \times \mathbf{B}}{R_{\mathrm{C}}^2}.$$ (10.13)

This drift is known as the curvature drift. It may be noted that the field configuration sketched in Figure 10.4 does not satisfy $\nabla \times \mathbf{B} = 0$. In vacuum, a curved magnetic field should have some non-uniformities associated with it. If this is so, then some gradient drift is present in combination with the curvature drift (see Exercise 10.2). Like the gradient drift, the curvature drift is also in opposite directions for opposite charges and can give rise to a current by making the electrons and ions drift in opposite directions.

10.3 Magnetic mirrors

Let us consider a magnetic configuration symmetric around a field line in such a way that the strength of the magnetic field varies as we move along the central field line. Figure 10.5 shows such a magnetic configuration in which we want to study the motion of a charged particle.

We use cylindrical coordinates with the z axis along the central field line. Although the predominant component of the magnetic field is B_z, there is a small B_r component which can be found out from the condition $\nabla \cdot \mathbf{B} = 0$, which here gives

$$\frac{1}{r}\frac{\partial}{\partial r}(rB_r) + \frac{\partial B_z}{\partial z} = 0.$$ (10.14)

We neglect the small radial variation of $\partial B_z / \partial z$ in the neighbourhood of the central axis and take it to be constant there. Then B_r in the central region can be readily obtained by integrating (10.14):

$$rB_r = -\int_0^r r'\frac{\partial B_z}{\partial z}\,dr' = -\frac{1}{2}r^2\frac{\partial B_z}{\partial z}$$

so that

$$B_r = -\frac{1}{2}r\frac{\partial B_z}{\partial z}.$$ (10.15)

Figure 10.5 A magnetic mirror configuration.

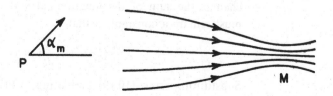

For a particle gyrating around the central field line, the z component of the Lorentz force is

$$F_z = -\frac{q}{c} u_\theta B_r.$$

The θ-velocity $u_\perp$ for positively (negatively) charged particles is in the negative (positive) θ direction so that we can write $\mp u_\perp$ for u_θ. Substituting for B_r from (10.15),

$$F_z = \mp \frac{q}{2c} u_\perp r_{\mathrm{L}} \frac{\partial B_z}{\partial z} = -\mu \frac{\partial B_z}{\partial z}, \tag{10.16}$$

where

$$\mu = \pm \frac{q}{2c} u_\perp r_{\mathrm{L}} = \frac{\frac{1}{2} m u_\perp^2}{B} \tag{10.17}$$

on using (10.2) to eliminate r_{L}. If we eliminate $u_\perp$ instead of r_{L}, then we get

$$\mu = \pm \frac{\omega_{\mathrm{c}}}{2\pi} q \cdot \pi r_{\mathrm{L}}^2 \cdot \frac{1}{c}. \tag{10.18}$$

We can interpret $-(\omega_{\mathrm{c}}/2\pi)q$ as the current associated with the gyrating charge q. We also know that the magnetic moment associated with a current encircling an area is obtained by multiplying the current with the area (see, for example, Panofsky and Phillips 1962, §7-11; Jackson 1975, §5.6). Hence it is clear from (10.18) that μ is the *magnetic moment* of the gyrating particle.

The z component of motion of the charged particle is

$$m \frac{du_\parallel}{dt} = F_z = -\mu \frac{\partial B_z}{\partial z}$$

from (10.16). The rate of change of the longitudinal kinetic energy is given by

$$\frac{d}{dt} (\tfrac{1}{2} m u_\parallel^2) = u_\parallel m \frac{du_\parallel}{dt} = -\mu \frac{dB}{dt}, \tag{10.19}$$

where we have substituted dB/dt for $u_\parallel(\partial B/\partial z)$. It may be noted that the kinetic energy is a scalar, and we are using terms like 'longitudinal' or 'transverse' kinetic energy merely to mean the kinetic energy associated with longitudinal or transverse motion. Since the kinetic energy of a charged particle moving in a static magnetic field cannot change, the sum of the longitudinal and transverse kinetic energies must remain a constant so that

$$\frac{d}{dt} (\tfrac{1}{2} m u_\parallel^2) + \frac{d}{dt} (\tfrac{1}{2} m u_\perp^2) = 0.$$

Substituting from (10.19) and using (10.17) to write μB for $\tfrac{1}{2} m u_\perp^2$, we have

$$-\mu\frac{dB}{dt} + \frac{d}{dt}(\mu B) = 0,$$

from which

$$\frac{d\mu}{dt} = 0. \tag{10.20}$$

This means that the magnetic moment μ defined by (10.17) is conserved during the motion of the guiding centre. The transverse kinetic energy $\frac{1}{2}mu_\perp^2$ then has to increase when the particle moves into regions of stronger B. The transverse kinetic energy, however, can never exceed the total kinetic energy. Therefore, when the particle reaches a region of sufficiently strong B where the transverse kinetic energy becomes equal to the total kinetic energy, it is not possible for the particle to penetrate further into regions of stronger magnetic field. The only possibility is that the particle gets reflected back. A region of increasing magnetic field as sketched in Figure 10.5 can thus act as a reflector and is known as the *magnetic mirror*.

It is obvious that a charged particle moving along the symmetry axis in Figure 10.5 is unaffected by magnetic forces and hence is not reflected by the magnetic mirror. If a magnetic mirror has a maximum magnetic field B_m at the point M as shown in Figure 10.5, then charged particles from the point P are expected to penetrate through the mirror if they have velocity vectors sufficiently close to the symmetry axis, say within a critical angle α_m. We can estimate this critical angle by using the fact that a charged particle at P with the velocity vector inclined to the central axis at the critical angle α_m is expected to be reflected from the region of maximum magnetic field B_m. Let u_0 be the velocity amplitude of the particle so that its starting transverse velocity at P is

$$u_{\perp 0} = u_0 \sin \alpha_m. \tag{10.21}$$

If B is the magnetic field at P, then the constancy of μ given by (10.17) implies

$$\frac{u_{\perp 0}^2}{B} = \frac{u_0^2}{B_m}, \tag{10.22}$$

since the transverse velocity at the time of reflection has to equal u_0. We find from (10.21) and (10.22) that

$$\sin^2 \alpha_m = \frac{B}{B_m}. \tag{10.23}$$

We can consider a cone at P around the central axis with the opening angle α_m. Such a cone is called the *loss cone*, because particles with velocity vectors lying within that cone will pass through the mirror and be lost from the system.

10.4 Formation of the Van Allen belt

In the previous two sections, we have presented some of the most important results of the plasma orbit theory. We now want to discuss how the combination of these results helps us to understand the motion of a charged particle in a complicated magnetic field.

Let us first consider a region with two magnetic mirrors at the two ends, as shown in Figure 10.6(a). Unless a charged particle has its velocity vector very close to the axis of the system, it is reflected back and forth between the mirrors, thereby remaining trapped within the region. Hence this type of magnetic configuration is called a *magnetic bottle*. Such magnetic bottle configurations are often used in the laboratory to confine charged particles (Rose and Clark 1961, Chapter 15).

We now look at the magnetic configuration outside a spherical dipole shown in Figure 10.6(b). Since field lines come together near the poles, we expect the poles to act as magnetic mirrors. Hence the region between the points P and Q can be thought of as a magnetic bottle which is curved. A charged particle may remain trapped in this magnetic bottle, while moving back and forth between the polar regions. But this is not the whole story. A charged particle gyrating around a curved field line is subject to curvature drift. Since the magnetic field is falling with radius, this non-uniformity gives rise to a gradient drift as well. It is easy to see that both the curvature drift and the gradient drift will produce a motion of the guiding centre in the

Figure 10.6 (a) A magnetic bottle configuration. (b) A dipolar field with a belt of trapped charged particles.

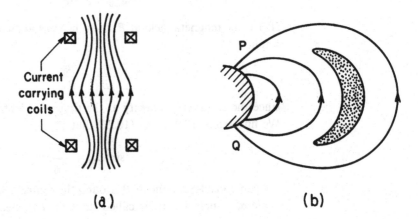

Current carrying coils

(a)   (b)

azimuthal direction. Hence the guiding centre, while moving to and fro between the polar regions, also goes around the polar axis of the spherical body. Figure 10.1 shows such motion of a charged particle along with the trajectory of the guiding centre as computed by Alfvén (1950).

Apart from giving an idea as to how the results of the guiding centre theory can be put together, the above discussion should make it clear that it is possible for charged particles to remain trapped in the Earth's magnetosphere. Such a region of trapped charged particles was actually discovered by Van Allen's group (Van Allen *et al.* 1958) with instruments borne on spacecrafts. This belt of charged particles is known as the *Van Allen belt*. We have pointed out in §10.2.1–2 that the gradient and the curvature drifts of electrons and ions are in opposite directions. This gives rise to an azimuthal current through this belt called the *ring current*.

10.5 Cosmic rays. Particle acceleration in astrophysics

It was established by the balloon flight experiments of Hess (1912) that the Earth is exposed to some ionizing rays coming from above the Earth's atmosphere. It was later ascertained that these *cosmic rays* are actually not rays, but highly energetic charged particles—mostly electrons and light nuclei. A question of fundamental importance is to determine if cosmic rays are something local existing in the neighbourhood of the Earth and the solar system, or if they fill up the whole Galaxy or even the whole Universe. Most astrophysicists at present hold the view that the cosmic rays are a galactic phenomenon. These charged particles are accelerated within our Galaxy and remain confined within it by the galactic magnetic field. For an analysis of the arguments in support of this view, we recommend the reader to the book by Longair (1994, §20.5). We should mention that such energetic charged particles are believed to exist in other galaxies as well. One way of proving the existence of energetic charged particles in extragalactic systems is to search for synchrotron radiation—a type of radiation given out by energetic charged particles spiralling around magnetic fields (see, for example, Rybicki and Lightman 1978, Chapter 6; Longair 1994, Chapter 18). Radio telescopes have discovered synchrotron radiation in many extragalactic systems, implying that the acceleration of charged particles to very high energies must be a fairly ubiquitous process in the astrophysical Universe. There is evidence for particles with energies as high as 10^{20} eV. For comparison, remember that the rest mass energy of a nucleon is of the order of 10^9 eV.

The study of cosmic rays played a curious role in the development of the twentieth century physics. Until the time of World War II, cosmic rays were studied mainly by particle physicists. These were the days before the gigantic particle accelerators when cosmic rays were the only source of high-energy particles. Several new particles were first discovered from cosmic ray studies. During this period, cosmic rays were rarely discussed in the astrophysical literature. Since the existence of cosmic rays could not be reconciled with the theoretical ideas of astrophysics in that period, most astrophysicists pretended as if these vexing particles did not exist. The famous theory of Fermi (1949) discussed below finally paved the way for bringing cosmic rays within the fold of astrophysics. Fermi's theory came exactly at the time when particle physicists felt that they had learnt what was to be learnt from cosmic rays and started losing interest in them. From then onwards, the study of cosmic rays became an integral part of modern astrophysics.

Figure 10.7 shows the spectrum of cosmic ray electrons at the top of the Earth's atmosphere. From energies of the order of about 10^3 MeV to energies of the order of 10^6 MeV, the spectrum can be fitted quite well with a power law

$$N(E)dE \propto E^{-2.66}dE. \tag{10.24}$$

From the study of the synchrotron radiation coming from extragalactic sources, it can be inferred that the electrons in those sources also must have a power-law distribution with an index close to 2.5, in fair agreement with what is observed in cosmic ray measurements. The aim of any particle acceleration theory is to explain the origin of this power-law spectrum with the observed index.

We now summarize the main ideas put forward by Fermi (1949). The interstellar clouds are known to carry magnetic fields. Since the field lines must spread out when coming outside from the clouds, the surfaces of the clouds should act as magnetic mirrors and reflect charged particles. Just as a ball picks up energy on being hit by a bat, Fermi (1949) visualized that charged particles can be accelerated by being hit repeatedly by the moving magnetic clouds. Although a particle gains energy in a 'head-on' collision with a moving cloud, there can also be 'trailing' collisions in which energy is lost. Hence we have to show that the particle on average gains energy in collisions. To understand the basic physics, let us consider a simple model of only one-dimensional motions of clouds and guiding centres. So all the collisions can neatly be divided into head-on and trailing collisions. We present a Newtonian treatment of the problem here. Since the energetic particles are relativistic, one should actually use a relativistic treatment. We refer the reader to Longair (1994, §21.3) for the rela-

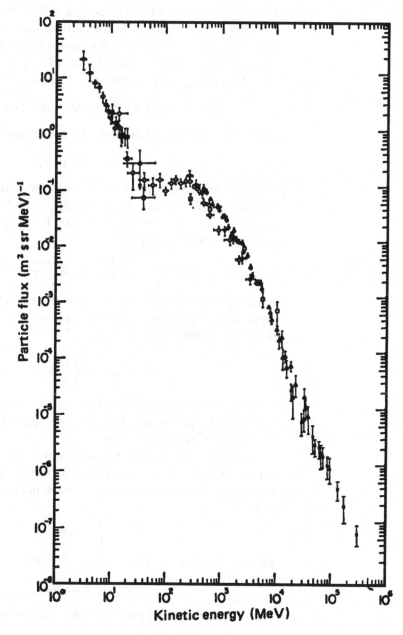

Figure 10.7 The spectrum of cosmic ray electrons at the top of the Earth's atmosphere. From Meyer (1969). (©Annual Reviews Inc. Reproduced with permission from *Annual Reviews of Astronomy and Astrophysics*.)

tivistic treatment along with a clear discussion of several other aspects of the problem.

Let us consider the clouds to move with velocity U in one dimension, i.e. half of the clouds are moving in one direction and the other half moving in the opposite direction. Let a particle of initial velocity u undergo a head-on collision with a cloud. The initial velocity seen from the rest frame of the cloud is $u + U$. If the collision is elastic, it would appear from this frame that the particle also bounces back in the opposite direction with the same magnitude of velocity $u + U$. In the observer's frame, this reflected velocity appears to be $u + 2U$. Hence the energy gain according to the observer is

$$\Delta E_+ = \tfrac{1}{2}m(u + 2U)^2 - \tfrac{1}{2}mu^2 = 2mU(u + U). \tag{10.25}$$

The energy loss in a trailing collision can similarly be shown to be

$$\Delta E_- = -2mU(u - U). \tag{10.26}$$

The probability of head-on collisions is proportional to the relative velocity $u + U$, whereas the probability of trailing collisions is proportional to the relative velocity $u - U$. The average energy gain in a collision is therefore equal to

$$\Delta E_{\text{ave}} = \Delta E_+ \frac{u + U}{2u} + \Delta E_- \frac{u - U}{2u} = 4mU^2 \tag{10.27}$$

on using (10.25) and (10.26). We now write down the corresponding expression for average energy gain in a relativistic treatment derived in Longair (1994, §21.3). It is

$$\Delta E_{\text{ave}} = 4 \left(\frac{U}{c} \right)^2 E. \tag{10.28}$$

It is easy to see that (10.28) reduces to (10.27) in the non-relativistic limit on putting $E = mc^2$. Readers good at special relativity may try to derive (10.28) (Exercise 10.4).

The main point to note in (10.28) is that the average energy gain is proportional to the energy. Hence the energy of a particle suffering repeated collisions is expected to increase, obeying an equation of the form

$$\frac{dE}{dt} = \alpha E, \tag{10.29}$$

where α is a constant. The solution of (10.29) is

$$E(t) = E_0 \exp(\alpha t), \tag{10.30}$$

where E_0 is the initial energy. If all particles started with the same initial energy E_0, then a particle acquires energy E after remaining confined in the acceleration region for time

$$t = \frac{1}{\alpha} \ln \left(\frac{E}{E_0} \right).$$ (10.31)

We expect particles to be continuously lost from the acceleration region. If τ is the mean confinement time, then the probability that the confinement time of a particle is between t and $t + dt$ is

$$N(t)dt = \frac{\exp(-t/\tau)}{\tau} dt.$$ (10.32)

This is exactly like the kinetic theory result of finding the probability that the time between two collisions for a particle is in the range t to $t + dt$ and is discussed in any elementary textbook presenting kinetic theory (see, for example, Reif 1965, §12.1; Saha and Srivastava 1965, §3.30) Now a particle with confinement time between t, $t + dt$ would acquire the energy between E, $E + dE$. Substituting for t from (10.31) and for dt from (10.29), we are led from (10.32) to

$$N(E)dE = \frac{\exp \left[-\frac{1}{\alpha \tau} \ln \left(\frac{E}{E_0} \right) \right]}{\tau} \cdot \frac{dE}{\alpha E},$$

from which

$$N(E) \propto E^{-(1 + \frac{1}{\alpha \tau})}.$$ (10.33)

We thus end up with a power-law spectrum.

This theory of Fermi (1949), although somewhat heuristic in nature and based on several ad hoc assumptions, gives us a clue as to how a power-law spectrum may arise. There are, however, very big gaps in the theory. Since it is not so straightforward to estimate α and τ, the index of the power law cannot be calculated easily. Furthermore, there is no indication in the theory why this index should be close to some universal value in different astrophysical systems. We also see from (10.28) that the average energy gain is proportional to $(U/c)^2$. Since the clouds are moving at non-relativistic speeds, this is a very small number and the acceleration process is quite inefficient. Because of this quadratic dependence on U, this process is referred as the *second-order Fermi acceleration*.

If we could somehow arrange that only head-on collisions take place, then the acceleration process would be much more efficient. For $u \gg U$, it follows from (10.25) that the energy gain will depend linearly on U rather than quadratically. The acceleration resulting from such a situation is called the *first-order Fermi acceleration*. But is it possible for this to happen in Nature? It was pointed out by several authors in the late 1970s (Axford, Leer and Skadron 1977; Krymsky 1977; Bell 1978; Blandford and Ostriker 1978) that shock waves produced in supernova explosions may provide sites for the first-order Fermi acceleration. Magnetic irregularities are expected on both sides of the advancing shock wave. It is possible that a charged particle is trapped at the shock front and reflected repeatedly from magnetic irregularities on both sides. Such collisions are always head-on and lead to much more efficient acceleration compared to Fermi's original proposal of acceleration by moving interstellar clouds. We again refer the reader to Longair (1994, §21.4) for a detailed discussion of this theory. Although many questions are still unanswered, acceleration in supernova shocks seems at present to be the most promising mechanism for producing cosmic rays.

Exercises

10.1   Consider a charged particle moving in a uniform magnetic field $\mathbf{B} = B\hat{\mathbf{e}}_z$. Write down the x, y and z components of the equation of motion. Solve these equations to show that the charged particle moves in a helical trajectory.

10.2   Suppose a region of *free space* has a cylindrically symmetric magnetic field $\mathbf{B} = B_\theta(r)\hat{\mathbf{e}}_\theta$. Show that the guiding centre of a gyrating charged particle has a combination of curvature and gradient drifts given by

$$\mathbf{u}_{\mathrm{gc}} = -\frac{cm}{q}\frac{\mathbf{R}_C \times \mathbf{B}}{R_C^2 B^2}\left(u_\parallel^2 + \frac{1}{2}u_\perp^2\right),$$

where the different symbols have the same meanings as in §10.2.2, $u_\parallel$ being the component of velocity in the direction of the magnetic field and $u_\perp$ being the component transverse to it.

10.3   Consider an electron in the Earth's Van Allen belt moving with 10% of the speed of light. If the Earth's magnetic field near the surface is of the order of 0.3 G, make a very rough estimate (say, correct within a factor of 10 or so) of the time the electron takes in going around the Earth once. How will you justify the use of the guiding centre approximation in this problem?

10.4   (For those who are good in special relativity.) Consider the one-dimensional problem of a relativistic particle being reflected from a set of reflectors moving with speeds either U

or $-U$. Using special relativity, show that the average energy gained per collision is given by (10.28).

10.5 The total power radiated per unit time by a charge q moving with acceleration a is given by

$$P = \frac{2q^2 a^2}{3c^3}.$$

Estimate the strength of the magnetic field such that an electron moving at 1% of the speed of light in that field would radiate away an appreciable amount of kinetic energy during one period of gyration. Is it justified to neglect the radiation reaction in the equation of motion (10.1) while considering motions of electrons in astrophysical magnetic fields?

11 Dynamics of many charged particles

11.1 Basic properties of plasmas

After discussing the motion of a single charged particle in the previous chapter, we now start studying the dynamics of many charged particles. The term *plasma* to denote a collection of ions and electrons was coined by Langmuir (1928). While discussing the ionized gas produced in the discharge tube, Langmuir (1928) wrote: "We shall use the name *plasma* to describe this region containing balanced charges of ions and electrons." The easiest way of obtaining a plasma is to heat a gas to sufficiently high temperature such that several atoms break into ions and electrons. For a gas in thermodynamic equilibrium, the degree of ionization at a given pressure and temperature can be obtained from the Saha ionization equation (Saha 1920) derived in many standard textbooks (see, for example, Mihalas 1978, §5-1; Rybicki and Lightman 1979, §9.5). Here we merely state the Saha equation without proof. We write it down for the ionization of hydrogen gas. This form of Saha equation will hold for other gases as well if we do not consider the possibility of multiple ionization due to the loss of more than one electron from the atom. The degree of ionization x is given by

$$\frac{x^2}{x-1} = \frac{(2\pi m_e)^{3/2}}{h^3} \frac{(\kappa_B T)^{5/2}}{p_{gas}} \exp(-\frac{\chi}{\kappa_B T}), \qquad (11.1)$$

where p_{gas} is the total gas pressure, m_e is the mass of the electron and χ is the ionization energy, other symbols having the usual meanings.

It is clear from the Saha equation that the degree of ionization in a gas increases with the increase in temperature. The interiors of main-sequence stars are hot enough to ensure that the gases there exist as highly ionized plasmas. It is to be noted that the Saha equation holds only for plasmas which are in thermodynamic equilibrium.

The regions of ionized hydrogen around the early-type stars known as the H II regions are almost completely ionized by the ultraviolet radiation coming from the central stars and are not in thermodynamic equilibrium. The ionization levels inside the H II regions are not determined by the Saha equation, but by the flux of ultraviolet photons from the central stars. It is estimated that more than 99% of the material in the Universe exists in the plasma state, even though the Saha equation may not be applicable to some of this plasma.

If we consider a volume of plasma large enough to contain many charged particles, we expect that volume to be close to charge-neutral. Any charge imbalance would produce strong electrostatic forces to restore the charge-neutrality. We will see later that charge imbalances may exist only over a short distance (called the Debye length) or for a short period of time (the inverse of the so-called plasma frequency). What makes the macroscopic behaviour of plasmas so different from that of the neutral gases, however, is the fact that an electromotive force applied to a plasma can drive large currents by making the electrons and ions move in opposite directions. Volumes of plasma can therefore sustain large currents in spite of being nearly charge-neutral. One often says that a plasma is a *quasi-neutral gas*.

Since many charged particles in a plasma can interact simultaneously through long-range electromagnetic interactions, there can be many phenomena in plasmas which are caused by collective interactions of many particles. The electrostatic oscillations discussed in the next chapter is an example of such collective behaviour. We shall see that each charged particle moves due to the electrostatic field produced by all the particles around it. Different charged particles moving in this way, however, can give rise to coherent oscillations. In the next section, we shall introduce the *plasma parameter*, which will provide a quantitative way of estimating the importance of long-range interactions.

Suppose we consider the motion of one charged test particle in a plasma. When no other particle is close enough to cause an appreciable deflection in the trajectory of our test particle, we can assume the electromagnetic field experienced by this test particle to be a smooth field caused by the long-range electromagnetic interactions of many distant particles. However, when another charged particle comes close enough to our test particle to cause a deflection by the Coulomb interaction, we have to treat that as a collision. We saw in §3.4 that collisions in neutral gases play crucial roles in keeping the systems close to thermodynamic equilibrium. We expect that collisions in plasmas also would play similar roles. But collisions between char ̧ed particles are much harder to treat than collisions between neutral particles. Hence the mathematical theory of plasmas is still in a much less satisfactory stage of development.

If a plasma is weakly ionized, then a charged particle is more likely to have a collision with a neutral atom rather than another charged particle. A collision between a charged particle and a neutral atom takes place only when they are very close and usually produces large deflections. Hence such collisions are fairly similar to collisions between neutral particles and the Boltzmann equation can be applied to weakly ionized plasmas. In Chapter 13 we shall present some discussion on the collisional processes in weakly ionized plasmas.

For highly ionized plasmas, in contrast, the Coulomb collision between charged particles is the important collisional process. Whenever two charged particles interact in such a way that the trajectory of at least one particle (the lighter particle, if the two particles are not alike) shows a 'noticeable' deflection, we may regard it as a collision. Defined in this fashion, most collisions between charged particles would take place when the separations between them are typically larger than the separations for neutral particle collisions and a majority of such collisions would produce small deflections in the trajectories of the particles. Hence the velocity of a charged particle would change appreciably more likely due to the succession of many small collisions rather than due to one large collision. Figure 11.1 shows the typical trajectory of a charged particle in a plasma. It is clear that this trajectory is very different from the trajectory of a particle in a neutral gas as shown in Figure 2.1. We shall give an outline as to how one tries to develop a kinetic theory of a gas of charged particles, which is much more complicated than the kinetic theory of neutral gases.

As in the case of neutral gases, eventually we attempt to develop continuum fluid models of plasmas. We saw in Chapter 3 that a hydrodynamic model could be developed by closing the moment equations

Figure 11.1 The typical trajectory of a charged particle in a plasma.

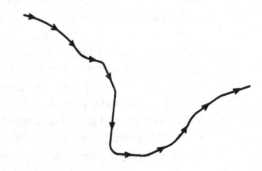

if the neutral gas was close to thermodynamic equilibrium within local regions. We even calculated departures from the thermodynamic equilibrium, which enabled us to estimate the transport coefficients. A similar presentation for plasmas is beyond the scope of this elementary textbook. We shall develop fluid models of plasmas merely on the basis of the *assumption* that the plasmas under consideration are close to thermodynamic equilibrium within local regions. Properties of a plasma in complete thermodynamic equilibrium can be investigated by employing the powerful techniques of statistical mechanics through the manipulation of the partition function. But a systematic derivation of the continuum fluid model of a plasma which is spatially inhomogeneous and hence not in complete thermodynamic equilibrium—especially in the presence of a strong magnetic field—is a formidable problem. One can perhaps give more justification for the fluid models than we provide here. But a completely satisfactory derivation of fluid models for inhomogeneous magnetized plasmas is not available at the present time. We shall see that our fluid models often give fairly reliable results in agreement with experiments, but we do not have a good understanding of their limitations. We do not always understand why and when the models break down. We shall make more comments on these in appropriate places.

There is one important topic in the study of plasmas which is completely left out of this book. According to classical electrodynamics, an accelerated charged particle emits electromagnetic radiation (see, for example, Panofsky and Phillips 1962, Chapter 20; Jackson 1975, Chapter 14). Hence, whenever the velocity of a charged particle in the plasma changes (in magnitude or direction), some electromagnetic radiation comes out of the plasma. When two charged particles collide with each other and change their velocities, the type of radiation coming out is known as *bremsstrahlung* (Rybicki and Lightman 1979, Chapter 5). On the other hand, when a relativistic charged particle spirals around a magnetic field, we get *synchrotron radiation* (Rybicki and Lightman 1979, Chapter 6). Although a study of the radiation processes in plasmas is tremendously important in astrophysics, it requires theoretical concepts and techniques which are very different from the concepts and techniques developed in this book. Hence a discussion of radiation processes would have been quite a bit outside the main line of development in this book. We recommend the reader to the excellent graduate textbook by Rybicki and Lightman (1979) for an introduction to the radiation processes important in astrophysics. The propagation of electromagnetic waves in plasmas, discussed in §12.2 and §12.5, will be the only topic in this book pertaining to the subject of interaction between plasmas and electromagnetic radiation.

11.2 Debye shielding. The plasma parameter

To understand how good the assumption of charge-neutrality is, let us consider the charge separation produced by introducing a charge Q inside a plasma. The theory of electrostatic screening of a charge was first developed by Debye and Hückel (1923) in their study of electrolytic solutions. If n_e is the number density of electrons and n_i that of ions (assumed singly ionized), then the charge density at a point in space is given by $(n_i - n_e)e$. Denoting the electrostatic potential by ϕ, the Poisson equation is

$$\nabla^2 \phi = -4\pi(n_i - n_e)e. \tag{11.2}$$

If the plasma is in thermodynamic equilibrium and n is the number density of ions or electrons far away from the charge Q, then we expect

$$n_i = n \exp\left(-\frac{e\phi}{\kappa_B T}\right), \qquad n_e = n \exp\left(\frac{e\phi}{\kappa_B T}\right). \tag{11.3}$$

Substituting in (11.2),

$$\nabla^2 \phi = 4\pi n e \left[\exp\left(\frac{e\phi}{\kappa_B T}\right) - \exp\left(-\frac{e\phi}{\kappa_B T}\right)\right].$$

We now expand in Taylor series and neglect terms quadratic or higher-order in $(e\phi/\kappa_B T)$, which are expected to be small in usual circumstances. This gives

$$\nabla^2 \phi = \frac{\phi}{\lambda_D^2}, \tag{11.4}$$

where

$$\lambda_D = \left(\frac{\kappa_B T}{8\pi n e^2}\right)^{1/2} \tag{11.5}$$

is known as the Debye length. The potential around the charge Q satisfies (11.4). The appropriate solution is

$$\phi = Q\frac{\exp(-r/\lambda_D)}{r}. \tag{11.6}$$

It thus appears that the effect of the charge is screened beyond a distance λ_D. A plasma can therefore be considered charge-neutral when distances larger than the Debye length are considered.

Although the electric field of a charged particle in principle extends to infinity, the influence of a charged particle in a plasma is effectively felt to a distance λ_D, i.e. within a volume of the order λ_D^3 called the Debye volume. Hence the number of particles on which this charged particle would exert an influence is of order $n\lambda_D^3$. We have already mentioned in §11.1 that a plasma can exhibit collective phenomena arising out of mutual interactions of many charged particles. The number $n\lambda_D^3$ gives a measure of the number of particles which can interact simultaneously. The inverse of this number is known as the *plasma parameter*

$$g = \frac{1}{n\lambda_D^3} = \frac{(8\pi)^{3/2}e^3n^{1/2}}{(\kappa_B T)^{3/2}} \qquad (11.7)$$

on substituting from (11.5). When the plasma parameter g is smaller, there is more collective interaction in the plasma. It is to be noted that g is smaller for smaller n. So the number of particles interacting collectively is more for a low-density plasma. In such a plasma, the Debye shielding is less effective so that the Debye volume λ_D^3 is much larger. Hence, even if the number of particles per unit volume is less, the total number of particles in the Debye volume is larger.

The average distance between the particles of a plasma is of the order $n^{-1/3}$. Therefore the average potential energy of electrostatic interaction between a pair of nearby particles is of the order $e^2n^{1/3}$. Hence the ratio of average potential to kinetic energy is

$$\frac{\langle\text{P.E.}\rangle}{\langle\text{K.E.}\rangle} \approx \frac{e^2n^{1/3}}{\kappa_B T} \propto g^{2/3}. \qquad (11.8)$$

Another interpretation of the plasma parameter, therefore, is that it is a measure of the potential energy of interactions compared to the kinetic energy. When g is small (as for a low-density plasma), the interaction amongst particles is weak, but a large number of particles interact simultaneously. On the other hand, a larger g implies few particles interacting collectively, but interacting strongly. The limit of small g is referred to as the *plasma limit*. Most textbooks on plasmas take the smallness of the plasma parameter g as a condition for the definition of a plasma.

11.3 Different types of plasmas

From the expressions (11.5) and (11.7), it is clear that, apart from the mathematical and physical constants, the characteristics of a plasma are determined by the number density n of charged particles and the temperature T. Now these two physical variables n and T are found to vary over many orders of magnitude in the different types of plasma known to us. This is clearly seen in Figure 11.2, where different plasma systems are indicated in a plot of n against T. Since the temperature of a plasma is often denoted in eV, the top axis of the figure gives the temperature scale in eV. We shall see in §12.2 that the characteristic frequency of a plasma, known as the plasma frequency, depends on the number density n. The values of the plasma frequency are indicated on the right axis. The plasma frequency for the Earth's ionosphere is about 10^7 Hz lying in the radio range, whereas the plasma frequency for the electron gas in a metal is about 10^{15} Hz which is in the ultraviolet. For the convenience of the readers, Appendix E gives the expressions for various plasma parameters after putting the numerical values of mathematical and physical constants.

Figure 11.2 Different plasma systems indicated on a plot of the number density n of charged particles against the temperature T.

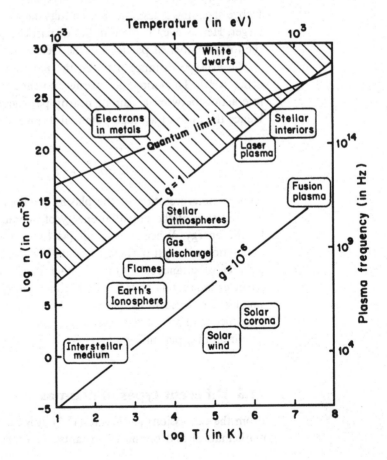

The straight lines corresponding to $g = 1$ and $g = 10^{-6}$ are indicated in Figure 11.2. We pointed out in the previous section that one often takes the condition $g < 1$ as a part of the definition of a plasma. The region excluded by this definition is shaded. It should be clear from the figure that many plasma systems of interest to us have $g < 1$, and the theoretical results which we shall derive in the following chapters should hold for them. For the region above the line labelled as 'quantum limit', the condition (1.1) is not satisfied and the wave packets of the particles overlap with each other. Hence the electrons inside a metal or a white dwarf star cannot be regarded as classical particles. It would appear from Figure 11.2 that these electron gases would have very high g and the usual results of plasma physics would not apply to them at all. However, the expression (11.7) for g does not hold for these systems, since we tacitly assumed a classical distribution while applying (11.3) in our derivation. To derive the Debye length for a quantum system, we have to replace (11.3) by expressions appropriate for a Fermi–Dirac distribution (see, for example, Kittel 1995, pp. 280–3; Ashkroft and Mermin 1976, pp. 340–2). On calculating the Debye length in this way, it is found that $g = (n\lambda_{\mathrm{D}}^3)^{-1}$ for these systems is not as large as one would expect on the basis of (11.7). Hence, *some results* of plasma physics are applicable to a system like the electron gas in a metal.

According to (11.8), one implication of small g is that the potential energy of interactions amongst nearby particles is small compared to the kinetic energy, and the plasma can be treated as a perfect gas, i.e. a gas of non-interacting particles with the properties listed in §6.1. This holds even for material at the centres of stars, where the density may be higher than the density of water. Matter of such density does not behave as a perfect gas at terrestrial temperatures. Hence, in the early years of stellar research, it was not clear whether matter at the centre of a main-sequence star would behave as a perfect gas (Eddington 1926, §6–7). Only after it was realized that the perfect gas laws hold down to the centres of stars, rapid advances were made in the study of stellar structure. It may seem at first sight that the mutual interactions amongst electrons would be very important inside a metal or a white dwarf star (because of $g \gg 1$). However, in a system of Fermions with most states filled up to the Fermi energy, collisions amongst nearby particles are suppressed due to the fact that the particles may not have free states available for occupation after the collision (Kittel 1995, pp. 294–7; Ashkroft and Mermin 1976, pp. 346–8). Hence electrons inside a metal or a white dwarf star are often approximated as a perfect gas made up of non-interacting Fermions.

11.4 BBGKY hierarchy

We now discuss what is perhaps the most promising approach for developing a kinetic theory for systems like plasmas in which many-particle interactions are important. One strategy we have been following in most parts of the present book is to give full derivations of whatever we are discussing. An exception will be made now. The kinetic theory of plasmas is such a complex subject that a full treatment is beyond the scope of our book. On the other hand, it is so important that we want to give the reader at least an idea of the subject. Hence we present an outline of the subject without full derivations of all the results mentioned. Some of the results of plasma kinetic theory are applicable to stellar dynamics as well. Since collisions in both fully ionized plasmas and stellar systems are caused by inverse-square forces, the mathematical treatments of collisions in these two systems are quite similar. There is, however, one big difference between plasma kinetic theory and stellar dynamics, as already mentioned in §3.7. The gravitational attraction in a stellar system cannot be screened, and the large-scale gravitational field existing over the system makes an overall thermodynamic equilibrium impossible. On the other hand, the electrical forces in a plasma are screened beyond the Debye length and do not come in the way of the plasma having a proper thermodynamic equilibrium. Hence those results of plasma kinetic theory which explicitly depend upon the plasma being in thermodynamic equilibrium do not hold for stellar systems.

We had introduced the concept of an ensemble in §1.4. We now consider an ensemble for a system of N particles, which means that we have to think of a Γ-space (see §1.5) in which a configuration of the system is represented by a point. The many probable configurations would correspond to many points in the Γ-space making up the ensemble. Let $\rho_N(\mathbf{x}_1, \mathbf{u}_1, \ldots, \mathbf{x}_N, \mathbf{u}_N, t)$ be the density of the ensemble points. We now want to find the joint probability $f_K(\mathbf{x}_1, \mathbf{u}_1, \ldots, \mathbf{x}_K, \mathbf{u}_K, t)$ that the particles 1 to K have coordinates $(\mathbf{x}_1, \mathbf{u}_1)$, ..., $(\mathbf{x}_K, \mathbf{u}_K)$ at time t. This can be obtained from the ensemble density by integrating over the coordinates of the remaining particles $K + 1$ to N, i.e.

$$f_K(\mathbf{x}_1, \mathbf{u}_1, \ldots, \mathbf{x}_K, \mathbf{u}_K, t) = V^K \int d\mathbf{x}_{K+1} d\mathbf{u}_{K+1} \ldots d\mathbf{x}_N d\mathbf{u}_N \, \rho_N, \quad (11.9)$$

where V is the physical volume which the N particles can occupy. The factor V^K is introduced here to ensure proper normalization. We know from §1.4 that ρ_N satisfies Liouville's equation (1.4). By suitably integrating Liouville's equation, one can obtain the following equation for f_K:

$$\frac{\partial f_K}{\partial t} + \sum_{r=1}^{K} \mathbf{u}_r \cdot \nabla_{\mathbf{x}_r} f_K + \sum_{r=1}^{K} \sum_{\substack{s=1 \\ s \neq r}}^{K} \frac{\mathbf{F}_{rs}}{m} \cdot \nabla_{\mathbf{u}_r} f_K$$

$$+ \frac{(N-K)}{V} \sum_{r=1}^{K} \int d\mathbf{x}_{K+1} d\mathbf{u}_{K+1} \frac{\mathbf{F}_{r,K+1}}{m} \cdot \nabla_{\mathbf{u}_r} f_{K+1} = 0, \quad (11.10)$$

where $\mathbf{F}_{rs}$ is the force exerted by the s-th particle on the r-th particle. In order to avoid any confusion, we have indicated all the summations explicitly rather than using the summation convention. For a very clear derivation of (11.10) from Liouville's equation, we refer the reader to Nicholson (1983, §4.3). We see from (11.10) that the equation for f_K involves f_{K+1}. Hence it follows that the equation for f_1 will involve f_2, the equation for f_2 will involve f_3, and so on. This hierarchy of equations goes by the name *BBGKY hierarchy* after the scientists who developed this formalism (Bogoliubov 1946; Born and Green 1949; Kirkwood 1946; Yvon 1935).

The first function $f_1(\mathbf{x}_1, \mathbf{u}_1, t)$ is clearly the usual probability distribution function. The next function $f_2(\mathbf{x}_1, \mathbf{u}_1, \mathbf{x}_2, \mathbf{u}_2, t)$ is the joint probability that the particle 1 has the coordinates $(\mathbf{x}_1, \mathbf{u}_1)$ and the particle 2 the coordinates $(\mathbf{x}_2, \mathbf{u}_2)$ simultaneously at time t. If the particle 1 does not influence the particle 2 in any way, then this joint probability would simply be $f_1(\mathbf{x}_1, \mathbf{u}_1, t) f_1(\mathbf{x}_2, \mathbf{u}_2, t)$. If, however, the two particles attract each other, then the probability of their being found close by is enhanced. On the other hand, if they repel each other, then this probability is reduced. Hence we can in general write

$$f_2(\mathbf{x}_1, \mathbf{u}_1, \mathbf{x}_2, \mathbf{u}_2, t) = f_1(\mathbf{x}_1, \mathbf{u}_1, t) f_1(\mathbf{x}_2, \mathbf{u}_2, t) + g(\mathbf{x}_1, \mathbf{u}_1, \mathbf{x}_2, \mathbf{u}_2, t), \quad (11.11)$$

where $g(\mathbf{x}_1, \mathbf{u}_1, \mathbf{x}_2, \mathbf{u}_2, t)$ is the two-particle correlation function. The above expansion can be written more compactly as

$$f_2(1, 2) = f_1(1) f_1(2) + g(1, 2), \quad (11.12)$$

where we have abbreviated by simply writing the number r in the place of $(\mathbf{x}_r, \mathbf{u}_r)$ and have not indicated the t dependences explicitly. Proceeding the same way, the next function can be written as

$$f_3(1, 2, 3) = f_1(1) f_1(2) f_1(3) + f_1(1) g(2, 3) + f_1(2) g(1, 3)$$
$$+ f_1(3) g(1, 2) + h(1, 2, 3), \quad (11.13)$$

where $h(1, 2, 3)$ is the three-particle correlation function. Expansions of the form (11.12) and (11.13) are known as the *Mayer cluster expansion* (Mayer and Mayer 1940, Chapter 13).

When a plasma is in thermodynamic equilibrium, one can apply the principles of statistical mechanics (through the Gibbs ensemble approach) to calculate various correlation functions and other quantities of interest. We shall not discuss this approach here, but refer the interested reader to Chapter 2 of Krall and Trivelpiece (1973). It can be shown that $f_1(\mathbf{x}_1, \mathbf{u}_1, t)$ for an equilibrium plasma is simply the Maxwellian distribution

$$f_{\mathrm{M}}(\mathbf{u}_1) = \left(\frac{m}{2\pi\kappa_{\mathrm{B}}T}\right)^{3/2} \exp\left(-\frac{mu_1^2}{2\kappa_{\mathrm{B}}T}\right). \tag{11.14}$$

The two-particle correlation function turns out to be

$$g(\mathbf{x}_1, \mathbf{u}_1, \mathbf{x}_2, \mathbf{u}_2) = -\frac{q_1 q_2}{\kappa_{\mathrm{B}}T} \frac{\exp\left(-|\mathbf{x}_2 - \mathbf{x}_1|/\lambda_{\mathrm{D}}\right)}{|\mathbf{x}_2 - \mathbf{x}_1|} f_{\mathrm{M}}(\mathbf{u}_1) f_{\mathrm{M}}(\mathbf{u}_2),$$

if the plasma is in thermodynamic equilibrium. As expected, this correlation function is negative when the two charges q_1 and q_2 are alike, whereas it is positive when the charges have opposite signs. The real non-trivial problem is to calculate all these various correlation functions for non-equilibrium plasmas with spatial inhomogeneities and macroscopic motions.

To have an idea how one handles non-equilibrium plasmas, let us look at the equation for $f_1(\mathbf{x}_1, \mathbf{u}_1, t)$ which one obtains from the general BBGKY hierarchy equation (11.10). Writing $N/V = n_0$, we get

$$\frac{\partial f_1(1)}{\partial t} + \mathbf{u}_1 \cdot \nabla_{\mathbf{x}_1} f_1(1) + n_0 \int d\mathbf{x}_2 d\mathbf{u}_2 \frac{\mathbf{F}_{12}}{m} \cdot \nabla_{\mathbf{u}_1} [f_1(1)f_1(2) + g(1,2)] = 0, \tag{11.15}$$

where we have substituted from (11.12). We thus see that we need a knowledge of $g(1,2)$ in order to calculate $f_1(1)$. One can obtain an equation for $g(1,2)$ from the next equation of the BBGKY hierarchy (i.e. from the equation for $f_2(1,2)$). But this equation involves the next correlation function $h(1,2,3)$. Hence we have to truncate at some correlation function in order to have a closed set of equations.

The simplest truncation is to ignore $g(1,2)$. Then (11.15) gives

$$\frac{\partial f_1(1)}{\partial t} + \mathbf{u}_1 \cdot \nabla_{\mathbf{x}_1} f_1(1) + \frac{\mathbf{F}}{m} \cdot \nabla_{\mathbf{u}_1} f_1(1) = 0, \tag{11.16}$$

where

$$\mathbf{F} = n_0 \int d\mathbf{x}_2 d\mathbf{u}_2 \, \mathbf{F}_{12} f_1(2) \tag{11.17}$$

is the smooth force acting on the particle 1 due to the long-range interactions of all the other particles. Since we have neglected $g(1,2)$

which arises due to interactions between neighbouring particles, **F** does not include the forces exerted during collisions. Equation (11.16) is known as the *Vlasov equation* (Vlasov 1945) and is one of the most important equations in plasma physics. It is sometimes also referred to as the collisionless Boltzmann equation because of its identical appearance with the Boltzmann equation without the collision term. One should, however, keep in mind that there are subtle differences between the Boltzmann equation and the Vlasov equation at a conceptual level. The Boltzmann equation is for particles without long-range interactions and the force **F** is usually some external force. The collisions are well-defined sharp events and can be neglected by setting the collision integral equal to zero. On the other hand, the Vlasov equation deals with particles with long-range interactions so collisions are more difficult to define. Since $g(1, 2)$ captures the essence of interactions between nearby particles, the neglect of $g(1, 2)$ in a sense implies the neglect of collisions. But the long-range interactions due to a smooth distribution of particles are taken into account through **F** as defined by (11.17). Because of these reasons, we shall not use the term collisionless Boltzmann equation for the Vlasov equation.

We expect that a non-equilibrium plasma left to itself will relax to an equilibrium with the Maxwellian distribution. This cannot be shown by the Vlasov equation where two-particle interactions are neglected by neglecting $g(1, 2)$. Hence an arbitrary distribution function f_1 obeying the Vlasov equation will not relax to a Maxwellian distribution. It is therefore essential to include $g(1, 2)$ if we want to study plasma relaxation through collisions. As mentioned already, $g(1, 2)$ captures the effect of collisions between particles. To calculate $g(1, 2)$, we have to keep the next equation in the BBGKY hierarchy. If we neglect the three-particle correlation $h(1, 2, 3)$, then we get closed equations. But these equations are still too difficult to handle in all generality. Impressive progress has been made in studying plasmas which are spatially homogeneous, though they may not have the Maxwellian distribution in the beginning. In other words, we have to deal with distribution functions $f_1(\mathbf{u}_1, t)$ independent of position. For such plasmas, a formal solution for $g(1, 2)$ can be written down and, on substituting it in (11.15), one gets a closed equation for $f_1(\mathbf{u}_1, t)$. This equation is called the *Lenard–Balescu equation* (Lenard 1960; Balescu 1960), which we shall not discuss here. We recommend Montgomery and Tidman (1964, Chapters 6–7) or Nicholson (1983, Chapter 5) for a systematic treatment of the subject. If we begin with an arbitrary initial distribution function $f_1(\mathbf{u}_1, t)$ and advance it according to the Lenard–Balescu equation, it is found that $f_1(\mathbf{u}_1, t)$ asymptotically approaches the Maxwellian distribution. We may mention that the Lenard–Balescu equation is a special case of what is known as the Fokker–Planck equation (Fokker 1914; Planck 1917). Whenever we have a system of many particles moving under forces which produce mostly small deflections in the trajectories, the effect of collisions is a diffusion in the velocity space and the Fokker–Planck equation is a natural outcome for such systems. One is led to the Fokker–Planck equation in the theory of Brownian motion or in collisional stellar dynamics, where we expect the trajectories to be like that shown in Figure 11.1. A very elementary and incomplete discussion of the Fokker–Planck equation is given in §11.6.

To summarize, a system of particles with long-range interactions can be treated by the BBGKY hierarchy with the Mayer cluster expansion. To close the equations, we have to neglect the K-particle correlation functions beyond a certain K. By neglecting the two-particle correlation function, we get the Vlasov equation. One of the limitations of this equation is that it will not make a plasma relax to a Maxwellian distribution, since we effectively neglect collisions by neglecting the two-particle correlation function. To study plasma relaxation, we have to go to one order higher in the equations.

11.5 From the Vlasov equation to the two-fluid model

If the Vlasov equation cannot even describe the relaxation of a plasma to equilibrium, of what use is it then? We now want to argue that the Vlasov equation is a tremendously useful equation in spite of its limitations. Suppose we consider the disturbance of a plasma from the equilibrium situation. One example is the study of waves excited in a plasma in thermodynamic equilibrium. The waves are expected to make the distribution functions of ions and electrons slightly different from the Maxwellian distribution characteristic of a plasma in equilibrium. It is true that the Vlasov equation cannot explain why the distribution functions are Maxwellian. If, however, the wave frequency is sufficiently high so that collisions can be neglected during an oscillation period of the wave, then the perturbations to the distribution function produced by the wave can be handled by the Vlasov equation. We have seen in Chapter 3 that a hydrodynamic description can be introduced for a system having the distribution function close to the Maxwellian. Hence we expect that a hydrodynamic description should be possible for the plasma we are considering. We now discuss how the Vlasov equation leads to the so-called two-fluid model of the plasma, in which a plasma is regarded as an inter-penetrating mixture of a negatively charged fluid of electrons and a positively charged fluid of ions.

Since the Vlasov equation cannot handle collisions, the two-fluid model developed below will be appropriate for collisionless plasma processes. It may be pointed out that the term 'collisionless' is used in plasma physics and stellar dynamics in somewhat different senses. The collisional relaxation time of a typical galaxy is larger than the age of the Universe. Such systems are referred to as *collisionless stellar systems* in which collisions have never played an important role *since the formation of the system*. On the other hand, here we shall be dealing with plasmas which are in thermodynamic equilibrium presumably due to collisional relaxation. But collisions can be neglected while studying some of the processes we are interested in.

We can introduce distribution functions $f_i(\mathbf{x}, \mathbf{u}, t)$ and $f_e(\mathbf{x}, \mathbf{u}, t)$ for ions and electrons respectively, both of which should satisfy the Vlasov equation. If the force $\mathbf{F}$ is purely electromagnetic, then we can write the Vlasov equation (11.16) as

$$\frac{\partial f_l}{\partial t} + \mathbf{u} \cdot \nabla f_l + \frac{q_l}{m_l} \left(\mathbf{E} + \frac{\mathbf{u}}{c} \times \mathbf{B} \right) \cdot \nabla_\mathbf{u} f_l = 0, \qquad (11.18)$$

where the subscript l can be i or e. It is obvious that $q_i = e$, $q_e = -e$, whereas m_i and m_e stand for the masses of ions and electrons respectively. If we multiply (11.18) by any quantity $\chi(\mathbf{u})$ and integrate over $\mathbf{u}$, then the derivation presented in §2.5 holds and we shall end up with equation (2.37) with n_l and m_l written for n and m. Although we now have a velocity-dependent force, it is to be noted that

$$\frac{\partial F_i}{\partial u_i} = 0,$$

because the nature of the vector product $\mathbf{u} \times \mathbf{B}$ is such that the i-th component of the force does not depend on u_i. Hence we write (2.37) as

$$\frac{\partial}{\partial t}(n_l \langle \chi \rangle) + \frac{\partial}{\partial x_i}(n_l \langle u_i \chi \rangle) - \frac{n_l}{m_l} \left\langle F_i \frac{\partial \chi}{\partial u_i} \right\rangle = 0. \qquad (11.19)$$

It is easy to see that taking $\chi = 1$ gives us the continuity equation

$$\frac{\partial n_l}{\partial t} + \nabla \cdot (n_l \mathbf{v}_l) = 0, \qquad (11.20)$$

where

$$\mathbf{v}_l = \langle \mathbf{u} \rangle_l$$

is the average velocity of the particles of type l. If we put $\chi = m_l u_j$, then we can manipulate the equations exactly as in §3.1 and §3.3. Assuming the pressure to be a scalar, we then arrive at the Euler equation as in §3.3. The Euler equation for both the ion fluid and the electron fluid is

$$m_l n_l \left[\frac{\partial \mathbf{v}_l}{\partial t} + (\mathbf{v}_l \cdot \nabla)\mathbf{v}_l \right] = -\nabla p_l + q_l n_l \left(\mathbf{E} + \frac{\mathbf{v}_l}{c} \times \mathbf{B} \right), \qquad (11.21)$$

where p_l is the pressure associated with the particles of type l. It is to be noted that this equation follows also from the macroscopic considerations presented in §4.1.1 by noting that the volume force acting on the charge $q_l n_l$ per unit volume is $q_l n_l (\mathbf{E} + \mathbf{v} \times \mathbf{B}/c)$.

Chapter 3 presents a detailed discussion of how to close the set of equations obtained from the moments of the Boltzmann equation by relating the pressure to other thermodynamic variables in the case of neutral gases. This procedure works if the system is close to thermodynamic equilibrium. For plasmas close to thermodynamic equilibrium, we expect that it should be possible to relate p_l to other variables. We shall not get into a detailed discussion of this point as we did in Chapter 3 for neutral gases. In most of the cases of interest, p_l can be taken to have one of the forms

$$p_l = 0, \qquad (11.22a)$$

$$p_l = \kappa_{\mathrm{B}} n_l T_l, \qquad (11.22b)$$

or

$$p_l = C n_l^\gamma. \qquad (11.22c)$$

When the random thermal motions are not important in a particular situation, we can use (11.22a) known as the *cold plasma approximation*. If each species of particles can be assumed to have a constant temperature T_l (T_i need not be equal to T_e always), then (11.22b) is applicable. On the other hand, (11.22c) with a constant C is the appropriate relation for adiabatic processes. In the presence of a strong magnetic field, the thermodynamic properties of the plasma can be very different in directions parallel and perpendicular to the magnetic field. A theory for such a system was developed by Chew, Goldberger and Low (1956).

Our aim is to construct a complete dynamical theory for the two-fluid model. Apart from the electromagnetic fields, the other variables appearing in the equations (11.20–11.22) are n_l, p_l and $\mathbf{v}_l$. Noting that $\mathbf{v}_l$ is a vector and l has two possibilities, we have altogether ten scalar variables. It is easy to see that the number of equations is the same. If we count six additional scalars for the different components of $\mathbf{E}$ and

B fields, then we have altogether sixteen scalars. To have a full dynamical theory, we have to combine (11.20–11.22) with equations for the electromagnetic fields. The appropriate equations for the electromagnetic fields are Maxwell's equations. Remembering that the charge density is given by $(n_i - n_e)e$ and the current density by $(n_i v_i - n_e v_e)e$, Maxwell's equations are

$$\nabla \cdot \mathbf{E} = 4\pi(n_i - n_e)e, \tag{11.23}$$

$$\nabla \cdot \mathbf{B} = 0, \tag{11.24}$$

$$\nabla \times \mathbf{B} = \frac{4\pi}{c}(n_i v_i - n_e v_e)e + \frac{1}{c}\frac{\partial \mathbf{E}}{\partial t}, \tag{11.25}$$

$$\nabla \times \mathbf{E} = -\frac{1}{c}\frac{\partial \mathbf{B}}{\partial t}. \tag{11.26}$$

It may be noted that usual electrodynamics textbooks distinguish between **E** and **D**, **H** and **B**. This distinction is not useful for plasmas. When we have well-defined conductors placed in a medium, it makes sense to distinguish between charges and currents on the conductor from the induced polarization charges and currents in the medium. By combining the polarization charge in the medium with the usual electric field, one then arrives at the concept of the electric displacement **D** (see, for example, Panofsky and Phillips 1962, §2-1; Jackson 1975, §4.3). Such considerations are not appropriate for a plasma where we normally do not distinguish between conductors and the medium. In the next chapter, however, we shall see that one often introduces a 'dielectric constant' for the plasma, which is defined somewhat differently from the usual definition of dielectric constant in elementary electrostatics.

Equations (11.20–11.26) constitute the full dynamical theory of the two-fluid model. Since there are sixteen scalar variables, it seems at first sight to be a complicated theory. We shall see in the next chapter that applying this theory is actually much simpler than what we may expect by looking at the full set of equations.

11.6 Fokker–Planck equation

We have already pointed out in §11.4 that one has to go beyond the Vlasov equation to incorporate the two-particle interactions in the plasma. In most cases, such interactions produce small deflections and hence small changes in particle velocities. We mentioned that the evolution of the distribution function in such a situation is given by the Lenard–Balescu equation, which is a special case of the Fokker–Planck equation (Fokker 1914; Planck 1917) applicable to systems of particles moving under stochastic forces that keep producing small changes in particle velocities. We give a very heuristic presentation of the Fokker–Planck equation, which holds in different forms for such diverse systems as Brownian particles in a liquid or a cluster of stars. A rigorous treatment of the subject is beyond the scope of this book and can be found in the famous review article of Chandrasekhar (1943a).

Let us consider a distribution function independent of space so that we can write $f(\mathbf{v}, t)$ as a function of velocity and time only. We now think of some initial situation in which all the particles have nearly equal velocities. This means that the tips of all the velocity vectors for different particles lie within a small volume d^3v of the velocity space. In the presence of stochastic forces producing small changes of velocities, the tips of many velocity vectors would diffuse out of this small volume d^3v after some time. Hence we expect $f(\mathbf{v}, t)$ to satisfy a diffusion equation in the velocity space, which we can write as

$$\frac{\partial f}{\partial t} = D_v \nabla_v^2 f. \tag{11.27}$$

This is exactly like the usual diffusion equation, except that we are considering diffusion in the velocity space and the operator ∇_v^2 appropriate for the velocity space appears in the place of the usual ∇^2. The coefficient D_v is related to the statistical properties of the stochastic forces.

This is usually not the whole story. Suppose we think of Brownian particles in a liquid, on which stochastic forces are exerted by the molecules of the liquid. The velocity vectors of the Brownian particles certainly keep on diffusing in the velocity space. But there is another effect as well. A Brownian particle moving with velocity $\mathbf{v}$ through the liquid experiences a drag opposing the motion. The resulting deceleration can be written as a negative acceleration of the form

$$\mathbf{a}(\mathbf{v}) = -\gamma \mathbf{v}, \tag{11.28}$$

where γ is the friction coefficient. We now want to find how this acceleration affects the distribution function. Let us forget about the diffusion in the velocity space for the time being. We shall put it back later. If there is no diffusion, the effect of the acceleration $\mathbf{a}(\mathbf{v})$ can be incorporated simply by writing an equation of continuity in the velocity space. In the usual equation of continuity (3.51), we replace the space coordinate by velocity and the velocity by acceleration.

This gives

$$\frac{\partial f}{\partial t} + \nabla_{\mathbf{v}} \cdot (f\mathbf{a}) = 0. \tag{11.29}$$

To treat both the acceleration $\mathbf{a}(\mathbf{v})$ and the diffusion in the velocity space, we have to combine (11.27) and (11.29) so that

$$\frac{\partial f}{\partial t} = \nabla_{\mathbf{v}} \cdot (f\gamma\mathbf{v}) + D_{\mathbf{v}}\nabla_{\mathbf{v}}^2 f, \tag{11.30}$$

where we have used (11.28) to substitute for $\mathbf{a}$. Equation (11.30) is the usual form of the Fokker–Planck equation.

To obtain the Fokker–Planck equation for a particular system, the main task is to calculate the friction coefficient γ and the diffusion coefficient $D_{\mathbf{v}}$ from the interactions appropriate for that particular system. It is obvious that Brownian particles in a liquid would have a non-zero friction coefficient γ. But do we expect such friction to be present in systems like a plasma or a star cluster where the particles interact through long-range forces? Chandrasekhar (1943b) derived the famous result that such a friction term should indeed arise in a stellar system. Let us qualitatively explain why this should be so. Similar arguments can be given for plasmas as well. Suppose a star has moved from point P to point Q as shown in Figure 11.3. While passing from P to Q, the star attracted the surrounding stars towards itself. Hence we expect the number density of stars around PQ to be slightly larger than that ahead of Q. The star at Q, therefore, experiences a net gravitational attraction in the backward direction (i.e. in the direction of QP). This important effect is known as the *dynamical friction*.

If the system is not homogeneous in space then, instead of obtaining the Fokker–Planck equation in the three-dimensional velocity space as

Figure 11.3 An illustration of dynamical friction. A star has moved from P to Q creating a region of enhanced density of surrounding stars behind it.

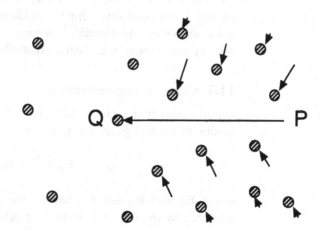

in (11.30), one has to obtain it in the six-dimensional space comprising of position and velocity coordinates. Binney and Tremaine (1987, §8.3) derive the Fokker–Planck equation for a collisional stellar dynamic system. Nicholson (1983, Chapter 5) obtains the Lenard–Balescu equation through the BBGKY hierarchy and shows that it is a special form of the Fokker–Planck equation. A particularly simple situation occurs if the distribution function depends on the magnitude v of velocity rather on the velocity $\mathbf{v}$, in addition to being spatially homogeneous. Then we write the distribution function as $f(v, t)$, which satisfies the scalar version of (11.30):

$$\frac{\partial f}{\partial t} = \frac{1}{v^2}\frac{\partial}{\partial v}(v^3 \gamma f) + D_v \frac{1}{v^2}\frac{\partial}{\partial v}\left(v^2\frac{\partial f}{\partial v}\right). \qquad (11.31)$$

If the coefficients γ and D_v are constant, then an arbitrary initial distribution function eventually relaxes to a Maxwellian. We shall not prove this general result. But it is easy to see that the relaxed asymptotic solution of (11.31) is a Maxwellian. To obtain the asymptotic, time-independent solution, we put the R.H.S. of (11.31) equal to zero. The solution of the resulting equation is

$$f \propto \exp\left(-\frac{\gamma}{2D_v}v^2\right). \qquad (11.32)$$

Thus we would have a Maxwellian distribution, provided

$$\frac{\gamma}{D_v} = \frac{m}{\kappa_B T}. \qquad (11.33)$$

Since the coefficients γ and D_v both ultimately arise from the same inter-particle interactions, it is not surprising for them to have such a relation. In the case of Brownian particles in a liquid, one can argue that such a relation should actually exist between γ and D_v (see, for example, Reif 1965, §15.7–15.12). Demonstrating the same for a plasma is more difficult. Using the Fokker–Planck equation approach, it was shown by MacDonald, Rosenbluth and Chuck (1957) that the velocity distribution in a plasma eventually relaxes to a Maxwellian.

11.7 Concluding remarks

Let us consider a system of N charged particles. The equation of motion of the r-th particle is given by

$$m_r\frac{d\mathbf{u}_r}{dt} = q_r\left[\sum_s \mathbf{E}_{rs} + \frac{\mathbf{u}_r}{c}\times\left(\mathbf{B}_{\text{ext}} + \sum_s \mathbf{B}_{rs}\right)\right],$$

where $\mathbf{E}_{rs}$ and $\mathbf{B}_{rs}$ are respectively the electric and magnetic field produced by the s-th particle at the position of the r-th particle. The possibility of an external magnetic field $\mathbf{B}_{ext}$ is indicated, although any external electric field is assumed to have been shielded in the interior of the system. The most complete account of the dynamics of this system is obtained by solving the equations of motion of all the N particles simultaneously. As this is not possible for realistic values of N, we develop less complete models of the system with the hope that these simpler models will provide us all the information we may wish to have. For a plasma, it appears that there is no simpler model which is applicable to *all* situations. Any approach short of solving the equations of motion of the N particles has its limitations. Since no unique simpler model serves all purposes, we have to develop different approaches for different problems, causing considerable confusion for the beginner.

The most obvious simpler model is to introduce distribution functions for the positive and negative charges. Apart from the theoretical problem that the long-range interactions amongst the particles make it very hard to derive sufficiently general equations for the distribution functions that encompass all situations, we have phenomena like confinement of particles in magnetic mirrors or particle acceleration which require the single-particle orbit theory approach for their understanding. So even the distribution function approach may not be suited for all purposes.

Techniques of statistical mechanics can be used to show that a plasma in thermodynamic equilibrium must have the Maxwellian distribution—a fact which is amply confirmed by laboratory experiments. We have seen that neutral gases admit of hydrodynamic models when the distribution functions are close to the Maxwellian. Hence we expect similar hydrodynamic models to hold for plasmas which have distribution functions close to the Maxwellian. However, the systematic establishment of such models for plasmas is a formidable theoretical problem not yet solved adequately. For a neutral gas, the Chapman–Enskog procedure enables us to calculate the departure from the Maxwellian distribution. This leads to a rigorous derivation of the hydrodynamic equations and gives an understanding when they may break down. It is much harder to follow a similar procedure for plasmas. Although the concept of a mean free path is not very useful when trajectories of particles change through many small deflections, one can define relaxation time τ as the average time required for the velocity of a particle to change appreciably. We shall discuss in Chapter 13 how to estimate the relaxation time τ. Using the relaxation time, we may be tempted to write down the BGK equation (3.29) for plasmas and calculate the departure from the Maxwellian from it. This

approach is not very reliable for plasmas, especially in the presence
of strong magnetic fields. To see the limitations of this approach, let
us look at (3.33) giving the departure from the Maxwellian distribu-
tion. One term in the expression is $\tau \mathbf{u} \cdot \nabla f^{(0)}$. This is the difference in
the value of $f^{(0)}$ between the points $\mathbf{x}$ and $\mathbf{x} + \tau \mathbf{u}$. This term arises
from the tacit assumption that a particle with velocity $\mathbf{u}$ traverses a
distance $\tau \mathbf{u}$ between collisions and thereby carries information from
the point $\mathbf{x}$ to the point $\mathbf{x} + \tau \mathbf{u}$. A charged particle in a plasma does
not traverse a straight distance $\tau \mathbf{u}$ in the relaxation time. Especially,
if there is a strong magnetic field, the charged particle moves in a
circular trajectory. Hence the approach of §3.4 is not very reliable for
inhomogeneous plasmas with magnetic fields, and the derivation of
the hydrodynamic equations is a much more complicated problem.

Since it is so difficult to derive hydrodynamic models of plasmas
rigorously, we do not have a very clear understanding when they
break down. For a neutral gas, the hydrodynamic model holds if the
mean free path λ is much smaller than the length scale L. A volume
of the neutral gas of size larger than λ but small compared to L can
be regarded as a genuine fluid element, because most particles inside
this volume remain inside for a sufficiently long time due to the many
collisions and therefore the volume retains a physical identity. In a
magnetized plasma, charged particles revolve around magnetic field
lines and a typical particle may spend a considerable amount of time
in a region of size of the order of the gyroradius r_L. Hence, if the length
scale of the plasma is much larger than the gyroradius r_L, we may
expect the hydrodynamic model to work. It is, however, a formidable
task to translate this qualitative idea into rigorous mathematics, and
it has still not been possible to do this completely satisfactorily. There
is ample experimental evidence that strong magnetic fields do indeed
make plasmas behave like hydrodynamic fluids. For many astrophysical plasma systems with magnetic fields, we expect fluid models to
hold even though collisions may be unimportant in the system.

If the scale of spatial inhomogeneities is sufficiently large, neutral
gases are adequately modelled by hydrodynamic equations. In the case
of a plasma, however, even when the system is spatially homogeneous,
the hydrodynamic description may be inadequate due to more subtle
reasons. The hydrodynamic description is based on the moments of
the distribution function f like $\rho = m \int f \, d^3u$, $\rho \mathbf{v} = m \int f \mathbf{u} \, d^3u$, and
so on. The hydrodynamic model works only when a finite number
of such low-order moments are sufficient to provide all the essential
information about the system. If the the distribution function f has
some unusual features, then a few low-order moments may not carry

all the necessary information and we may lose important physics by restricting ourselves to hydrodynamic equations. An example of this will be provided in the next chapter when we consider Landau damping. We shall see that the two-fluid model predicts that electrostatic waves in collisionless plasmas have no damping. On applying the Vlasov equation, however, we find subtle effects depending on the detailed pathologies of the distribution function out in the Maxwellian tail. These effects are not reflected in the first few low-order moments. Only by studying the distribution function fully, do we find out that the system has damping.

Since there is no unique simple model of a plasma which encompasses all situations, we have to familiarize ourselves with different models applied to different situations. What makes life particularly hard is that we are often unable to provide clear guidelines when a model works and when it does not work. Hence the best strategy is to develop an intuition about plasma physics by working out some examples of different models. This is done in the next few chapters. We are often like the blind men who visited an elephant and gave different descriptions of the elephant depending on the part of the elephant a particular blind man touched. Often a beginning student is completely disconcerted by this state of affairs. Ever since Galileo and Newton figured out that falling apples and planets are governed by simple laws, physicists have been on the lookout for simple laws to make sense out of diverse phenomena. Most of the elementary physics textbooks emphasize those topics for which simple and elegant solutions are possible. Hence a physics student raised on textbook problems often feels at a loss when encountering a system like a plasma which stubbornly refuses to admit of an all-encompassing simple model. We should keep in mind that there is no fundamental reason why everything in Nature should be governed by simple laws. There are systems and processes in Nature which are inherently complicated. The growth of a plant from a seed is an example of such a process. A plasma also appears to be an inherently complicated system, and we have to learn to live with this fact.

Exercises

11.1   Estimate the Debye lengths for the different types of plasmas shown in Figure 11.2.

11.2   Read about the BBGKY hierarchy from a book like Montgomery and Tidman (1964) or Nicholson (1983), and work out all the algebra to derive the basic BBGKY equation (11.10) from Liouville's equation.

12 Collisionless processes in plasmas

12.1 Introduction

In this chapter, we shall mainly study what happens when a plasma in thermodynamic equilibrium is slightly disturbed. If the effect of collisions can be neglected in studying the evolution of the disturbance, then we call it a *collisionless process*. In stellar dynamics, one often studies systems with relaxation times larger than the age of the Universe so that collisions have *never* been important in the system. Here, however, we shall consider plasmas which are close to thermodynamic equilibrium presumably as a result of collisions. But collisions happen to be unimportant in the particular processes we are going to look at.

If a plasma is close to thermodynamic equilibrium, then hydrodynamic models may be applicable as we pointed out in §11.5. We shall mostly use the two-fluid model which was developed in the previous chapter. Only §12.4 will provide an example of how calculations can be done with the help of the Vlasov equation. We shall see that the Vlasov equation can give rise to a conclusion different from what we get by using the two-fluid model due to rather subtle reasons.

We shall mainly consider waves of high frequencies (so that collisions are unlikely to occur during a period of the wave), as such waves are the most important examples of collisionless processes. For low-frequency processes (with frequency $\ll$ the plasma frequency ω_p defined in §12.2), we shall see in later chapters that charge separation can be neglected and the plasma may be regarded as a single fluid. We shall begin developing the one-fluid or *magnetohydrodynamic* model of the plasma from the next chapter. For the high-frequency waves discussed in this chapter, however, charge separation has to be taken into account and we employ the two-fluid model of regarding the plasma as an inter-penetrating mixture of a positively charged ion fluid and a negatively charged electron fluid. A plasma can sustain a whole 'zoo' of wave modes, which should be apparent to any person who has ever taken a look at any intermediate-level textbook on plasma physics. We shall discuss only the most important types of waves—especially those which astrophysicists are likely to encounter in their work. Apart from the high-frequency waves discussed in this chapter, we shall consider some low-frequency plasma waves in §14.5 using the MHD approximation.

Since ions are much heavier than electrons, they move much more sluggishly compared to electrons when subjected to high-frequency electromagnetic fields. While considering the response of the plasma to such fields, one does not introduce a large error if the motions of ions are neglected. All the calculations presented in this chapter neglect the motions of ions and consider only the electron fluid. Calculating the small correction due to the motions of ions is given as Exercise 12.1. Since we shall be dealing with the electron fluid only, we keep the notation simpler by denoting the number density of electrons and its perturbation as n_0 and n_1 respectively, without putting an extra subscript e.

12.2 Electromagnetic oscillations in cold plasmas

We begin our study of the response of a plasma to electromagnetic perturbations by assuming the plasma to be *cold*—a terminology introduced in §11.5 meaning that the pressure can be neglected when considering the dynamics of the plasma. The effect of pressure will be explored in the next two sections. We consider perturbations in a uniform homogenous plasma. The number density of the electron fluid can be written as $n_0 + n_1$. Let $\mathbf{v}_1$, $\mathbf{E}_1$ and $\mathbf{B}_1$ respectively denote the velocity of the electron fluid, the electric field and the magnetic field. The subscript 1 implies that the unperturbed values of all these quantities are zero.

If we linearize the equation of motion of the electron fluid as given by (11.21), it is straightforward to see that we get

$$m_e n_0 \frac{\partial \mathbf{v}_1}{\partial t} = -e n_0 \mathbf{E}_1. \tag{12.1}$$

It is to be noted that the magnetic force $\mathbf{v}_1 \times \mathbf{B}_1$ is of the second order in perturbed quantities and hence is neglected. We further obtain from (11.25) and (11.26) that

$$\nabla \times \mathbf{B}_1 = -\frac{4\pi}{c} n_0 e \mathbf{v}_1 + \frac{1}{c} \frac{\partial \mathbf{E}_1}{\partial t}, \tag{12.2}$$

$$\nabla \times \mathbf{E}_1 = -\frac{1}{c} \frac{\partial \mathbf{B}_1}{\partial t}. \tag{12.3}$$

Since equations (12.1–12.3) give the evolutions of the perturbation variables v_1, E_1 and B_1, it is possible to derive all the characteristics of the electromagnetic oscillations from these equations. One notes that the number density perturbation n_1 does not appear in the linearized perturbation equations for the cold plasma.

Let us assume that the time dependence of all the perturbed quantities is of the form $\exp(-i\omega t)$ so that we can everywhere replace $\partial/\partial t$ by $-i\omega$. We then get from (12.1)

$$v_1 = \frac{e}{i\omega m_e} E_1. \tag{12.4}$$

On substituting this for v_1 in (12.2), we have

$$\nabla \times B_1 = -\frac{i\omega}{c} \epsilon E_1, \tag{12.5}$$

where

$$\epsilon = 1 - \frac{\omega_p^2}{\omega^2} \tag{12.6}$$

is known as the *dielectric constant* of a plasma, ω_p being given by

$$\omega_p^2 = \frac{4\pi n_0 e^2}{m_e} \tag{12.7}$$

and being known as the *plasma frequency*. In the usual notation of electromagnetism textbooks, for a dielectric medium which does not conduct electricity, we have

$$\nabla \times B = \frac{1}{c} \frac{\partial D}{\partial t}$$

with the electric displacement $D = \epsilon E$. If we further substitute $-i\omega$ for $\partial/\partial t$ in the above equation, it takes the form (12.5). Because of this formal analogy, ϵ appearing in (12.5) is called the dielectric constant of the plasma. Substituting the values of electronic mass and charge in (12.7), the frequency $f_p = \omega_p/2\pi$ corresponding to the circular frequency ω_p is found to be

$$f_p \approx 10^4 \sqrt{n_0} \text{ Hz}, \tag{12.8}$$

where n_0 has to be given in cm^{-3}. One may remember (12.8) as rule of thumb for a quick estimate of the plasma frequency. The plasma frequencies for the various types of plasmas are indicated on the right axis of Figure 11.2. On taking a time derivative of (12.5) and using (12.3), we end up with

$$\frac{\omega^2}{c} \epsilon E_1 = c\nabla \times (\nabla \times E_1). \tag{12.9}$$

Since our background plasma is homogeneous, we look for solutions of the perturbed quantities which are sinusoidal in space. In other words, we assume all perturbations to be of the form $\exp(i\mathbf{k}.\mathbf{x} - i\omega t)$. On substituting in (12.9), we get

$$\mathbf{k} \times (\mathbf{k} \times \mathbf{E}_1) = -\frac{\omega^2}{c^2}\left(1 - \frac{\omega_p^2}{\omega^2}\right)\mathbf{E}_1, \qquad (12.10)$$

where we have substituted for ϵ from (12.6). Without any loss of generality, we can choose our z axis in the direction of the propagation vector $\mathbf{k}$, i.e. we write $\mathbf{k} = k\hat{\mathbf{e}}_z$. On substituting this in (12.10), we obtain the following matrix equation

$$\begin{pmatrix} \omega^2 - \omega_p^2 - k^2c^2 & 0 & 0 \\ 0 & \omega^2 - \omega_p^2 - k^2c^2 & 0 \\ 0 & 0 & \omega^2 - \omega_p^2 \end{pmatrix} \begin{pmatrix} E_{1x} \\ E_{1y} \\ E_{1z} \end{pmatrix} = \begin{pmatrix} 0 \\ 0 \\ 0 \end{pmatrix}. \qquad (12.11)$$

It is clear from (12.11) that the x and y directions are symmetrical, as we expect. The z direction, being the direction along $\mathbf{k}$, is distinguishable. This indicates that we may have two physically distinct types of wave modes. They are discussed below. The existence of these two wave modes in the plasma was first recognized in the classic paper of Tonks and Langmuir (1929).

12.2.1 Plasma oscillations

One solution of the matrix equation (12.11) is

$$E_{1x} = E_{1y} = 0, \qquad \omega^2 = \omega_p^2. \qquad (12.12)$$

Here the electric field is completely in the direction of the propagation vector $\mathbf{k}$, and it follows from (12.1) that all the displacements are also in the same direction. We also note that the group velocity $(\partial\omega/\partial k)$ is zero. We therefore have a non-propagating longitudinal oscillation with its frequency equal to the plasma frequency ω_p. Such oscillations are known as *plasma oscillations*. They are often called *Langmuir oscillations*, after Langmuir who was the pioneer in the study of these oscillations (Langmuir 1928; Tonks and Langmuir 1929). We shall see in the next section that these oscillations become propagating waves when a non-zero electron pressure is included.

It is not difficult to understand the physical nature of these oscillations. Against a background of nearly immobile and hence uniformly distributed ions, there will be alternate layers of compression and rarefaction of the electron gas (unless $\mathbf{k} = 0$ so that the wavelength is infinite). The electrostatic forces arising out of such a charge imbalance drive these oscillations. These oscillations were discovered experimentally by Penning (1926) and then theoretically explained by Tonks and Langmuir (1929).

12.2.2 Electromagnetic waves

The only other possible solution of the matrix equation (12.11) is

$$E_{1z} = 0, \qquad \omega^2 = \omega_p^2 + k^2 c^2. \tag{12.13}$$

This clearly corresponds to a transverse wave. It is actually nothing but the ordinary electromagnetic wave modified by the presence of the plasma. If $\omega \gg \omega_p$, then we are led to limiting relation $\omega^2 = k^2 c^2$, which is the usual dispersion relation for electromagnetic waves in the vacuum. In other words, if the frequency of the wave is too high, even the electrons, which are much more mobile than the ions, are unable to respond sufficiently fast so that the plasma effects are negligible.

The phase velocity for this wave is

$$v_{ph} = \frac{\omega}{k} = \frac{c}{\sqrt{1 - \omega_p^2/\omega^2}}, \tag{12.14}$$

which is greater than the speed of light. This, however, is not a violation of special relativity, because no physical signal travels at this velocity. The group velocity

$$v_{gr} = \frac{d\omega}{dk} = c\sqrt{1 - \frac{\omega_p^2}{\omega^2}} \tag{12.15}$$

is less than the speed of light. It is also to be noted from (12.13) that if $\omega < \omega_p$, then k becomes imaginary so that the wave is evanescent. If an electromagnetic wave of frequency ω is sent towards a volume of plasma with a plasma frequency ω_p greater than ω (if $\omega < \omega_p$), then the electromagnetic wave is not able to pass through this plasma and the only possibility is that it is reflected back.

We have indicated in Figure 11.2 that the plasma frequency of the ionosphere is in the radio-frequency range. Hence radio waves with lower frequencies are reflected from the ionospheric plasma, making it possible to communicate by radio waves between faraway points on the Earth's surface. Electron gases in metals have plasma frequencies in the ultraviolet. Hence visible light cannot propagate through metals and is usually reflected from a metallic surface. But metals become transparent in the ultraviolet. This was discovered by Wood (1933) and explained by Zener (1933). The plasma frequency of alkali metals calculated from the electron number density agrees quite well with the frequency beyond which the metals become transparent (Kittel 1995, pp. 274–5).

The refractive index of a medium is given by

$$\mu = \frac{c}{v_{\text{ph}}} = \frac{ck}{\omega}. \tag{12.16}$$

It follows from (12.14) that the refractive index of a plasma is

$$\mu^2 = 1 - \frac{\omega_p^2}{\omega^2}. \tag{12.17}$$

Well before Thomson (1897) discovered the electron, Lorentz (1878) developed a theory of the refractive index assuming that atoms contain harmonically bound charges inside. In the limit of free charges, this theory leads to the expression (12.17) (see Exercise 12.2).

12.3 Warm plasma waves

We now want to show that the plasma oscillations introduced in §12.2.1 become propagating waves when we take the electron pressure to be non-zero. We shall now use the adiabatic relation (11.22c) for the electron pressure rather than using (11.22a) as done for cold plasmas.

We linearize the Euler equation (11.21) for the electron fluid, which now has one additional pressure perturbation term compared to (12.1) and is

$$m_e n_0 \frac{\partial \mathbf{v}_1}{\partial t} = -e n_0 \mathbf{E}_1 - \frac{\gamma p_0}{n_0} \nabla n_1. \tag{12.18}$$

In the previous section, it was not necessary to consider the equation of continuity, since we had a full set of equations (12.1–12.3) which did not involve n_1. Now we linearize the equation of continuity (11.20), which gives

$$\frac{\partial n_1}{\partial t} + n_0 \nabla \cdot \mathbf{v}_1 = 0. \tag{12.19}$$

From (11.23), we further have

$$\nabla \cdot \mathbf{E}_1 = -4\pi e n_1. \tag{12.20}$$

We thus have three equations (12.18–12.20) involving the three perturbation variables n_1, $\mathbf{v}_1$ and $\mathbf{E}_1$. To proceed further, we assume all these variables to vary in space and time as $\exp[i(kx - \omega t)]$. The x component of (12.18) becomes

$$-i\omega m_e n_0 v_{1x} = -e n_0 E_{1x} - i\frac{\gamma p_0}{n_0} k n_1, \tag{12.21}$$

whereas (12.19) and (12.20) give

$$-i\omega n_1 + i n_0 k v_{1x} = 0, \tag{12.22}$$

$$i k E_{1x} = -4\pi e n_1. \tag{12.23}$$

On combining (12.21–12.23), we obtain the dispersion relation

$$\omega^2 = \omega_{\mathrm{p}}^2 + k^2 \frac{\gamma p_0}{m_e n_0}, \tag{12.24}$$

where ω_{p} is the plasma frequency defined in (12.7).

We now estimate the value of γ, which was introduced in (6.4). Here we are considering one-dimensional motions of the electrons, and the collisionless condition implies that the energy of these motions remains completely decoupled from energies due to motions in the other directions. Hence, to find γ in the present situation, we have to take c_V corresponding to the internal energy of motion in one dimension only. According to the equipartition principle (see, for example, Saha and Srivastava 1965, §3.26–3.27; Reif 1965, §7.5–7.6), the general expression for c_V corresponding to motions in N dimensions is $(N/2)R$. Substituting this in (6.4), we have

$$\gamma = \frac{N+2}{N}.$$

For longitudinal electrostatic waves, we have to take $N = 1$ so that $\gamma = 3$.

Writing $p_0 = \kappa_{\mathrm{B}} n_0 T$, we have from (12.24):

$$\omega^2 = \omega_{\mathrm{p}}^2 + k^2 \frac{3\kappa_{\mathrm{B}} T}{m_{\mathrm{e}}}, \tag{12.25}$$

It is easy to see that the group velocity is

$$v_{\mathrm{gr}} = \frac{d\omega}{dk} = \frac{3\kappa_{\mathrm{B}} T}{m_{\mathrm{e}}} \frac{k}{\omega}.$$

The waves are therefore propagating as long as the temperature is non-zero.

12.4 Vlasov theory of plasma waves

We now want to show how the plasma waves obtained in the previous section can be studied with the help of the Vlasov equation. Apart from illustrating how the Vlasov equation can be applied to solve actual problems, it will throw light on some subtle effects not captured in the two-fluid model. This analysis was first carried out in a famous paper by Landau (1946).

We write the distribution function for the electrons as

$$f_e(\mathbf{x}, \mathbf{u}, t) = f_0(\mathbf{u}) + f_1(\mathbf{x}, \mathbf{u}, t),$$ (12.26)

where $f_0(\mathbf{u})$ is the unperturbed Maxwellian distribution as given in (11.14) and $f_1(\mathbf{x}, \mathbf{u}, t)$ is the perturbation due to the presence of the wave. Substituting this in the Vlasov equation (11.18) and linearizing, we get

$$\frac{\partial f_1}{\partial t} + \mathbf{u} \cdot \nabla f_1 - \frac{e}{m_e} \mathbf{E}_1 \cdot \nabla_{\mathbf{u}} f_0 = 0.$$ (12.27)

We have not kept the magnetic field, because we are anticipating longitudinal electrostatic waves which do not produce magnetic fields. We take all the perturbed quantities to vary as $\exp[i(kx - \omega t)]$ as in the previous section. Such a spatial variation would imply that the electric field $\mathbf{E}_1$ would be in the x direction. Hence, from (12.27),

$$(-i\omega + iku_x)f_1 = \frac{e}{m_e} E_{1x} \frac{\partial f_0}{\partial u_x}.$$ (12.28)

Noting that the number density perturbation n_1 is obtained by integrating f_1 over the velocity space, (12.20) can be written as

$$ikE_{1x} = -4\pi e \int f_1 \, d^3u.$$ (12.29)

On substituting into (12.29) the expression of f_1 obtained from (12.28), we get

$$ikE_{1x} = -\frac{4\pi e^2}{m_e} E_{1x} \int \frac{\partial f_0/\partial u_x}{(-i\omega + iku_x)} d^3u.$$ (12.30)

Let us now write

$$f_0(\mathbf{u}) = n_0 \hat{f}_0(u_x)\hat{f}_0(u_y)\hat{f}_0(u_z),$$ (12.31)

where n_0 is the unperturbed number density and $\hat{f}_0(s)$ is the one-dimensional Maxwellian

$$\hat{f}_0(s) = \left(\frac{m_e}{2\pi\kappa_B T}\right)^{1/2} \exp\left(-\frac{m_e s^2}{2\kappa_B T}\right).$$ (12.32)

Remembering that $\int \hat{f}_0(u_y)du_y = 1$, $\int \hat{f}_0(u_z)du_z = 1$ (see Appendix B) and cancelling E_{1x} from the two sides, (12.30) gives us

$$1 = \frac{\omega_p^2}{k^2} \int_{-\infty}^{\infty} \frac{\partial \hat{f}_0(u_x)/\partial u_x}{u_x - \omega/k} du_x.$$ (12.33)

If we can carry out the integration over the dummy variable u_x, then we shall get the dispersion relation connecting ω with k.

Carrying out the integration in (12.33) happens to be a rather tricky business because of the pole at $u_x = \omega/k$. If ω and k are both real, then this pole is on the real axis. But do we expect this to be so? Let us consider a sinusoidal disturbance in space so that k is real. If the nature of the system is such that this disturbance evolves in time in a sinusoidal fashion as well, then ω is also real. But if the disturbance either damps or grows in time, then ω must have an imaginary part. Hence finding the correct position of the pole and the correct contour of integration is not straightforward. As was shown by Landau (1946), the most satisfactory way of proceeding is to treat it as an initial value problem where we study the evolution of an initial perturbation by carrying out a Laplace transformation in time. The readers are referred to Krall and Trivelpiece (1973, §8.3–8.6) for a rigorous treatment of the problem. It is found from the rigorous treatment that ω has a small imaginary part such that the pole is situated sightly off the real axis (although very close to it) and we get the correct answer by evaluating the integral in (12.33) along the Landau contour shown in Figure 12.1. We now present the results that we get by carrying out the integration along this contour without trying to justify this contour any further.

The integral in (12.33) can now be substituted by the principal value of the integral in addition to $i\pi$ times the residue at the pole due to the small semicircular path, i.e.

$$ 1 = \frac{\omega_p^2}{k^2} \left[P \int_{-\infty}^{\infty} \frac{\partial \hat{f}_0(u_x)/\partial u_x}{u_x - \omega/k} du_x + i\pi \left. \frac{\partial \hat{f}_0}{\partial u_x} \right|_{u_x = \omega/k} \right]. \qquad (12.34) $$

The principal value can be evaluated readily if the phase speed ω/k of the wave is much larger than the typical thermal velocities in warm plasma, i.e. if $\omega/k \gg u_x$. We can then use the binomial expansion

$$ -\frac{1}{u_x - \omega/k} = \frac{1}{\omega/k} + \frac{u_x}{(\omega/k)^2} + \frac{u_x^2}{(\omega/k)^3} + \frac{u_x^3}{(\omega/k)^4} + \cdots . \qquad (12.35) $$

Figure 12.1 Landau contour for evaluating the integral in (12.33).

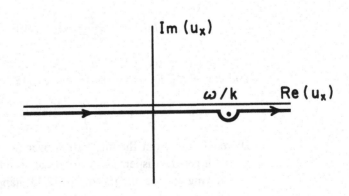

On substituting this in (12.34) and using (12.32), we find that only the odd powers of u_x in (12.35) contribute to the integral. Using the results quoted in Appendix B to work out the integrals, (12.34) leads to

$$1 = \frac{\omega_p^2}{k^2}\left[\frac{k^2}{\omega^2} + 3\frac{\kappa_B T}{m_e}\frac{k^4}{\omega^4} + i\pi \left.\frac{\partial \hat{f}_0}{\partial u_x}\right|_{u_x=\omega/k}\right],$$

from which

$$\omega^2 = \omega_p^2 + 3\frac{\kappa_B T}{m_e}k^2\frac{\omega_p^2}{\omega^2} + i\pi\frac{\omega_p^2\omega^2}{k^2}\left.\frac{\partial \hat{f}_0}{\partial u_x}\right|_{u_x=\omega/k}. \tag{12.36}$$

Since the second and third terms on the R.H.S. of (12.36) are small compared to the first term, we shall not introduce a large error if we substitute ω_p^2 in the place of ω^2 in those terms. Then, apart from an additional imaginary term, (12.36) becomes the same as the dispersion relation (12.25) derived in the previous section. On taking the square root of (12.36) with the help of the binomial theorem, we have

$$\omega = \omega_p\left[1 + \frac{3}{2}\lambda_{De}^2 k^2 + i\frac{\pi}{2}\frac{\omega_p^2}{k^2}\left.\frac{\partial \hat{f}_0}{\partial u_x}\right|_{u_x=\omega/k}\right], \tag{12.37}$$

where

$$\lambda_{De}^2 = \frac{\kappa_B T}{4\pi n_0 e^2}. \tag{12.38}$$

It is clear the imaginary part is negative (implying damping) or positive (implying instability) depending on the sign of $\partial \hat{f}_0/\partial u_x$ at $u_x = \omega/k$.

12.4.1 Landau damping

If $\hat{f}_0(u_x)$ is given by (12.32), then it follows from (12.37) that

$$\mathrm{Im}(\omega) = -\sqrt{\frac{\pi}{8}}\frac{\omega_p}{(k\lambda_{De})^3}\exp\left[-\frac{1}{2}(k\lambda_{De})^{-2} - \frac{3}{2}\right]. \tag{12.39}$$

It is easy to see that the negative $\mathrm{Im}(\omega)$ corresponds to damping. This *Landau damping* is seen to be small if the wavelength is much larger than the Debye length (see Exercise 12.3).

When Landau (1946) first discovered this damping, it came as a surprise. Normally one thinks of damping as a dissipative process and hence expects it to be present only in systems where collisions can convert a part of the wave energy into random thermal energy. Damping in a collisionless system seems, at first sight, quite mystifying. Since damping means that a part of the wave energy is converted into some other form, one asks the question where, in a collisionless system, could the energy have gone. Figure 12.2(a) shows the distribution function $\hat{f}_0(u_x)$ indicating the point where u_x is equal to the phase velocity ω/k of the plasma wave. Since the slope of $\hat{f}_0(u_x)$ is negative, we conclude that there are more particles moving slightly slower than the phase velocity than there are particles moving slightly faster than the phase velocity. Now the plasma wave corresponds to a sinusoidal electrostatic potential propagating at the phase velocity $u_x = \omega/k$. The troughs of this electrostatic potential would capture charged particles moving with low relative speed such that the captured particles thereafter tend to travel with the potential trough. Particles near a trough with u_x slightly less than ω/k are therefore accelerated to move with the trough. On the other hand, particles with u_x slightly more than ω/k are decelerated. For a negative slope of the distribution function at the phase velocity ω/k, there are more particles which are accelerated than which are decelerated. The wave therefore puts a net amount of energy in the particles so that there is a loss of wave energy. If we could follow the nonlinear evolution of the distribution function, then it would look after some time as shown in Figure 12.2(b). Such a process in which a wave exchanges energy with the particles in the plasma is called a *wave–particle interaction*. Landau damping is one of the most striking examples of wave–particle interaction.

12.4.2 Two-stream instability

Let us consider a stream of plasma with a non-zero average velocity impinging on a plasma at rest (i.e. with zero average velocity). The distribution function of the combined system would be as shown in Figure 12.3. Consider a plasma wave with phase velocity ω/k as indicated. Since the slope of the distribution function is positive at that point, it follows from (12.37) that $\text{Im}(\omega)$ is positive so that we

Figure 12.2 The basis of Landau damping. (a) The initial Maxwellian distribution with the phase velocity ω/k indicated in the tail. (b) The eventual distribution function of the electrons after the nonlinear evolution due to Landau damping.

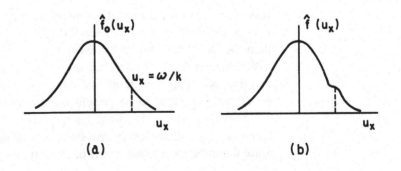

have a growing wave rather than a damped one. In other words, the system has an instability such that the energy of the relative motion between the plasma streams is fed into a plasma wave of appropriate phase velocity. This is known as the *two-stream instability*. Because of this instability, a stream of plasma impinging on another plasma cannot pass through it freely. The passage of the stream is halted, giving rise to plasma waves in the process.

This is a remarkable illustration of long-range interactions in plasmas. When considering low-density plamsas, we may naively think that two inter-penetrating plasma streams would pass through each other smoothly if collisions are unimportant. But this is not possible due to the two-stream instability arising out of the long-range interactions. Even a low-density plasma can act almost like a solid obstacle, preventing another plasma stream from penetrating too far inside.

12.5 Electromagnetic wave propagation parallel to magnetic field

We have already noted that there can be magnetic fields inside plasmas, although stationary electric fields cannot exist over distances larger than the Debye length. The propagation of electromagnetic waves in unmagnetized plasmas was studied in §12.2. We now want to show how the results are modified by the presence of a magnetic field. This topic is of great interest in astrophysics and space science, where we often have to consider electromagnetic wave propagation in magnetized plasmas. The general problem of electromagnetic waves propagating along a direction at an arbitrary angle to the magnetic field is very complicated. The general theory was worked out by Appleton (1927)

Figure 12.3 A double-humped distribution function representing two streams with ω/k indicated at a point where $d\hat{f}_0/du_x$ is positive.

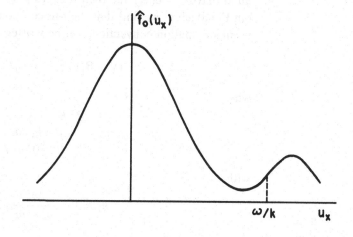

and Hartree (1931), who were studying the propagation of radio waves in the ionosphere. Here we give a flavour of the subject by considering the propagation of an electromagnetic wave parallel to the magnetic field in the plasma. The problem of an electromagnetic wave propagating perpendicular to the magnetic field is given as Exercise 12.5. We assume the plasma to be cold (i.e. $p_e = 0$) to keep the equations more manageable.

As in §12.2, we choose the z axis in the direction of propagation so that the propagation vector can be written as $\mathbf{k} = k\hat{\mathbf{e}}_z$. The unperturbed magnetic field in the plasma is taken as $\mathbf{B}_0 = B_0\hat{\mathbf{e}}_z$. The linearized perturbation equations can be written down the same way as in §12.2. Since we shall be considering all the perturbed quantities to have time dependences of the form $\exp(-i\omega t)$, we write down the linearized perturbation equations by replacing the time derivatives by $-i\omega$. On linearizing the equation of motion of the electron fluid (11.21), we get

$$-i\omega m_e \mathbf{v}_1 = -e\left(\mathbf{E}_1 + \frac{\mathbf{v}_1}{c} \times \mathbf{B}_0\right). \tag{12.40}$$

Apart from replacement of the time derivative by $-i\omega$, the only other difference between (12.1) and (12.40) is the inclusion of the Lorentz force term due to the unperturbed magnetic field. The equations (12.2) and (12.3) remain the same. Let us write them down with $-i\omega$ replacing the time derivatives:

$$\nabla \times \mathbf{B}_1 = -\frac{i\omega}{c}\mathbf{E}_1 - \frac{4\pi}{c}n_0 e\mathbf{v}_1, \tag{12.41}$$

$$\nabla \times \mathbf{E}_1 = \frac{i\omega}{c}\mathbf{B}_1. \tag{12.42}$$

We can find expressions for v_{1x}, v_{1y} and v_{1z} from (12.40) (keeping in mind that $\mathbf{B}_0 = B_0\hat{\mathbf{e}}_z$) and then substitute them in (12.41). On carrying out the algebra, we find that the i-th component of $\nabla \times \mathbf{B}_1$, using the usual summation convention, can be written as

$$(\nabla \times \mathbf{B}_1)_i = -\frac{i\omega}{c}\epsilon_{ij}E_{1j}, \tag{12.43}$$

where

$$\epsilon_{ij} = \begin{pmatrix} h_1 & ih_2 & 0 \\ -ih_2 & h_1 & 0 \\ 0 & 0 & h_3 \end{pmatrix} \tag{12.44}$$

with

$$h_1 = 1 - \frac{\omega_p^2}{\omega^2 - \omega_c^2}, \tag{12.45}$$

$$h_2 = \frac{\omega_c}{\omega} \frac{\omega_p^2}{\omega^2 - \omega_c^2},$$ (12.46)

$$h_3 = 1 - \frac{\omega_p^2}{\omega^2}.$$ (12.47)

Here ω_p is the plasma frequency given by (12.7) and ω_c is the cyclotron frequency $B_0 e/m_e c$ as in (10.2). On comparing (12.43) with (12.5), we see that the dielectric constant ϵ_{ij} has now become a second-rank tensor rather than a scalar as in the case of unmagnetized plasmas. The special case of unmagnetized plasmas can be obtained on putting $\omega_c = 0$ in our analysis. This would make ϵ_{ij} diagonal with all the diagonal elements equal to h_3. We further note that h_3 is equal to the scalar dielectric constant given by (12.6) so that our analysis reduces to the analysis of §12.2 on putting $\omega_c = 0$.

Taking the curl of (12.42), we get

$$[\nabla \times (\nabla \times \mathbf{E}_1)]_i = \frac{\omega^2}{c^2} \epsilon_{ij} E_{1j}.$$ (12.48)

Replacing ∇ by $ik\hat{e}_z$ in (12.48), we are led to the matrix equation

$$\begin{pmatrix} 1 - \frac{h_1}{n^2} & -\frac{ih_2}{n^2} & 0 \\ \frac{ih_2}{n^2} & 1 - \frac{h_1}{n^2} & 0 \\ 0 & 0 & -\frac{h_3}{n^2} \end{pmatrix} \begin{pmatrix} E_{1x} \\ E_{1y} \\ E_{1z} \end{pmatrix} = \begin{pmatrix} 0 \\ 0 \\ 0 \end{pmatrix},$$ (12.49)

where

$$n^2 = \frac{c^2 k^2}{\omega^2}.$$ (12.50)

This can be compared with the matrix equation (12.11). It is straightforward to show that (12.49) reduces to (12.11) if $\omega_c = 0$.

One easy solution of the matrix equation (12.49) is

$$E_{1x} = E_{1y} = 0, \qquad h_3 = 0.$$ (12.51)

From the expression (12.47) of h_3, we see that this is nothing but the usual plasma oscillation given by (12.12). Since the displacements are in the same direction as the unperturbed magnetic field (i.e. in z direction), the electron motions are unaffected by this magnetic field so that the plasma oscillation is not modified by the presence of the parallel magnetic field.

We are here interested in electromagnetic waves rather than the plasma oscillation. Hence we consider the other possible solution of (12.49) given by

$$E_{1z} = 0, \quad \begin{pmatrix} 1 - \frac{h_1}{n^2} & -\frac{ih_2}{n^2} \\ \frac{ih_2}{n^2} & 1 - \frac{h_1}{n^2} \end{pmatrix} \begin{pmatrix} E_{1x} \\ E_{1y} \end{pmatrix} = \begin{pmatrix} 0 \\ 0 \end{pmatrix}.$$ (12.52)

Setting the determinant equal to zero, we have

$$\left(1 - \frac{h_1}{n^2}\right)^2 - \frac{h_2^2}{n^4} = 0,$$

of which the solution is

$$n^2 = h_1 \pm h_2. \tag{12.53}$$

The plus and minus signs here correspond to the two possible eigenmodes. We shall now discuss these two eigenmodes separately.

First let us take the positive sign in (12.53). It is easy to see from (12.52) that we must have

$$E_{1x} - iE_{1y} = 0$$

in this case. The wave eigenfunction then has the form

$$\mathbf{E}_L = (\hat{\mathbf{e}}_x - i\hat{\mathbf{e}}_y) \exp[i(k_L z - \omega t)]. \tag{12.54}$$

This corresponds to a left circularly polarized wave, and we have written k_L for its wavenumber. The expression for k_L can be found by substituting (12.45), (12.46) and (12.50) in (12.53):

$$k_L = \frac{\omega}{c} \left[1 - \frac{\omega_p^2}{\omega(\omega + \omega_c)}\right]^{1/2}. \tag{12.55}$$

On the other hand, if we choose the minus sign in (12.53), then (12.52) leads to

$$E_{1x} + iE_{1y} = 0.$$

The corresponding eigenfunction

$$\mathbf{E}_R = (\hat{\mathbf{e}}_x + i\hat{\mathbf{e}}_y) \exp[i(k_R z - \omega t)] \tag{12.56}$$

represents a right circularly polarized wave of which the wavenumber is

$$k_R = \frac{\omega}{c} \left[1 - \frac{\omega_p^2}{\omega(\omega - \omega_c)}\right]^{1/2}. \tag{12.57}$$

From (12.55) and (12.57), it is clear that the two eigenmodes have different phase and group velocities. Hence, when we consider electromagnetic waves propagating parallel to a magnetic field in a plasma, we have two possible modes corresponding to left and right circularly polarized waves propagating at different speeds. It is easy to see that any electromagnetic wave of frequency ω propagating parallel to the magnetic field can be represented as a superposition of these two modes. Figure 12.4 shows the relation between the frequency and wavenumber as given by (12.55) and (12.57), for the case $\omega_p > \omega_c$. It is obvious from (12.55) that a real value of k_L is possible only when ω is larger than a value ω_1 which satisfies $\omega_1(\omega_1 + \omega_c) = \omega_p^2$. For the right circularly polarized wave, however, it follows from (12.57) that two branches are possible. One branch corresponds to $\omega < \omega_c$ for which k_R is real. The other branch of real k_R occurs if ω is larger than some value ω_2 satisfying $\omega_2(\omega_2 - \omega_c) = \omega_p^2$.

12.5.1 Faraday rotation

Let us now consider the propagation of a linearly polarized electromagnetic wave along the magnetic field. Such a linearly polarized wave is obtained simply by adding the expressions (12.54) and (12.56) of left and right circularly polarized waves, i.e.

$$\mathbf{E} = \mathbf{E}_L + \mathbf{E}_R. \tag{12.58}$$

On substituting from (12.54) and (12.56) into (12.58), we have

$$E_x = [\exp(ik_R z) + \exp(ik_L z)] \exp(-i\omega t), \tag{12.59}$$

$$E_y = i[\exp(ik_R z) - \exp(ik_L z)] \exp(-i\omega t). \tag{12.60}$$

We note that $E_y = 0$ at $z = 0$, which means that the wave is polarized along the x axis at $z = 0$. If θ be the angle of inclination of the plane

Figure 12.4 The graphical representation of the dispersion relation for an electromagnetic wave propagating parallel to a magnetic field in the plasma.

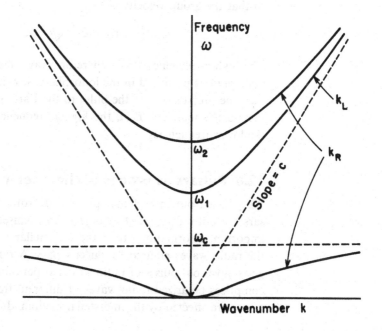

of polarization with respect to the x axis at any arbitrary z, then

$$\tan \theta = \frac{E_y}{E_x} = i\frac{1 - \exp[i(k_L - k_R)z]}{1 + \exp[i(k_L - k_R)z]}$$

$$= \tan \frac{k_L - k_R}{2} z$$

from which

$$\theta = \frac{k_L - k_R}{2} z. \tag{12.61}$$

The linear dependence of θ on z implies that the plane of polarization rotates with a uniform angular velocity as the wave propagates along the z direction.

12.5.2 Whistlers

During World War I, some radio operators working with audio-frequency waves (≈ 10 kHz) occasionally detected short-duration signals of whistling sound in which the frequency would start high and finish low. The correct explanation for these so-called *whistlers* was given by Storey (1953). In this very low-frequency limit ($\omega \ll \omega_c, \omega_p$), (12.57) gives

$$k_R = \frac{\omega_p}{c} \sqrt{\frac{\omega}{\omega_c}}$$

so that the group velocity is

$$v_{gr} = 2\frac{c}{\omega_p} \sqrt{\omega_c \omega}. \tag{12.62}$$

The higher-frequency waves therefore travel faster. If an electromagnetic pulse is produced in the ionosphere, say due to a lightning flash, then the propagation of the pulse to the Earth's surface is affected by the Earth's magnetic field, the higher frequencies reaching ahead of the lower frequencies.

12.6 Pulsars as probes of the interstellar medium

We pointed out in §9.4 that pulsars are rotating neutron stars. Pulsars are not only interesting objects by themselves, they also provide important information about the interstellar medium through which the radio waves emitted by pulsars pass. A pulsar, as its name suggests, gives out pulses of radio waves at periodic intervals. Each pulse comprises polarized radio waves of different frequencies. These radio waves are affected by the interstellar medium during their propagation to reach us. As seen in §12.2.2, waves with different frequencies travel with different group velocities. If the interstellar medium has a component of magnetic field parallel to the propagation direction, then the planes of polarization of different waves suffer Faraday rotation depending on frequency, as shown in §12.5.1. The pulse therefore has a spread in the arrival times of waves with different frequencies, their planes of polarization showing a spread as well. The plasma effects are noticeable for long-wavelength radio waves, but are completely negligible at optical frequencies (see Exercise 12.6).

The propagation time of a radio wave with frequency ω is given by

$$t_{\mathrm{p}} = \int_0^L \frac{dl}{v_{\mathrm{gr}}} = \frac{1}{c} \int_0^L \left(1 + \frac{\omega_{\mathrm{p}}^2}{2\omega^2} \right) dl, \qquad (12.63)$$

where we have substituted for the group velocity from (12.15) and made a binomial expansion in the small quantity $\omega_{\mathrm{p}}^2/\omega^2$. On substituting for ω_{p}^2 from (12.7), we get

$$t_{\mathrm{p}} = \frac{L}{c} + \frac{2\pi e^2}{m_{\mathrm{e}} c \omega^2} \int_0^L n_{\mathrm{e}}\, dl. \qquad (12.64)$$

The spread in the arrival times of waves with different frequencies is then given by

$$\frac{dt_{\mathrm{p}}}{d\omega} = -\frac{4\pi e^2}{m_{\mathrm{e}} c \omega^3} \int_0^L n_{\mathrm{e}}\, dl. \qquad (12.65)$$

From the spread in arrival times, one can then calculate the quantity $\int_0^L n_{\mathrm{e}} dl = \langle n_{\mathrm{e}} \rangle L$, which is known as the *dispersion measure*. Since the electron density in the interstellar medium in the solar neighbourhood is about $\langle n_{\mathrm{e}} \rangle \approx 0.03$ cm^{-3}, one can estimate the distance of the pulsar from the dispersion measure.

To find out the amount of Faraday rotation for a radio wave with given frequency, we merely generalize (12.61) to an integral, i.e.

$$\theta_{\mathrm{p}} = \int_0^L \frac{k_{\mathrm{L}} - k_{\mathrm{R}}}{2}\, dl.$$

On substituting from (12.55) and (12.57) after making binomial expansions, we get

$$\theta_{\mathrm{p}} = \int_0^L \frac{\omega}{2c} \left[\frac{\omega_{\mathrm{p}}^2}{2\omega(\omega - \omega_{\mathrm{c}})} - \frac{\omega_{\mathrm{p}}^2}{2\omega(\omega + \omega_{\mathrm{c}})} \right] dl$$

$$= \int_0^L \frac{\omega_{\mathrm{p}}^2 \omega_{\mathrm{c}}}{2c\,\omega^2}\, dl, \qquad (12.66)$$

where we have neglected ω_{c}^2 compared to ω^2 in the denominator.

Let us substitute for ω_p^2 from (12.7) and substitute $\omega_c = B_\parallel e/m_e c$ for the cyclotron frequency, where $B_\parallel$ is the component of magnetic field along the line of sight which is responsible for the Faraday rotation. Then (12.66) becomes

$$\theta_p = \frac{2\pi e^3}{m_e^2 c^2 \omega^2} \int_0^L n_e B_\parallel \, dl. \tag{12.67}$$

The spread in the planes of polarization is then obtained by differentiating this expression, i.e.

$$\frac{d\theta_p}{d\omega} = -\frac{4\pi e^3}{m_e^2 c^2 \omega^3} \int_0^L n_e B_\parallel \, dl. \tag{12.68}$$

By analyzing the polarization of pulsar signals, one can then find the quantity $\int_0^L n_e B_\parallel \, dl$.

Exercises

12.1   Consider a cold plasma in which the ions have charge $q_i = Ze$ and mass m_i. Without neglecting the motions of the ions, show that the plasma oscillations have the frequency

$$\omega = \omega_p \sqrt{1 + \frac{Z m_e}{m_i}},$$

where ω_p is given by (12.7). Estimate the percentage of error we make by neglecting the motions of ions.

12.2   Suppose all the electrons in a medium are harmonically bound inside atoms with some damping present. Then the linearized equation of motion is

$$m_e \left(\frac{\partial \mathbf{v}_1}{\partial t} + \gamma \mathbf{v}_1 + \omega_0^2 \mathbf{x}_1 \right) = -e\mathbf{E}_1$$

instead of (12.1). Here $\mathbf{x}_1$ is the displacement from the mean position, ω_0^2 the spring constant and γ the damping constant. Show that the refractive index of the medium is given by

$$\mu^2 = 1 - \frac{\omega_p^2}{\omega^2 - \omega_0^2 + i\gamma\omega}.$$

Note that this leads to (12.17) in the limit of free undamped electrons.

It may be pointed out that the effective electric field at the molecular level becomes slightly different from $\mathbf{E}$ when the surrounding medium is polarized. This effect has to be taken into account in a more accurate theory of refractive index (see, for example, Born and Wolf 1980, §2.3.2–2.3.4).

12.3 Consider warm plasma waves with wavelengths (i) $10\lambda_D$ and (ii) $25\lambda_D$, where λ_D is the Debye length. By what factors will they be damped due to Landau damping during one period?

12.4 Work out all the algebra for the derivation given in §12.5.

12.5 Consider a plasma with a uniform magnetic field $\mathbf{B}_0 = B_0\hat{\mathbf{e}}_z$, through which an electromagnetic wave is propagating with its propagation vector $\mathbf{k} = k\hat{\mathbf{e}}_x$ perpendicular to the magnetic field. Show that there are two possible electromagnetic wave modes:

(a) one mode (called the *ordinary wave*) with displacements in the z direction (i.e. $\parallel \mathbf{B}_0$) having the dispersion relation

$$\omega^2 = \omega_p^2 + k^2 c^2,$$

where ω_p is as defined in (12.7);

(b) one mode (called the *extraordinary wave*) with displacements in the xy plane (i.e. $\perp \mathbf{B}_0$) having the dispersion relation

$$\frac{c^2 k^2}{\omega^2} = 1 - \frac{\omega_p^2}{\omega^2} \frac{\omega^2 - \omega_p^2}{\omega^2 - \omega_p^2 - \omega_c^2},$$

where $\omega_c = B_0 e / m_e c$.

Note that the second mode is not fully transverse.

12.6 Suppose a pulsar is at a distance of about 100 pc (1 pc $\approx 3 \times 10^{18}$ cm) from us. Assume the electron density and the line-of-sight component of magnetic field in the intervening space to be about 0.03 cm^{-3} and 10^{-6} G respectively. For a radio wave of frequency $\omega = 10^9$ s^{-1}, find the time delay introduced by plasma effects and the Faraday rotation. What would have been the values of these quantities for visible light?

13 Collisional processes and the one-fluid model

13.1 Collisions and diffusion in weakly ionized plasmas

In the previous chapter, we neglected collisions amongst plasma particles. Now we want to look at the effects of collisions. We mentioned in Chapter 11 that treating collisions between charged particles is a complicated problem. Before we take up that subject, let us first consider a weakly ionized plasma in which electrons and ions mostly collide with neutral atoms. These collisions are more well defined, like collisions in a neutral gas, and hence can be treated more easily.

Even in the case of weakly ionized plasmas, we can think of electrons and ions as making up one fluid each. These fluids are now intermixed with a much denser fluid made up of neutral atoms. Let us consider the simple situation of the neutral fluid at rest so that we do not have to consider the equation of motion of the neutral fluid explicitly. Let $\mathbf{v}_e$ and $\mathbf{v}_i$ be the velocities of the electron fluid and the ion fluid respectively with respect to the background neutral fluid. If collisions were negligible, then $\mathbf{v}_e$ or $\mathbf{v}_i$ would have satisfied (11.21). We now consider how (11.21) will have to be modified due to collisions. If the electrons or ions have some relative velocity with respect to the neutrals, then we expect that collisions with the neutrals would hinder this relative velocity. If the typical collision time is τ_l, then the collisions are expected to produce a deceleration of order $-\mathbf{v}_l/\tau_l$. We can treat the effects of collisions heuristically by putting such a deceleration term in (11.21) so that

$$m_l n_l \frac{d\mathbf{v}_l}{dt} = -\nabla p_l + q_l n_l \left(\mathbf{E} + \frac{\mathbf{v}_l}{c} \times \mathbf{B} \right) - m_l n_l v_{c,l} \mathbf{v}_l, \tag{13.1}$$

where we have used the collision frequency $v_{c,l} = \tau_l^{-1}$ instead of τ_l.

A discussion of motions in the presence of a magnetic field will be taken up in the next section. Let us now consider a situation without **B**. It follows from (13.1) that either an electric field **E** or a pressure gradient ∇p_l can induce motions of the charged fluids. Since such motions are braked by collisions, one can think of a steady state. On putting the time derivative in (13.1) equal to zero in addition to $\mathbf{B} = 0$, we find that the velocity in the steady state is

$$\mathbf{v}_l = \frac{1}{m_l n_l \nu_{c,l}}(q_l n_l \mathbf{E} - \kappa_B T \nabla n_l), \tag{13.2}$$

where we have assumed the isothermal condition so that the temperature T can be taken outside the derivative after setting $p_l = \kappa_B T n_l$. The particle number density flux is given by

$$\Gamma_l = n_l \mathbf{v}_l = \pm \mu_l n_l \mathbf{E} - D_l \nabla n_l, \tag{13.3}$$

where

$$\mu_l = \frac{e}{m_l \nu_{c,l}} \tag{13.4}$$

and

$$D_l = \frac{\kappa_B T}{m_l \nu_{c,l}}. \tag{13.5}$$

are known as the *mobility* and the *diffusion coefficient* respectively. The positive or negative sign before μ in (13.3) has to be used depending on whether we are dealing with ions or electrons. It may be noted that we are assuming the ions to be singly charged with charge e.

Let us consider a weakly ionized plasma in which the density of charged particles varies from place to place. If there is an imbalance between the number densities of electrons and ions, it would lead to a very strong electric field. We therefore expect $n_i \approx n_e$, which we write as n. If n varies in space, then charged particles diffuse from regions of higher n to regions of lower n. In order to avoid charge separation, we must also have the number density fluxes Γ_i and Γ_e equal to each other. Such a diffusion process is known as the *ambipolar diffusion*, in which the ions and electrons diffuse jointly against a background of neutral atoms.

Setting $\Gamma_i = \Gamma_e = \Gamma$ and using (13.3),

$$\mu_i n \mathbf{E} - D_i \nabla n = -\mu_e n \mathbf{E} - D_e \nabla n,$$

from which

$$\mathbf{E} = \frac{D_i - D_e}{\mu_i + \mu_e} \frac{\nabla n}{n}. \tag{13.6}$$

Such an electric field would be set up inside the plasma in order to prevent the electrons and ions from separating. On substituting for $\mathbf{E}$ from (13.6) into (13.3), we have

$$\Gamma = \mu_i \frac{D_i - D_e}{\mu_i + \mu_e} \nabla n - D_i \nabla n,$$

which can be written as

$$\Gamma = -D_a \nabla n, \tag{13.7}$$

where

$$D_a = \frac{\mu_i D_e + \mu_e D_i}{\mu_i + \mu_e} \tag{13.8}$$

is the *coefficient of ambipolar diffusion*.

13.2 Diffusion across magnetic fields

One method of keeping a plasma confined in a region is to apply a strong magnetic field. The charged particles then gyrate around magnetic field lines and remain confined within a region of space of the order of the gyroradius. If, however, we are considering a weakly ionized plasma, then collisions with neutral atoms cause the charged particles to be scattered and produce a diffusion. Let us now calculate this diffusion for the electron fluid. As in the previous chapter, we simplify the notation by throwing away the subscript e referring to the electron fluid.

Let us consider a central region where many electrons are gyrating around magnetic field lines and focus our attention on the steady state such that electrons diffuse out of the central region at a steady rate. Putting the time derivative in (13.1) to zero, we get

$$-en\left(\mathbf{E} + \frac{\mathbf{v}}{c} \times \mathbf{B}\right) - \kappa_B T \nabla n - m_e n v_c \mathbf{v} = 0.$$

Eventually the electron density in the central region will decrease due to the particles diffusing away. But we can assume a steady state over time scales short compared to this time for density decrease. As we saw in §13.1, an electric field may arise to keep electrons and ions from separating out. Hence we kept $\mathbf{E}$ in addition to $\mathbf{B}$ in the above equation. If we choose the z axis in the direction of the magnetic field $\mathbf{B}$, then the x and y components of this equation are

$$v_x = -\mu E_x - \frac{D}{n}\frac{\partial n}{\partial x} - \frac{\omega_c}{v_c}v_y, \tag{13.9}$$

$$v_y = -\mu E_y - \frac{D}{n}\frac{\partial n}{\partial y} + \frac{\omega_c}{v_c}v_x, \tag{13.10}$$

where μ and D are as defined in (13.4) and (13.5), ω_c being the electron cyclotron frequency Be/m_ec (see (10.2)). Note that the subscript c in v_c and ω_c stand for 'collision' and 'cyclotron' respectively. On substituting for v_x in (13.10) from (13.9), we obtain

$$v_y(1 + \omega_c^2\tau^2) = -\mu E_y - \frac{D}{n}\frac{\partial n}{\partial y} - \omega_c^2\tau^2 c\frac{E_x}{B} - \omega_c^2\tau^2 c\frac{\kappa_B T}{Be}\frac{1}{n}\frac{\partial n}{\partial x}, \quad (13.11)$$

where $\tau = v_c^{-1}$. We note that $-c(E_x/B)$ is nothing but the y component of the $\mathbf{E} \times \mathbf{B}$ drift defined in (10.8). The last term in (13.11) corresponds to what is known as the *diamagnetic drift*. This arises whenever we have a non-uniform plasma in a magnetic field and can be understood from Figure 13.1. The magnetic field is directed into the page so that all the electrons are gyrating in the plane of the page. We show a box with a higher electron density toward the left side. More electrons pass in the downward direction through this box so that we obtain a drift perpendicular to the density gradient. Substituting for v_y in (13.9) from (13.10) gives

$$v_x(1 + \omega_c^2\tau^2) = -\mu E_x - \frac{D}{n}\frac{\partial n}{\partial x} + \omega_c^2\tau^2 c\frac{E_y}{B} + \omega_c^2\tau^2 c\frac{\kappa_B T}{Be}\frac{1}{n}\frac{\partial n}{\partial y}. \quad (13.12)$$

We can combine (13.11) and (13.12) together in the form

$$\mathbf{v}_\perp = -\mu_\perp\mathbf{E} - D_\perp\frac{\nabla n}{n} + \frac{\mathbf{v}_E + \mathbf{v}_D}{1 + v_c^2/\omega_c^2}, \quad (13.13)$$

where $\mathbf{v}_E$ and $\mathbf{v}_D$ are the $\mathbf{E} \times \mathbf{B}$ and the diamagnetic drifts respectively, and

$$\mu_\perp = \frac{\mu}{1 + \omega_c^2\tau^2}, \quad (13.14)$$

$$D_\perp = \frac{D}{1 + \omega_c^2\tau^2}, \quad (13.15)$$

Figure 13.1 The origin of the diamagnetic drift due to a magnetic field directed into the page.

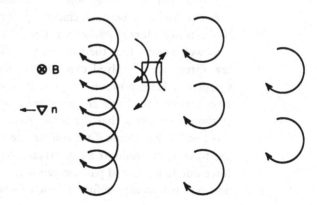

are the mobility and diffusion coefficients perpendicular to the magnetic field. It is easy to see that these coefficients parallel to the magnetic field remain unaffected. A magnetic field therefore makes these coefficients anisotropic.

If collisions are very efficient, in the sense of the collision frequency being much larger than the cyclotron frequency, then $\omega_c^2 \tau^2 \ll 1$ and we see from (13.14) and (13.15) that isotropy is restored. Let us now consider the opposite limit of the collisions being infrequent ($\tau \gg \omega_c^{-1}$) so that the effects of the magnetic field are pronounced, then we neglect 1 compared to $\omega_c^2 \tau^2$ and obtain from (13.5) and (13.15) that

$$D_\perp = \frac{\kappa_B T \nu_c}{m_e \omega_c^2}. \tag{13.16}$$

On comparing (13.16) with (13.5), we notice something interesting. In the absence of the magnetic field, the diffusion coefficient as given by (13.5) *decreases* with collision frequency ν_c. On the other hand, when the effect of the magnetic field is dominant, the perpendicular diffusion *is enhanced* by collisions as seen from (13.15). Collisions therefore play very different roles depending on whether there is a magnetic field in the plasma or not. When there is no magnetic field, collisions only impede the free streaming of charged particles and the transport coefficients diminish with more frequent collisions. In the presence of a magnetic field, however, it is due to the collisions that the transport across magnetic field lines becomes possible. In the absence of collisions, the charged particles would keep on gyrating in the same place.

13.3 Collisions in a fully ionized plasma

We now turn to a fully ionized plasma within which we have to consider collisions between charged particles. The force of interaction between two charged particles varies as the inverse square of distance. It is well known that the two-body problem with the inverse-square law force can be solved exactly (see, for example, Goldstein 1980, Chapter 3). Here, however, we present an approximate treatment which gives much of the basic physics more simply than the exact treatment. It should be noted that the total cross-section in the exact solution turns out to be infinite if we assume the inverse-square dependence of the force to be valid at all distances. Within a plasma, however, the force due to a charged particle gets screened beyond the Debye length and one has to take refuge to such facts in the rigorous analysis in order to keep the cross-section finite. Here we follow a more heuristic approach.

If the impact parameter in a collision is large, the deflection produced becomes small. We would regard only those encounters as *collisions* in which the deflection is sufficiently large. Let r_0 be the limiting impact parameter for which the deflection is sufficiently large to make the change of momentum of a particle comparable to the original momentum of the particle. Let us make an estimate of r_0. If u be the typical relative velocity between the particles, then r_0/u is the time during which the particles are close enough to make the interaction strongest. Since the strongest interaction is about e^2/r_0^2, the impulse is of the order

$$\frac{e^2}{r_0^2} \cdot \frac{r_0}{u} \approx \frac{e^2}{r_0 u} \approx \Delta p,$$

where Δp is the change in momentum produced by the impulse and we want it to be of order p or $m_e u$ for the limiting impact parameter r_0. Therefore

$$r_0 \approx \frac{e^2}{m_e u^2}, \tag{13.17}$$

where we are using the mass of the electron, because it is easier to deflect electrons and any event causing a large deflection of an electron would be regarded as a collision. Only when the impact parameter is less than r_0, the deflection is large enough for the event to be counted as a collision. We can therefore take πr_0^2 as the collision cross-section. A particle moving with velocity u undergoes collisions in unit time with those particles which lie within a cylinder of volume $\pi r_0^2 u$. If n is the particle number density, the collision frequency is

$$\nu_c \approx \pi r_0^2 n u \approx \frac{\pi n e^4}{m_e^2 u^3} \tag{13.18}$$

on substituting for r_0 from (13.17). Writing $u \approx (\kappa_B T/m_e)^{1/2}$ for the typical thermal velocities, we get from (13.18) that

$$\nu_c \approx \frac{\pi n e^4}{m_e^{1/2} (\kappa_B T)^{3/2}}. \tag{13.19}$$

Although we have swept several subtle points under the rug, this expression gives the correct dependence of the collision frequency on the various physical quantities. A more complete analysis adds in front of the expression a numerical factor not too much larger than unity. We refer the reader to Spitzer (1962, Chapter 5) for a detailed discussion of the subject.

13.4 Towards a one-fluid model

When we consider phenomena with length scales much larger than the Debye length and time scales much larger than the inverse of the plasma frequency, the charge separation in the plasma can be neglected. In such a situation, a fully ionized plasma can be regarded as a single fluid. When considering phenomena with large time scales, the collisions cannot be neglected. Hence let us first discuss how collisions can be incorporated in the two-fluid model for a fully ionized plasma, in a way similar to what we did for the weakly ionized plasmas in §13.1. Then we shall proceed to develop the one-fluid model. We simplify our treatment by assuming that the ions have the charge $+e$. It is fairly straightforward to generalize to a plasma in which the ions have the charge Ze so that the electron number density has to be taken Z times the ion number density in order to have charge neutrality.

Let us first consider the electron fluid. The collisions between electrons do not change the momentum of the electron fluid. Only when electrons collide with ions, some momentum is transferred from the electrons to the ions. If v_e and v_i are the velocities of the electron and ion fluids, then the electron fluid loses an average momentum $m_e(v_e - v_i)$ to the ion fluid in each electron–ion encounter. The equation of motion of the electron fluid is then similar to (13.1) with the relative velocity between the electron and ion fluids appearing in the collision term:

$$m_e n \frac{\partial v_e}{\partial t} = -\nabla p_e - ne \left(E + \frac{v_e}{c} \times B \right) - m_e n \nu_c (v_e - v_i). \qquad (13.20)$$

Here ν_c is the frequency of electron–ion collisions approximately given by (13.19). We shall be using the same number density n for the electron and the ion fluids, since we are interested in the situation in which charge separation is negligible. It may also be noted that in (13.20) we have replaced the Lagrangian derivative dv_e/dt by the Eulerian derivative $\partial v_e/\partial t$. This essentially means that we are throwing away the term $(v_e \cdot \nabla)v_e$. Since this term is quadratic in v_e, it is certainly negligible when v_e is very small. But we may often be interested in situations when this is not so. Throwing away this term may not therefore seem very justifiable. We shall later make some comments on how to do things better.

Since the current density is given by

$$\mathbf{j} = ne(v_i - v_e), \qquad (13.21)$$

it is clear that the collision term in (13.20) is proportional to the current density so that we can write

$$m_e n \frac{\partial \mathbf{v}_e}{\partial t} = -\nabla p_e - ne\left(\mathbf{E} + \frac{\mathbf{v}_e}{c} \times \mathbf{B}\right) + ne\eta\mathbf{j}, \tag{13.22}$$

where

$$\eta = \frac{m_e v_c}{ne^2}. \tag{13.23}$$

To understand the physical significance of the coefficient η, let us consider a uniform unmagnetized plasma in a steady state with an electric field $\mathbf{E}$ driving a current $\mathbf{j}$. Then the only non-zero terms in (13.22) are

$$-ne\mathbf{E} + ne\eta\mathbf{j} = 0$$

so that

$$\mathbf{E} = \eta\mathbf{j}.$$

It is clear from this expression that η is the electrical resistivity of the plasma. On substituting for v_c from (13.19), (13.23) becomes

$$\eta \approx \frac{\pi m_e^{1/2} e^2}{(\kappa_B T)^{3/2}}. \tag{13.24}$$

For a fully ionized plasma with ions of charge Ze, Spitzer and Härm (1953) carried out a rigorous analysis to show that the resistivity is given by

$$\eta = \frac{\pi^{3/2} m_e^{1/2} Z e^2}{\gamma_E 2 (2\kappa_B T)^{3/2}} \ln \Lambda, \tag{13.25}$$

where γ_E is a numerical factor depending on Z, which turns out to be 0.582 for $Z = 1$, and

$$\Lambda = \frac{3}{2Ze^2}\left(\frac{\kappa_B^3 T^3}{\pi n}\right)^{1/2}. \tag{13.26}$$

The expression (13.25) for resistivity is often known as the *Spitzer resistivity*. On comparing the approximate expression (13.24) with the rigorous expression (13.25), we note that the approximate treatment gives the various factors correctly. Putting the values of various quantities in (13.25), we find the Spitzer resistivity of fully ionized hydrogen plasma to be

$$\eta = 7.3 \times 10^{-9} \frac{\ln \Lambda}{T^{3/2}} \quad \text{e.s.u.} \tag{13.27}$$

Let us now consider the equation of motion of the ion fluid. Since the momentum loss of the electron fluid is the momentum gain of the ion fluid, the collision term in the equation of the ion fluid should be equal to that for the electron fluid in (13.22) with an opposite sign, i.e.

$$m_i n \frac{\partial \mathbf{v}_i}{\partial t} = -\nabla p_i + ne \left(\mathbf{E} + \frac{\mathbf{v}_i}{c} \times \mathbf{B} \right) - ne\eta \mathbf{j}. \qquad (13.28)$$

We shall now combine (13.28) with (13.22) to obtain various relations for the single-fluid model.

The density ρ and the fluid velocity $\mathbf{v}$ in the single-fluid model are obviously given by

$$\rho = n(m_i + m_e), \qquad (13.29)$$

$$\mathbf{v} = \frac{m_i \mathbf{v}_i + m_e \mathbf{v}_e}{m_i + m_e}. \qquad (13.30)$$

On adding (13.22) and (13.28), we get

$$n \frac{\partial}{\partial t} (m_i \mathbf{v}_i + m_e \mathbf{v}_e) = \frac{ne(\mathbf{v}_i - \mathbf{v}_e)}{c} \times \mathbf{B} - \nabla(p_i + p_e).$$

Using (13.21), (13.29) and (13.30), this gives

$$\rho \frac{\partial \mathbf{v}}{\partial t} = \frac{1}{c} \mathbf{j} \times \mathbf{B} - \nabla p, \qquad (13.31)$$

where we have substituted

$$p = p_i + p_e. \qquad (13.32)$$

But is the single-fluid pressure equal to the sum of the pressures of the electron and the ion fluids? It should be clear from (3.6) that the pressure tensor arises from the random velocities of the particles around the average velocity. So, in a full treatment, the pressures of the electron fluid, the ion fluid and the single-fluid plasma should all be regarded as tensors and defined around the corresponding average velocities. If this is done, then the analysis can be done without throwing away terms like $(\mathbf{v}_e \cdot \nabla)\mathbf{v}_e$ as we have done in (13.20), and one eventually obtains an additional term $(\mathbf{v} \cdot \nabla)\mathbf{v}$ in the final equation which is not present in (13.31). This is given as Exercise 13.3. So (13.32) should be taken in the spirit of the approximation we are following in the present derivation.

We now multiply (13.22) by m_i and subtract it from (13.28) multiplied by m_e. This gives

$$m_i m_e n \frac{\partial}{\partial t} (\mathbf{v}_i - \mathbf{v}_e) = ne(m_i + m_e)\mathbf{E} + \frac{ne}{c}(m_e \mathbf{v}_i + m_i \mathbf{v}_e) \times \mathbf{B}$$
$$- m_e \nabla p_i + m_i \nabla p_e - (m_i + m_e)ne\eta \mathbf{j}. \qquad (13.33)$$

We now use (13.21), (13.29) and (13.30) in addition to noting that

$$m_e \mathbf{v}_i + m_i \mathbf{v}_e = m_i \mathbf{v}_i + m_e \mathbf{v}_e + (m_e - m_i)(\mathbf{v}_i - \mathbf{v}_e) = \frac{\rho}{n} \mathbf{v} + \frac{m_e - m_i}{ne} \mathbf{j}.$$

Then (13.33) leads to

$$\mathbf{E} + \frac{\mathbf{v}}{c} \times \mathbf{B} - \eta \mathbf{j} = \frac{1}{e\rho} \left[\frac{m_e m_i n}{e} \frac{\partial}{\partial t} \left(\frac{\mathbf{j}}{n} \right) + (m_i - m_e) \frac{\mathbf{j}}{c} \times \mathbf{B} + m_e \nabla p_i - m_i \nabla p_e \right].$$

(13.34)

This equation, known as the *generalized Ohm's law*, shows how the current density $\mathbf{j}$ is related to the electromagnetic fields and other quantities. At first sight, the equation may appear quite complicated. Luckily it rarely happens that we have to use the full equation. For most purposes, it is sufficient to keep only a few dominant terms in the equation. For example, an order-of-magnitude estimate shows that the term involving the time derivative of $\mathbf{j}$ is negligible compared to the term $\eta \mathbf{j}$ (with η given by (13.23)), if the time scale of evolution of the system is much larger than the typical collision time. Remembering further that $m_e \ll m_i$, (13.34) simplifies to

$$\mathbf{E} + \frac{\mathbf{v}}{c} \times \mathbf{B} - \eta \mathbf{j} = \frac{1}{ne} \left[\frac{\mathbf{j}}{c} \times \mathbf{B} - \nabla p_e \right].$$

(13.35)

Here the term involving $\mathbf{j} \times \mathbf{B}$ corresponds to the *Hall effect* (see, for example, Kittel 1995, pp. 164–6; Ashkroft and Mermin 1976, pp. 11–15). We, however, often deal with situations where the terms on the R.H.S. of (13.35) are small so that an even simpler equation suffices

$$\mathbf{j} = \sigma \left(\mathbf{E} + \frac{\mathbf{v}}{c} \times \mathbf{B} \right),$$

(13.36)

where

$$\sigma = \eta^{-1}$$

(13.37)

is the *electrical conductivity*. In all the discussions in the remainder of this book, we shall restrict ourselves only to situations where the simplified equation (13.36) holds. This simplified equation (13.36) is usually referred to as *Ohm's law*.

The single-fluid model of the plasma is known as *magnetohydrodynamics*, abbreviated as *MHD*. The equations (13.31) and (13.36) are going to play central roles in the development of MHD in the next chapter. The equations of continuity for the electron and ion fluids taken in the form (11.20) can readily be combined to give the one-fluid equation of continuity

$$\frac{\partial \rho}{\partial t} + \nabla \cdot (\rho \mathbf{v}) = 0.$$

(13.38)

We shall introduce and discuss the full set of MHD equations in the next chapter.

13.5 Transport phenomena in fully ionized plasmas

One of the aims of the plasma kinetic theory is to derive expressions for the transport coefficients like viscosity and thermal conductivity, which are expected to appear in the one-fluid model. These coefficients for neutral gases were calculated in §3.4. We first obtained an expression for the departure from the Maxwellian distribution and then this was used in the calculation of the transport coefficients. To obtain the departure from the Maxwellian, we used (3.29), in which the effect of collisions was incorporated through the BGK expression $-(f - f^{(0)})/\tau$. We pointed out in §11.7 that the BGK model is not very reliable for plasmas, especially in the presence of magnetic fields. One, therefore, needs a better model for treating collisions to obtain the departure from the Maxwellian in a plasma. A calculation of departures from the Maxwellian and the resulting transport coefficients happens to be much more complicated in the case of plasmas than in the case of neutral gases. While writing down the equations (11.21) or (11.22) for the fluid model, we had essentially neglected the transport processes. Since a proper calculation of plasma transport coefficients is beyond the scope of this book, we merely quote some results.

The first step is to calculate the departure from the Maxwellian by using some method of handling collisions. Several models have been suggested on different grounds to account for collisions in plasmas. In §11.4 we mentioned the Lenard–Balescu model, which, however, is applicable only to spatially homogenous plasmas and hence is not suitable for calculating transport coefficients. The model of Rosenbluth, MacDonald and Judd (1957) has been used by Shkarofsky, Bernstein and Robinson (1963) to study transport processes in plasmas. These calculations are very complicated. The main results are summarized in §13.4 of Montgomery and Tidman (1964). All the transport coefficients become highly anisotropic in the presence of a strong magnetic field, just as we saw the diffusion coefficient in a weakly ionized plasma to become anisotropic in §13.2. In the limit of a vanishing magnetic field, however, the coefficients are essentially the same as what one would obtain from the BGK model, apart from numerical factors of the order of unity. So we can use the expressions (3.36) or (3.40) for unmagnetized plasmas. Let us write the expression (3.36) for thermal conductivity again:

$$K = \frac{5}{2}\tau n \frac{\kappa_B^2 T}{m_e},\qquad(13.39)$$

where we have put the electronic mass in the place of mass, since it is the random motion of electrons in a plasma which is responsible for the heat conduction. We now replace τ by the inverse of ν_c as given by (13.19) to get the following approximate expression of thermal conductivity

$$K = \frac{5}{2} \frac{\kappa_B^{7/2} T^{5/2}}{\pi m_e^{1/2} e^4}. \tag{13.40}$$

It follows that the thermal conductivity K of a fully ionized, unmagnetized plasma goes as $T^{5/2}$.

From (13.23) the electrical conductivity, which is the inverse of η, is given by

$$\sigma = \frac{ne^2 \tau}{m_e}. \tag{13.41}$$

Combining (13.39) and (13.41), we get the elegant result

$$\frac{K}{\sigma T} = \frac{5}{2} \left(\frac{\kappa_B}{e} \right)^2, \tag{13.42}$$

which is independent of the uncertainties in estimating the collision time τ. Since some results for plasmas are applicable to electron gases in metals, we expect something like (13.42) to hold for metals as well. As the electrons in the metal obey the Fermi–Dirac statistics, we have to take that fact into account while calculating K. This leads to

$$\frac{K}{\sigma T} = \frac{\pi^2}{3} \left(\frac{\kappa_B}{e} \right)^2 \tag{13.43}$$

for metals (see, for example, Kittel 1995, pp. 166–8). It was experimentally discovered by Wiedemann and Franz (1853) that the ratio K/σ is nearly the same for all metals at a fixed temperature. Then Lorenz (1872) noted that this ratio is proportional to T. Soon after Thomson (1897) established the existence of electrons, Drude (1900) realized that electrons are responsible for both the thermal and the electrical conduction in metals, and gave the theoretical explanation of the Wiedemann–Franz–Lorenz experimental law. The quantity $K/\sigma T$ is often called the *Lorenz number*. The experimental values of the Lorenz number for different metals are listed in the Table on p. 168 of Kittel (1995), showing that the values are nearly constant and close to the theoretical value as given by (13.43).

The values of various transport coefficients for a fully ionized hydrogen plasma are given in Appendix E. These values are based on the expressions of transport coefficients conveniently collected in a brief paper by Chapman (1954).

13.6 Lorentz transformation of electromagnetic fields. The non-relativistic approximation

We now consider how electromagnetic fields are transformed when we go from one inertial frame to another, since a knowledge of this subject provides a deeper understanding of some equations in plasma theory. This is a standard topic covered in all graduate-level electrodynamics textbooks (Panofsky and Phillips 1962, §18.1–18.2; Jackson 1975, §11.9–11.10). Here we summarize the main results without derivation.

The components of the electromagnetic field can be combined into a second-rank covariant tensor to which the Lorentz transformation can be applied. This leads to the following results. Suppose **E** and **B** are the electromagnetic fields in a laboratory frame as observed by us. If a part of a plasma is moving with velocity **v**, we can go to that frame. The fields **E'** and **B'** in that frame with velocity **v** are related to **E** and **B** as follows

$$E'_\parallel = E_\parallel, \tag{13.44}$$

$$B'_\parallel = B_\parallel, \tag{13.45}$$

$$\mathbf{E}'_\perp = \gamma \left(\mathbf{E}_\perp + \frac{\mathbf{v}}{c} \times \mathbf{B}_\perp \right), \tag{13.46}$$

$$\mathbf{B}'_\perp = \gamma \left(\mathbf{B}_\perp - \frac{\mathbf{v}}{c} \times \mathbf{E}_\perp \right), \tag{13.47}$$

where

$$\gamma = \frac{1}{\sqrt{1 - v^2/c^2}} \tag{13.48}$$

is the Lorentz factor, and the subscripts $\parallel$ and $\perp$ refer to the directions parallel and perpendicular to **v**.

We now consider how these transformation laws can be simplified if the velocities in the plasma are non-relativistic. In mechanics problems, it is customary to neglect terms of the order of $|v|/c$ in the non-relativistic limit. Here, however, we have to be a little more careful. If it appears from a frame that a plasma has only a magnetic field, it follows from (13.46) that observers in most other frames see an electric field in the plasma in addition to the magnetic field. If we want to capture this effect, then we should keep terms of order $|v|/c$, but we can throw away terms of the order of v^2/c^2. This is an approximation we shall be using throughout. It is clear from (13.48) that the Lorentz factor γ approximates to 1. Then (13.44) and (13.46) can be combined to

$$\mathbf{E}' = \mathbf{E} + \frac{\mathbf{v}}{c} \times \mathbf{B}. \tag{13.49}$$

We now give a physical interpretation to Ohm's law (13.36). It is clear from (13.49) that Ohm's law can be written as

$$\mathbf{j} = \sigma\mathbf{E}'. \tag{13.50}$$

Anybody familiar with the covariant formulation of electrodynamics should be able to show easily that, to order $|v|/c$, the current density $\mathbf{j}$ does not change from one frame to another. Ohm's law therefore merely asserts that the current density within a small volume of plasma is proportional to the electric field as seen in the frame in which that volume is at rest. For a highly conducting plasma (i.e. for very large σ), the current density in (13.50) remains finite only if $\mathbf{E}'$ is very small. It then follows from (13.49) that

$$|E| \approx \frac{|v|}{c}|B|. \tag{13.51}$$

It may at first appear from (13.45) and (13.47) that the transformation law for the magnetic field to order $|v|/c$ would be

$$\mathbf{B}' = \mathbf{B} - \frac{\mathbf{v}}{c} \times \mathbf{E}.$$

It follows, however, from (13.51) that the $\mathbf{v} \times \mathbf{E}/c$ term is of order v^2/c^2 with respect to other terms. We therefore write

$$\mathbf{B}' = \mathbf{B}. \tag{13.52}$$

Equations (13.49) and (13.52) give the appropriate transformation laws for the electromagnetic fields between frames moving with non-relativistic speeds.

We now show that the displacement current has to be neglected when we throw away terms of order v^2/c^2. Let us look at the appropriate Maxwell's equation

$$\nabla \times \mathbf{B} = \frac{4\pi}{c}\mathbf{j} + \frac{1}{c}\frac{\partial \mathbf{E}}{\partial t}.$$

Taking l and t as the typical length and time scales,

$$\frac{\frac{1}{c}\frac{\partial \mathbf{E}}{\partial t}}{|\nabla \times \mathbf{B}|} \approx \frac{|E|/ct}{|B|/l} \approx \frac{|v|}{c}\frac{|E|}{|B|},$$

where we have taken the velocity scale to be of order l/t. From (13.51), we then have

$$\frac{\frac{1}{c}\frac{\partial \mathbf{E}}{\partial t}}{|\nabla \times \mathbf{B}|} \approx \frac{v^2}{c^2}.$$

Neglecting the displacement current term, we therefore write

$$\nabla \times \mathbf{B} = \frac{4\pi}{c}\mathbf{j}. \qquad (13.53)$$

In the MHD model, we always use this equation in this form without the displacement current. This equation implies that the current density $\mathbf{j}$ and the magnetic field $\mathbf{B}$ have a one-to-one relation between them. If the current density $\mathbf{j}$ is given, then we can find out the magnetic field $\mathbf{B}$. On the other hand, from a knowledge of the magnetic field $\mathbf{B}$, the current density $\mathbf{j}$ can be inferred.

We now combine (13.36) and (13.53) to write the electric field as

$$\mathbf{E} = \frac{c}{4\pi\sigma}\nabla \times \mathbf{B} - \frac{\mathbf{v}}{c} \times \mathbf{B}. \qquad (13.54)$$

This means that $\mathbf{E}$ can be obtained from $\mathbf{v}$ and $\mathbf{B}$. We have seen that the electric field within a certain volume of plasma is always negligible in the rest frame of that volume. Now we point out that the electric field in any frame can be calculated from the other variables. Hence we do not have to consider $\mathbf{E}$ as an independent variable in MHD.

When developing the hydrodynamic model of neutral fluids, we saw that two thermodynamic variables and the velocity field $\mathbf{v}(\mathbf{x},t)$ appear as the independent dynamical variables in the theory. It should be clear by now that the hydrodynamic model of plasmas requires one additional variable—the magnetic field $\mathbf{B}(\mathbf{x},t)$. Because of (13.54), $\mathbf{E}(\mathbf{x},t)$ is not a variable we need to consider. By now we have introduced most of the ingredients and assumptions which go into building the hydrodynamic one-fluid model of plasmas. In the next chapter, we put all these things together to figure out the basic equations of MHD and then proceed from there.

13.7 A brief note on pulsar magnetospheres

Consider a highly conducting plasma such that σ is very large. It follows from (13.36) that we must have

$$\mathbf{E} + \frac{\mathbf{v}}{c} \times \mathbf{B} \approx 0 \qquad (13.55)$$

to keep $\mathbf{j}$ finite. The magnitude of the electric field $\mathbf{E}$ is then as given by (13.51). This means that a large electric field may be present in a frame with respect to which the plasma is moving. Such an electric field is usually of no consequence, since the electric field in the rest frame of the plasma as given by (13.49) would vanish. Only if there is an appreciable electric field in the rest frame of the plasma, something drastic may happen. Is this possible in any astrophysical situation?

One important result of the Lorentz transformation is that the dot product $\mathbf{E} \cdot \mathbf{B}$ is invariant between frames. Hence, if $\mathbf{E} \cdot \mathbf{B} \neq 0$ in a frame, then it is not possible for the electric field to vanish in any frame including the rest frame of the plasma. If (13.55) holds, it is easy to see that $\mathbf{E} \cdot \mathbf{B} = 0$. We now consider an astrophysical example where this may not be the case.

We argued in §9.4 that a pulsar must be a rapidly rotating neutron star. If the neutron star has a magnetic field and the space around the neutron star is a vacuum, then it can be shown that $\mathbf{E} \cdot \mathbf{B} \neq 0$ outside the surface of the neutron star—an important result first pointed out by Goldreich and Julian (1969) soon after the discovery of pulsars. To be specific, let us assume that the magnetic field of the neutron star is of a dipolar nature. In the vacuum outside the neutron star, the magnetic field has to be a potential field so that we can write

$$\mathbf{B}^{\text{out}} = B_0 a^3 \left(\frac{\cos \theta}{r^3} \hat{\mathbf{e}}_r + \frac{\sin \theta}{2r^3} \hat{\mathbf{e}}_\theta \right), \qquad (13.56)$$

where a is the radius of the neutron star. This is the standard expression of a dipolar potential field. We are assuming that the reader is familiar with the mathematical theory of three-dimensional potential fields in spherical coordinates (see, for example, Panofsky and Phillips 1962, Chapter 5; Jackson 1975, Chapter 3). If there are no surface currents, then the magnetic field is continuous across the surface of the neutron star so that just inside the surface

$$\mathbf{B}^{\text{in}}|_{r=a} = B_0 \left(\cos \theta \, \hat{\mathbf{e}}_r + \frac{\sin \theta}{2} \hat{\mathbf{e}}_\theta \right). \qquad (13.57)$$

If the neutron star is rotating with angular velocity $\boldsymbol{\Omega}$, then the electric field just inside the surface can be obtained from (13.55) and (13.57) using $\mathbf{v} = \boldsymbol{\Omega} \times \mathbf{r}$. This gives

$$\mathbf{E}^{\text{in}}|_{r=a} = \frac{\Omega B_0 a \sin \theta}{c} \left(\frac{\sin \theta}{2} \hat{\mathbf{e}}_r - \cos \theta \, \hat{\mathbf{e}}_\theta \right). \qquad (13.58)$$

The tangential component of this electric field has to be equal to the tangential component of the electric field just outside the surface, which can be written as

$$E_\theta^{\text{out}}|_{r=a} = -\frac{\partial}{\partial \theta} \left(\frac{\Omega B_0 a \sin^2 \theta}{2c} \right) = \frac{\partial}{\partial \theta} \left[\frac{\Omega B_0 a}{3c} P_2(\cos \theta) \right]. \qquad (13.59)$$

The electric field in the vacuum around the neutron star is a potential field satisfying the boundary condition (13.59). It is easy to see that the appropriate electrostatic potential is

$$\phi = -\frac{\Omega B_0}{3c}\frac{a^5}{r^3}P_2(\cos\theta), \tag{13.60}$$

which corresponds to a quadrupolar field. The electric field is given by $\mathbf{E} = -\nabla\phi$. From (13.56) and (13.60), we obtain $\mathbf{E} \cdot \mathbf{B}$ outside the neutron star, which is

$$\mathbf{E} \cdot \mathbf{B} = -\frac{\Omega a}{c}\left(\frac{a}{r}\right)^7 B_0^2 \cos^3\theta. \tag{13.61}$$

A rough estimate shows that the electric force near the surface of the neutron star would be much stronger than the gravitational force (see Exercise 13.5). We would expect this electric force to lift charged particles from the surface and make them move along magnetic field lines. Goldreich and Julian (1969) correctly pointed out that it is not self-consistent to assume a vacuum around a rotating magnetized neutron star. The pulsar should rather be surrounded by a magnetosphere filled with charged particles, and the above analysis will have to be modified to avoid large unbalanced forces. Presumably, the complicated physical processes in the pulsar magnetosphere are responsible for the radio emission from the pulsar.

Exercises

13.1   If a collision term is included in the equation of motion for the electrons (the collision frequency being v_c) while treating the propagation of electromagnetic waves through plasmas, how is the expression (12.6) for the dielectric constant modified? Obtain the dispersion relation for the electromagnetic wave and derive an expression for the length of plasma which attenuates a propagating electromagnetic wave by 50%.

13.2   Consider an isothermal atmosphere of plasma in a constant gravitational field $\mathbf{g}$. Assuming the ions to be singly ionized, write down the force balance equations of the ion and the electron fluids. Show that they can be combined to give the usual hydrostatic equation and an electric field

$$\mathbf{E} = -\frac{m_i}{2e}\mathbf{g}$$

has to exist in the atmosphere to prevent charge separation.

13.3   Suppose we want to derive the one-fluid model from the two-fluid model without neglecting the $(\mathbf{v} \cdot \nabla)\mathbf{v}$ terms as done in (13.22) and (13.28). We take the pressure to be a tensor and write

$$P_{l,jk} = nm_l \langle (u_{l,j} - v_{l,j})(u_{l,k} - v_{l,k}) \rangle$$

for the *jk* component of the pressure tensor. The subscript $l = e$, i would correspond to the electron and the ion fluids respectively. As usual, $\mathbf{u}_l$ is the particle velocity for the species l and $\mathbf{v}_l = \langle \mathbf{u}_l \rangle$. So, apart from adding terms $(\mathbf{v_e} \cdot \nabla)\mathbf{v_e}$ and $(\mathbf{v_i} \cdot \nabla)\mathbf{v_i}$ in (13.22) and (13.28), we replace the pressure gradient terms by the appropriate gradients of the pressure tensor. Show that these two equations can be combined to give an additional term $(\mathbf{v} \cdot \nabla)\mathbf{v}$ in (13.31) apart from making the pressure a tensor given by

$$P_{jk} = nm_e \langle (u_{e,j} - v_j)(u_{e,k} - v_k) \rangle + nm_i \langle (u_{i,j} - v_j)(u_{i,k} - v_k) \rangle.$$

Note that P_{jk} is not exactly equal to the sum of $P_{e,jk}$ and $P_{i,jk}$. (Since i is used to denote the ion fluid here, we have not used it as a coordinate index!)

13.4   If the electric and magnetic fields are perpendicular to each other, the guiding centre of a charged particle moves with a constant speed given by (10.8). Show that the electric field is zero in the frame of the guiding centre and discuss the significance of this result.

13.5   Consider a rotating neutron star in a vacuum with electric and magnetic fields as calculated in §13.7. Assume the following values of various physical quantities for a typical neutron star: radius ≈ 10 km, mass $\approx 10^{33}$ g, magnetic field $\approx 10^{11}$ G, rotation period ≈ 1 s. Estimate the ratio of the electric and gravitational forces at the surface of the neutron star.

14 Basic magnetohydro-dynamics

14.1 The fundamental equations

We discussed in the previous chapter how the one-fluid or the MHD model of the plasma can be developed starting from microscopic considerations. It was not possible to give as thorough or as systematic a presentation of the subject as we did in Chapter 3, where the hydrodynamic model for neutral fluids was developed from the microscopic theory. We have not rigorously established the conditions under which the one-fluid model of a plasma holds. We saw in Chapter 3 that frequent collisions make a neutral gas behave like a continuous fluid. Collisions certainly help in establishing fluidlike behaviour. It was, however, mentioned in §11.7 that a strong magnetic field in a plasma can also keep charged particles confined within local regions for sufficient time, thereby giving rise to fluidlike behaviour even in the absence of collisions.

Between the microscopic model based on distribution functions and the macroscopic one-fluid model, there exists the intermediate two-fluid model of the plasma discussed in Chapters 11–13. This was referred to in Table 1.1 as the $2\frac{1}{2}$ level. When we consider phenomena in which electrons and ions respond differently (such as the propagation of electromagnetic waves through a plasma), the two-fluid model has to be applied rather than the MHD model. The MHD model is applicable only when charge separation is negligible. The condition for it is that the length scales should be larger than the Debye length and the time scales larger than the inverse of plasma frequency. We have further introduced the non-relativistic approximation of throwing away terms of the order v^2/c^2 in §13.6. At first sight, it may appear that a model which is valid only when so many conditions are satisfied must be very restrictive. We shall, however, see that the

MHD model can be applied to a wide class of problems involving both astrophysical and laboratory plasmas.

When we consider non-relativistic and slowly varying (i.e. the time scale $\gg$ inverse of plasma frequency) motions of plasmas under the action of mechanical and magnetic forces, the MHD model is the appropriate model to apply. The main attraction of this model is that it provides an elegant and clean dynamical theory. In addition to the usual hydrodynamic variables (the two thermodynamic variables and the velocity field $\mathbf{v}(\mathbf{x})$), the magnetic field $\mathbf{B}(\mathbf{x})$ has to be specified for a description of the plasma in the MHD model. It has been pointed out in §13.6 that the electric field $\mathbf{E}(\mathbf{x})$ is not an independent variable. We shall see that the basic equations of the MHD model tell us how all these variables evolve with time. We should keep in mind that the main limitation of the MHD model is that it cannot be applied to high-frequency phenomena which may involve charge separation (plasma oscillations or electromagnetic waves in plasmas).

We now discuss the governing equations of the MHD model. We write down the equations in a form which assumes that the various transport coefficients (such as viscosity or resistivity) are scalars. We saw in §13.2 that diffusion in the presence of a strong magnetic field may become anisotropic. The same is true for the other transport coefficients, which all have different values along and perpendicular to the magnetic field if a strong magnetic field is present. The condition for isotropy is that the collision frequency has to be higher than the cyclotron frequency. We write the equations in forms which are valid in this limit. It is, however, not very difficult to generalize these equations to the situation of anisotropic transport coefficients, although we shall not do it here. We shall mostly consider the various transport coefficients to be constant in space as well, because that simplifies the equations further.

We saw in §4.1 and §5.1 that the hydrodynamic equations can be derived from macroscopic considerations, resulting in exactly the same equations which were obtained from kinetic theory. We now discuss how the MHD equations follow from macroscopic considerations. The equation of continuity

$$\frac{\partial \rho}{\partial t} + \nabla \cdot (\rho \mathbf{v}) = 0 \qquad (14.1)$$

remains the same. We now have to add the magnetic body force to the Navier–Stokes equation (5.10), which gives, in a rather straightforward manner,

$$\frac{\partial \mathbf{v}}{\partial t} + (\mathbf{v} \cdot \nabla)\mathbf{v} = \mathbf{F} - \frac{1}{\rho}\nabla p + \frac{1}{\rho c}\mathbf{j} \times \mathbf{B} + \nu \nabla^2 \mathbf{v}. \qquad (14.2)$$

We gave a partial derivation of this equation from microscopic considerations in §13.4. That partial derivation should give an idea how this equation arises from microscopic physics, although a full rigorous derivation involves some subtleties. The viscosity term, for example, was not introduced in §13.4. We, however, expect that the macroscopic considerations presented in §5.1 should hold for a plasma which behaves like a continuous fluid, and the form of the viscosity term should therefore be the same. Substituting for $\mathbf{j}$ in (14.2) from (13.53), we have

$$\frac{\partial \mathbf{v}}{\partial t} + (\mathbf{v} \cdot \nabla)\mathbf{v} = \mathbf{F} - \frac{1}{\rho}\nabla p + \frac{1}{4\pi\rho}(\nabla \times \mathbf{B}) \times \mathbf{B} + \nu \nabla^2 \mathbf{v}. \qquad (14.3)$$

From the vector identity (A.6), we get

$$(\nabla \times \mathbf{B}) \times \mathbf{B} = (\mathbf{B} \cdot \nabla)\mathbf{B} - \nabla \left(\frac{B^2}{2}\right). \qquad (14.4)$$

Hence (14.3) can also be written as

$$\frac{\partial \mathbf{v}}{\partial t} + (\mathbf{v} \cdot \nabla)\mathbf{v} = \mathbf{F} - \frac{1}{\rho}\nabla \left(p + \frac{B^2}{8\pi}\right) + \frac{(\mathbf{B} \cdot \nabla)\mathbf{B}}{4\pi\rho} + \nu \nabla^2 \mathbf{v}. \qquad (14.5)$$

It is clear from this that the magnetic force introduces a pressure $B^2/8\pi$. The other magnetic term $(\mathbf{B} \cdot \nabla)\mathbf{B}$ can be shown to be of the nature of a tension along magnetic field lines. To see this, let us write the i-th component of Lorentz force in (14.3) in the form

$$\left[\frac{1}{4\pi\rho}(\nabla \times \mathbf{B}) \times \mathbf{B}\right]_i = -\frac{1}{\rho}\frac{\partial}{\partial x_i}\left(\frac{B^2}{8\pi}\delta_{ij} - \frac{B_i B_j}{4\pi}\right) \qquad (14.6)$$

making use of (14.4) and noting that $\partial B_i/\partial x_i = \nabla \cdot \mathbf{B} = 0$. Now (14.3) can easily be put in the form of (4.8) with an additional term due to the Lorentz force, i.e.

$$\rho \frac{dv_i}{dt} = \rho F_i - \frac{\partial}{\partial x_j}(P_{ij} + \mathscr{M}_{ij}), \qquad (14.7)$$

where

$$\mathscr{M}_{ij} = \frac{B^2}{8\pi}\delta_{ij} - \frac{B_i B_j}{4\pi}. \qquad (14.8)$$

We have pointed out in §4.1.1 that the diagonal components of P_{ij} correspond to pressure and the off-diagonal components to viscous shear. In exactly the same way, the diagonal part $B^2/8\pi$ of $\mathscr{M}_{ij}$ can be interpreted as the magnetic pressure. The nature of the remaining part of the tensor $\mathscr{M}_{ij}$ becomes clear if we choose one of the coordinate axes, say the z axis, in the direction of the local magnetic field. Then, in that neighbourhood, (14.8) implies

$$\mathscr{M}_{ij} = \begin{pmatrix} B^2/8\pi & 0 & 0 \\ 0 & B^2/8\pi & 0 \\ 0 & 0 & -B^2/8\pi \end{pmatrix}. \tag{14.9}$$

We can write the zz component as

$$\mathscr{M}_{zz} = \frac{B^2}{8\pi} - \frac{B^2}{4\pi}.$$

The part $B^2/8\pi$ combines with the M_{xx} and M_{yy} components to give an isotropic pressure. The remaining part $-B^2/4\pi$ corresponds to excess negative pressure or tension in the z direction. Thus a magnetic field has a tension along field lines in addition to having an isotropic pressure associated with it. The part of $\mathscr{M}_{ij}$ remaining after subtracting $(B^2/8\pi)\delta_{ij}$ gives this tension along field lines.

The energy equation is still of the form (4.13) with the heat gain term $-\mathscr{L}$ now containing an Ohmic heating term j^2/σ added to $\nabla\cdot(K\nabla T)$ as given in (4.15). In the remaining parts of the book, we shall, however, mostly be concerned with topics in which compressibility does not play an important role. As pointed out in §4.2, the energy equation will not have to be considered in these situations. Since magnetic fields have energies and stresses associated with them, one may wonder whether these have to be included in the energy equation. It turns out that the work done by the magnetic stresses remains stored in the system in the form of magnetic energy, apart from the term j^2/σ corresponding to the conversion of magnetic energy to thermal energy (see Exercises 14.1 and 14.2). Hence only this heating term j^2/σ appears in the equation of internal thermal energy.

In order to have a full dynamical theory, we now require one more equation giving the time derivative of the magnetic field in addition to (14.1), (14.3) and the energy equation. Substituting from (13.54) in the Maxwell equation

$$\frac{\partial \mathbf{B}}{\partial t} = -c\nabla \times \mathbf{E},$$

we obtain

$$\frac{\partial \mathbf{B}}{\partial t} = \nabla \times (\mathbf{v} \times \mathbf{B}) + \lambda\nabla^2\mathbf{B}, \tag{14.10}$$

where

$$\lambda = \frac{c^2}{4\pi\sigma} \tag{14.11}$$

is called *magnetic diffusivity*. It may be noted that we have assumed the electrical conductivity σ to be spatially constant and hence taken it outside the spatial derivative in obtaining (14.10). The equation (14.10) is of central importance in MHD and is known as the *induction equation*. Some obvious consequences of this equation will be discussed in the next section.

If we compare the MHD model with the hydrodynamic model for neutral fluids, the differences are found be the magnetic force term added to the Navier–Stokes equation and the Ohmic heating term added to the energy equation—in addition to a whole new equation for the magnetic field, the induction equation (14.10). To summarize, a state in the MHD model is described by eight scalar variables: two thermodynamic quantities; three components of $\mathbf{v}(\mathbf{x})$; and three components of $\mathbf{B}(\mathbf{x})$. We also have eight basic equations: the continuity equation (14.1); three components of (14.3); the energy equation; and three components of (14.10). We thus have a full dynamical theory. In Chapters 14–16, we shall discuss only those topics which can be studied by the MHD model. A system satisfying the MHD equations is known as a *magnetofluid*. Apart from ionized gases, any liquid which is a good conductor of electricity behaves as a magnetofluid in the presence of magnetic fields. Mercury, the only metal to remain liquid at the room temperature, has often been used in experiments to verify predictions of the MHD equations.

14.2 Some consequences of the induction equation

Since the induction equation (14.10) is one completely new equation in the MHD model, we first look at this equation carefully before discussing other aspects of the MHD model. The first thing to note is that this equation is exactly analogous to equation (5.12) satisfied by vorticity. We now introduce a dimensionless number in exactly the same way we introduced the Reynolds number in §5.4. If B is the typical magnetic field, V the typical velocity and L the typical length scale, then the first term on the R.H.S. of (14.10) is of order VB/L and the second term of order $\lambda B/L^2$. On taking the ratio of these two terms, we get the dimensionless number

$$\mathscr{R}_{\mathrm{M}} = \frac{VB/L}{\lambda B/L^2} = \frac{LV}{\lambda} \tag{14.12}$$

known as the *magnetic Reynolds number*. Since $\mathscr{R}_{\mathrm{M}}$ is directly proportional to the size L of the system, it turns out to be much larger for astrophysical plasmas compared to laboratory plasmas.

Let us consider a hydrogen plasma of temperature 10^4 K. On using

(13.27) with $\ln \Lambda \approx 10$, we find from (14.11) that

$$\lambda = \frac{c^2 \eta}{4\pi} \approx 10^7 \text{cm}^2 \text{ s}^{-1}. \tag{14.13}$$

Taking this value of λ, we now estimate $\mathcal{R}_M$ for a laboratory system and for an astrophysical system. If we take $L \approx 10^2$ cm and $V \approx 10$ cm s^{-1} for a typical laboratory system, then we find from (14.12) that $\mathcal{R}_M \approx 10^{-4}$. Now let us consider granules or convection cells on the solar surface (see §7.3), which are very small objects by astrophysical standards. Using the typical values $L \approx 10^8$ cm and $V \approx 10^5$ cm s^{-1}, we find $\mathcal{R}_M \approx 10^6$. We therefore conclude that the magnetic Reynolds number is generally small ($\ll 1$) for laboratory systems and very large ($\gg 1$) for astrophysical systems.

If $\mathcal{R}_M \ll 1$, as in typical laboratory situations, then the second term on the R.H.S. of (14.10) is dominant and we may write

$$\text{Laboratory:} \quad \frac{\partial \mathbf{B}}{\partial t} \approx \lambda \nabla^2 \mathbf{B}. \tag{14.14}$$

On the other hand, in astrophysical systems where $\mathcal{R}_M \gg 1$, we may write

$$\text{Astrophysics:} \quad \frac{\partial \mathbf{B}}{\partial t} \approx \nabla \times (\mathbf{v} \times \mathbf{B}). \tag{14.15}$$

We want to emphasize that (14.14) and (14.15) should be taken as *extremely crude and simple-minded* approximations. The term $\nabla \times (\mathbf{v} \times \mathbf{B})$ cannot be neglected in many laboratory circumstances, whereas we shall see in the next chapter that the term $\lambda \nabla^2 \mathbf{B}$ can be *rather important* inside localized regions of astrophysical bodies due to some subtle reasons. The main point to note is that the magnetic fields behave very differently in the laboratory and in the astrophysical settings due to the dominance of two very different terms in the induction equation. Alfvén, who was the pioneer in the study of astrophysical magnetic fields, coined the phrase *cosmical electrodynamics* for the study of electromagnetic phenomena in astronomical systems as opposed to ordinary laboratory electrodynamics. This situation can be contrasted to the fact that the mechanical properties of laboratory-size and astronomical objects are not so different. The motions of stars are in many ways analogous to motions of tennis balls. But we have vast magnetic fields generated in stars and galaxies by interior plasma motions without any analogues in the laboratory environment. One is somewhat reminded of the fact that mechanics at the laboratory level and at the atomic level are very different, although the differences between electrodynamics at the laboratory level and at the astrophysical level are not as profound as the differences between classical mechanics and quantum mechanics.

We now point out some of the direct conclusions which can be drawn from (14.14) or (14.15). Since (14.14) is simply the vector diffusion equation, we conclude that a magnetic field within a laboratory plasma left to itself decays away. This result can be easily understood. By virtue of (13.53), magnetic fields are directly related to currents, and as currents in a system without a source of voltage dies away due to Ohmic dissipation, the magnetic field also must decay. Let us now look at (14.15). In the hypothetical limit of infinite conductivity (or zero resistivity), often known as the *ideal MHD limit*, we can write

$$\frac{\partial \mathbf{B}}{\partial t} = \nabla \times (\mathbf{v} \times \mathbf{B}). \qquad (14.16)$$

We now point out that in §4.6 we proved an important theorem for any arbitrary vector field $\mathbf{Q}$ satisfying an evolution equation

$$\frac{\partial \mathbf{Q}}{\partial t} = \nabla \times (\mathbf{v} \times \mathbf{Q}).$$

Since (14.16) is of this form, we can directly apply this theorem and conclude that

$$\frac{d}{dt} \int_S \mathbf{B} \cdot d\mathbf{S} = 0, \qquad (14.17)$$

where the surface integral can be thought to be over a surface made up of definite fluid elements and the Lagrangian time derivative implies that we are considering the variation in time while following the surface as the fluid elements making it are moving. A full discussion of the significance of this result is presented in §4.6. Physically we can say that the magnetic fields move with the fluid, just as we have found vortices to move with the fluid in §4.6. This result for magnetic fields was first pointed out by Alfvén (1942a), and it is therefore often called *Alfvén's theorem of flux-freezing.* Alfvén's theorem is the magnetic analogue of Kelvin's theorem for vorticity.

In astrophysical systems with high $\mathscr{R}_M$, we can therefore imagine the magnetic flux to be frozen in the plasma and to move with the plasma flows. Suppose we have straight magnetic field lines going through a plasma column as shown in Figure 14.1(a). If the plasma column is bent, then, in the high $\mathscr{R}_M$ limit, the magnetic field lines are also bent with it as shown in Figure 14.1(b). On the other hand, if one end of the plasma column is twisted as in Figure 14.1(c), then the magnetic field lines are also twisted. As a result of Alfvén's theorem of flux-freezing, the magnetic field in an astrophysical system can almost be regarded as a plastic material which can be bent, twisted or distorted by making the plasma move appropriately. This view of a magnetic field is radically different from that which we encounter in laboratory situations, where the magnetic field appears as something rather passive which we can switch on or off by sending a current through a coil. In the astrophysical setting, the magnetic field appears to acquire a life of its own.

We saw from (13.53) that there is a one-to-one correspondence between the magnetic field **B** and the current density **j**. We can therefore regard any one of them as our basic dynamical variable. There are obvious advantages of regarding **B** as the basic variable in the MHD model rather than **j**. If we know the initial configuration of the magnetic field and the nature of plasma flows, we can almost guess on the basis of Alfvén's theorem what the subsequent magnetic field configuration is going to be (as we saw in Figure 14.1). The human mind is more attuned to thinking geometrically rather than thinking analytically. We may be able to solve an equation describing a process, but only when we are able to make a mental picture of how the process proceeds, do we feel that we understand the process. The advantage of using the magnetic field as the basic variable in a high-$\mathscr{R}_M$ MHD system is that we can mentally visualize how the magnetic field evolves even without solving the equations. The current density **j** satisfies a more complicated equation, which can easily be obtained by taking the curl of (14.10), but which does not lead to such pictorial visualization of how the current density evolves. Regarding **B** as more basic than **j** is also a point of view which is contrary to our laboratory experience. In a laboratory, we usually send a current through a conductor and thereby produce a magnetic field. As soon as the current is switched off, the magnetic field goes away. Hence, in

Figure 14.1
Illustration of
flux-freezing. (a) A
straight column of
magnetic field. (b)
Magnetic
configuration after
bending the column.
(c) Magnetic
configuration after
twisting the column.

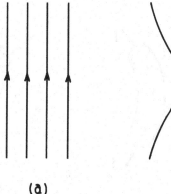
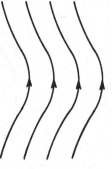

(a)　　　　　　(b)　　　　　　(c)

elementary electrodynamics, often the current is regarded as the more basic physical entity from which the magnetic field can be found.

Let us now consider an obvious corollary of Alfvén's theorem. Let P and Q be two fluid elements which lie on the same magnetic field line as shown in Figure 14.2(a). We consider a thin cylindrical surface around this field line. It is evident that the magnetic flux across this cylindrical surface is zero. After some time, the fluid elements P and Q take up positions P' and Q', whereas the fluid elements which made the previous cylindrical surface now make up a different cylindrical surface as shown in Figure 14.2(b). According to Alfvén's theorem, the magnetic flux through this new cylindrical surface should be zero, and this is possible only if the field line still passes along the axis of the cylindrical surface. This means that P' and Q' still lie on the same magnetic field line. In other words, if two fluid elements are connected by a field line, they will always remain connected by a field line in the limit of ideal MHD. We shall see in the next chapter that such connectivities introduce some topological constraints leading to far-reaching consequences.

When an astronomical object shrinks due to gravitational attraction, its magnetic field is expected to become stronger. If a is the radius of the equatorial cross-section of the body through which a magnetic field of order B is passing, then the magnetic flux linked with the equatorial plane is of order Ba^2. If the magnetic field is perfectly frozen, then this flux should remain an invariant during the contraction of the object. Some neutron stars are believed to have magnetic fields as high as 10^{12} G. Let us see if we can explain this magnetic field by assuming that the neutron star formed due to the collapse of an ordinary star of which the magnetic field got compressed. A star like the Sun has a radius of order 10^{11} cm, and the magnetic field near its pole is about 10 G. Since the radius of a typical neutron star is about 10^6 cm,

Figure 14.2 A sketch showing a cylindrical surface around the field line PQ and another cylindrical surface around $P'Q'$ made by the same fluid elements later.

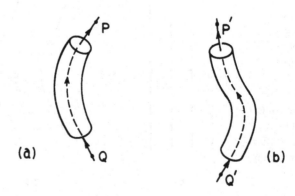

the equatorial area would decrease by a factor of 10^{10} if an ordinary star were to collapse to become the neutron star. If the magnetic flux remained frozen during this collapse, then the initial field of 10 G would finally become 10^{11} G, which is of the same order of magnitude as the magnetic fields of neutron stars.

14.3 Magnetohydrostatics

After developing the basic equations of hydrodynamics in Chapter 4, one of the first applications we considered was hydrostatic equilibrium. Here also we begin by looking at the static equilibrium solutions of MHD equations. For a static situation without any flow, (14.3) becomes

$$\rho \mathbf{F} - \nabla p + \frac{1}{4\pi}(\nabla \times \mathbf{B}) \times \mathbf{B} = 0. \tag{14.18}$$

The study of the solutions of this equation is known as *magnetohydrostatics*. It may be noted that a magnetic field **B** satisfying this equation is not in general a static solution of the induction equation (14.10). In the absence of motions, (14.10) becomes

$$\frac{\partial \mathbf{B}}{\partial t} = \lambda \nabla^2 \mathbf{B}$$

and we expect the magnetic field to decay away, eventually disrupting the force balance implied by (14.18). But it is still useful to consider the solutions of (14.18). In laboratory situations, often a magnetic field is maintained by driving external currents so that the field does not decay and the equilibrium satisfies (14.18). In the other limit of astrophysics with very large length and time scales, the decay of the magnetic field is often so slow that (14.18) can be taken as a condition of approximate equilibrium. If the various forces are not in balance (i.e. if (14.18) is not satisfied), then motions result with dynamical time scales usually much smaller than the decay time scale. Within the dynamical time scale, the various forces try to reach a balance satisfying (14.18) and then it takes a very long time for the magnetic field to decay. So there is a subtle difference between hydrostatics and magnetohydrostatics. Whereas hydrostatics deals with genuine static equilibria where we take note of all the hydrodynamic equations, magnetohydrostatics is concerned only with situations in which various forces balance.

Let us restrict ourselves to situations where mechanical body forces such as gravity are unimportant compared to the other forces. Then we put $\mathbf{F} = 0$ and (14.18) becomes

$$\nabla p = \frac{(\nabla \times \mathbf{B}) \times \mathbf{B}}{4\pi}, \tag{14.19}$$

which implies that the pressure has to be balanced by the magnetic stresses. A magnetic field satisfying (14.19) is called a *pressure-balanced field*. We now introduce a very important parameter known as the *plasma-β*, which is the ratio of gas pressure to magnetic pressure defined as

$$\beta = \frac{p}{B^2/8\pi}. \tag{14.20}$$

We often have to deal with low-β plasmas, within which the gas pressure is negligible compared to the magnetic pressure. In such a plasma, the magnetic stress cannot be balanced by the pressure gradient and the magnetic field has to adjust itself in such a fashion that the magnetic stress itself vanishes. The condition for this obviously is

$$(\nabla \times \mathbf{B}) \times \mathbf{B} = 0. \tag{14.21}$$

A magnetic field satisfying (14.21) is known as a *force-free field* (Lüst and Schlüter 1954).

We now consider some specific solutions of pressure-balanced and force-free fields in cylindrical geometry. Since many laboratory experiments are done with cylindrical columns of plasma, such solutions are of practical interest in addition to providing an insight into the nature of magnetohydrostatic equilibria. Equilibrium problems with less symmetry are more difficult to study. Two coordinates z and θ are ignorable in cylindrical symmetry. Often one has to deal with magnetohydrostatic systems in which only one coordinate is ignorable. In such a situation, one is led to what is known as the *Grad–Shafranov equation* (Grad and Rubin 1958; Shafranov 1958). The reader has been asked to derive this equation in Exercise 14.4.

14.3.1 Pressure-balanced plasma column

We begin by writing down (14.19) in cylindrical coordinates (r, θ, z) assuming cylindrical symmetry, i.e. assuming that nothing varies in θ or z directions. It follows trivially from $\nabla \cdot \mathbf{B} = 0$ that $B_r = 0$. So the magnetic field can be written as

$$\mathbf{B} = B_\theta(r)\hat{\mathbf{e}}_\theta + B_z(r)\hat{\mathbf{e}}_z. \tag{14.22}$$

Substituting this in (14.19), we obtain

$$\frac{d}{dr}\left(p + \frac{B_\theta^2 + B_z^2}{8\pi}\right) + \frac{B_\theta^2}{4\pi r} = 0 \tag{14.23}$$

on assuming the pressure $p(r)$ to be a function of r alone. This is the fundamental equation of magnetohydrostatics in cylindrical geometry.

We want to consider a situation in the which the magnetic field in the plasma column is produced by driving a current $\mathbf{j} = j(r)\hat{\mathbf{e}}_z$ along the axis of the column. It is straightforward to see that such a current would not produce any B_z and the only component B_θ is given by

$$\frac{1}{r}\frac{d}{dr}(rB_\theta) = \frac{4\pi}{c}j, \qquad (14.24)$$

which can be easily obtained from (13.53). In order to study the equilibrium of a plasma column due to a given current along its axis, we first have to find out B_θ from (14.24), and then we try to satisfy (14.23) with that B_θ, taking $B_z = 0$.

Let us first consider the case of a constant current density through the plasma column. If j in (14.24) is constant, then we get

$$B_\theta = \frac{2\pi}{c}jr.$$

Substituting this in (14.23) gives

$$\frac{dp}{dr} = -\frac{2\pi}{c^2}j^2r,$$

of which the solution is

$$p = p_0 - \frac{\pi j^2 r^2}{c^2}, \qquad (14.25)$$

where p_0 is the gas pressure at the centre of the column. Figure 14.3 shows the current density, the magnetic field component B_θ and the pressure as a function of radius. It is to be noted that the pressure falls

Figure 14.3 The profile of the azimuthal magnetic field B_θ and the pressure p inside a plasma column with a uniform current density j inside.

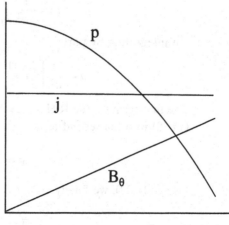

with radius, as we expect because of the pinching effect of the magnetic field. We also see from (14.25) that the pressure would become zero at a certain radius beyond which (14.25) ceases to hold, since a negative pressure is not physically admissible. If $r = a$ is the radius of the plasma column where the gas pressure falls to zero, we readily find from (14.25) that

$$a^2 = \frac{p_0 c^2}{\pi j^2}. \tag{14.26}$$

This raises the possibility that the pinching effect of the magnetic field may be able to confine a plasma column by making the pressure go to zero at a finite radius.

We now discuss the more general case of $j(r)$ being a function of r rather than a constant. If the plasma column is confined within a radius $r = a$ where the pressure goes to zero, then the total current is found from (14.24) to be

$$I = \int_0^a j(r) \cdot 2\pi r\, dr = \frac{c}{2} a B_\theta(a),$$

where $B_\theta(a)$ is the value of B_θ on the boundary surface of the plasma column given by

$$B_\theta(a) = \frac{2I}{ca}. \tag{14.27}$$

Putting $B_z = 0$, (14.23) can be written as

$$\frac{dp}{dr} = -\frac{1}{r^2} \frac{d}{dr} \left(\frac{r^2 B_\theta^2}{8\pi} \right). \tag{14.28}$$

The average pressure inside the plasma column is

$$\bar{p} = \frac{1}{\pi a^2} \int_0^a p(r) \cdot 2\pi r\, dr.$$

On integrating by parts,

$$\bar{p} = \frac{2}{a^2} \left[\frac{1}{2} r^2 p \Big|_0^a - \int_0^a \frac{1}{2} r^2 \frac{dp}{dr}\, dr \right].$$

The first term on the R.H.S. obviously vanishes. On substituting from (14.28) into the second term,

$$\bar{p} = \frac{B_\theta^2(a)}{8\pi}.$$

Using (14.27), we finally get

$$\bar{p} = \frac{I^2}{2\pi c^2 a^2}. \tag{14.29}$$

The significance of this equation is that it relates the average pressure $\bar{p}$ and radius a of a plasma column with the current I necessary to confine this plasma column. Often (14.29) is referred as the *Bennett pinch condition*, after Bennett (1934) who was the first person to study the static equilibrium of plasma columns with magnetic field.

14.3.2 Stability of plasma columns

Are the equilibrium configurations just discussed stable? Because, if they are not stable, then these configurations cannot be realized in laboratory experiments. A formal stability analysis for such configurations is fairly complicated, and we shall not get into that mathematical theory here. On physical considerations, however, one can argue that these configurations are actually *unstable*. Let us consider the two kinds of perturbations shown in Figures 14.4(a)–(b). Since magnetic field lines are crowded at the point P in Figure 14.4(a), it is obvious that the magnetic pressure is enhanced at that point and this additional pressure will push the plasma column in such a way as to enhance the kink perturbation. In other words, the type of kink perturbation shown in Figure 14.4(a), once initiated, starts growing so that the system is unstable. This is called the *kink instability*. It is also easy to see that B_θ at point Q in Figure 14.4(b) turns out to be larger than what B_θ just outside the column would have been if the column were not perturbed. The enhanced magnetic stress at Q will make the

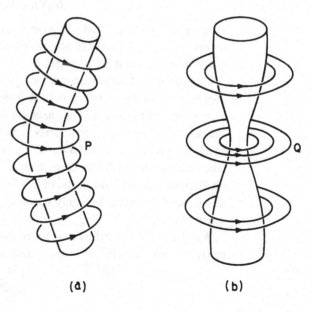

Figure 14.4 Unstable perturbations of a plasma column with azimuthal magnetic field. (a) Kink instability. (b) Sausage instability.

(a)   (b)

column still narrower there, triggering an instability. This is known as the *sausage instability*.

If there is a magnetic field along the axis of the plasma column (i.e. in the z direction), it tends to suppress both these types of instabilities. A kink perturbation bends the axial field lines, and the stress of the axial field clearly opposes the growth of the kink. In the case of a sausage perturbation, the axial field is compressed in the places where the column becomes narrower, and the enhanced pressure of the axial field opposes further narrowing of the column at that point. A rigorous analysis shows that $|B_z|$ has to be of the order of $|B_\theta|$ to suppress the instabilities. We shall discuss the relevance of cylindrical plasma configurations further in the next section.

14.3.3 Force-free fields

The force-free equation (14.21) implies that the curl of the magnetic field is in the direction of the magnetic field so that we can write

$$\nabla \times \mathbf{B} = \mu \mathbf{B}, \tag{14.30}$$

where μ is a scalar, which in principle can be a function of space. Some restriction on the spatial variation of μ follows on taking a divergence of (14.30), which gives

$$\nabla \cdot (\mu \mathbf{B}) = 0,$$

from which

$$\mathbf{B} \cdot \nabla \mu = 0. \tag{14.31}$$

This equation means that μ cannot vary along a magnetic field line. In general, μ can have different values on different field lines, but it has to be a constant on one field line.

The simplest case is to consider μ to be a constant. Since (14.30) then becomes a linear equation in $\mathbf{B}$, a magnetic field satisfying (14.30) with a constant μ is called a *linear force-free field*. A linear equation can in general be solved by a series expansion. Since (14.30) is a vector equation rather than a scalar equation, obtaining a general solution by series expansion is slightly complicated. This problem was first solved by Chandrasekhar and Kendall (1957). Here we shall not discuss this general solution, but only consider the solution with cylindrical symmetry.

Writing (14.30) in cylindrical coordinates assuming cylindrical symmetry (i.e. no variation of any quantity in θ or z directions) gives

$$-\frac{dB_z}{dr} = \mu B_\theta, \qquad \frac{1}{r}\frac{d}{dr}(rB_\theta) = \mu B_z. \tag{14.32}$$

Readers familiar with the properties of Bessel functions (see, for example, Arfken 1985, §11.1; Abramowitz and Stegun 1964, §9.1.27) should have no difficulty in verifying that the above equations are satisfied by

$$B_z = B_0 J_0(\mu r), \qquad B_\theta = B_0 J_1(\mu r). \tag{14.33}$$

This cylindrically symmetric solution appears as the first term in the general Chandrasekhar–Kendall expansion.

A force-free field with μ varying from one field line to another is known as a *nonlinear force-free field*. When we study such a field with cylindrical symmetry, it is more convenient to begin with (14.23) putting $p = 0$ rather than to begin with (14.32) taking μ to be a function of r. If we write

$$\frac{B_\theta^2 + B_z^2}{8\pi} = F(r), \tag{14.34}$$

then it is seen from (14.23) that

$$B_\theta^2 = -4\pi r \frac{dF}{dr}. \tag{14.35}$$

From (14.34) and (14.35), we can then write

$$B_z^2 = 8\pi \left(F + \frac{1}{2} r \frac{dF}{dr} \right). \tag{14.36}$$

The function $F(r)$ is often called the *generating function*, because the components of the magnetic field can at once be obtained from $F(r)$ on using (14.35) and (14.36). Certain conditions have to be satisfied by this generating function in order to ensure that B_θ and B_z turn out to be real. The function $F(r)$ has to be monotonically decreasing with r so that B_θ^2 given by (14.35) is everywhere positive. For B_z^2 given by (14.36) to be positive, $F(r)$ has to satisfy the condition

$$F + \frac{1}{2} r \frac{dF}{dr} \geq 0.$$

Given any generating function which satisfies these conditions, one can then at once obtain a nonlinear force-free field (see Exercise 14.5).

14.4 A note on fusion research

The huge energy needs of present-day civilization are mostly met by fossil fuels like coal and oil. Although experts may disagree on how long these fossil fuels will last, nobody thinks that they can supply the energy needs of the human race for more than a few centuries. It is, therefore, imperative to tap other sources of energy. One attractive possibility is the thermonuclear process, which generates energy in the interiors of stars. Deuterium happens to be the atom which seems most promising for this purpose. Since some molecules in sea-water have deuterium atoms in the place of hydrogen atoms, we have a huge source of deuterium atoms on the surface of the Earth. When two deuterium nuclei are brought together, they may fuse to produce a tritium or a helium nucleus, and a small fraction of mass is converted into energy during this process. Since deuterium nuclei are positively charged, normally two such nuclei would not come close to each other due to Coulomb repulsion. Only when two such nuclei approach each other with sufficiently high relative velocity, may it be possible to overcome the Coulomb barrier so that the thermonuclear reaction takes place. The easiest way of achieving this is to produce a sufficiently hot deuterium plasma. If the random velocities of the deuterium nuclei are sufficiently high due to the high temperature, then occasionally two nuclei may approach each other with high enough relative velocity to overcome the Coulomb barrier and lead to fusion.

Such a high-temperature deuterium plasma cannot be kept in ordinary material containers. A dense, hot plasma would burn the walls of the container. On the other hand, if the density and hence the total heat content of the hot plasma is low, then it would lose all its heat to the wall as soon as the plasma comes in contact with the walls. The only possibility is to keep the plasma confined by magnetic fields. We saw in the previous section that cylindrical magnetohydrostatic configurations are possible such that the plasma pressure falls to zero at a certain distance from the axis. One hopes that such magnetic configurations may keep the plasma confined, allowing the thermonuclear reactions to proceed (Rose and Clark 1961, Chapter 14). Just after World War II, several countries such as the U.S.A., the U.S.S.R. and the U.K. started large-scale projects to obtain energy by harnessing thermonuclear fusion. These developments provided a tremendous boost to plasma physics.

Most of these projects were initiated with the expectation that the commercial production of energy from fusion would be feasible within a few years—at most within a decade. Such hopes were quickly found to be premature. Confining plasmas for sufficiently long times turned out to be much more difficult than anybody anticipated. It is necessary to spend considerable energy in heating the plasma initially and setting up the system. Only if we are able get back a larger amount of energy as a result of fusion, can we start thinking of commercial applications. In order to get sufficient energy out of the system, the plasma has to remain confined for a sufficiently long time.

If n is the number density of the plasma and τ the time for which it remains confined, then $n\tau$ has to be larger than about 10^{16} cm^{-3} s in order to make the energy output larger than the energy input. This condition is known as the *Lawson criterion* (Lawson 1957). One way of satisfying this condition is to confine a low-density plasma for sufficient time with the help of magnetic fields. A typical plasma confinement experiment may have $n \approx 10^{15}$ cm^{-3} and $\tau \approx 0.1$ s, i.e. we may be two orders away from the Lawson criterion. The other approach is to produce sudden heat in a high-density material by an intense laser beam. Even though the confinement time τ would be very small, the high value of n may compensate for it. It has *not* been possible to satisfy the Lawson criterion in either approach as of today. So far, the magnetic confinement of low-density plasmas is the approach which is pursued more vigorously. The magnetohydrostatic equilibria in cylindrical geometry discussed in the previous section are very relevant for such experiments. In order to avoid the edges of the cylinder, one often makes a plasma device in the form of a torus, where essentially the cylinder is made to bend and close on itself. Figure 14.5 shows a *tokamak*, which appears one of the most promising plasma devices at the present time.

Why is it so difficult to confine plasmas? We have seen in §14.3.1–14.3.2 that a plasma column can be confined by sending a current along the axis, but the configuration is MHD unstable unless there is a strong magnetic field along the axis as well. When we set up a

Figure 14.5 The tokamak *Aditya* operating at the Institute for Plasma Research, Gandhinagar, India. Courtesy: Abhijit Sen.

plasma configuration, the first important condition is that it has to be MHD stable. Otherwise violent instabilities would set in immediately. Although the MHD instabilities are the most serious ones, there can be many other kinds of instabilities based on microscopic physics, which would plague an MHD stable configuration. As a result, the hot plasma in the core region diffuses to outer parts of the device much faster than one would expect. Much of present-day research in laboratory plasmas is aimed at understanding and suppressing these micro-instabilities. Whether we shall be able to provide the ultimate solution of mankind's energy problem may well depend on our success in this research.

14.5 Hydromagnetic waves

In Chapter 12, we considered several types of high-frequency waves in plasmas in which ions and electrons respond differently. Let us now consider perturbations in a plasma in which electrons and ions remain together so that the plasma can be regarded as a single fluid and the equations of MHD apply. We saw in §6.2 that perturbations in a uniform neutral gas give rise to acoustic waves. If there is a magnetic field in the plasma, then it can sustain waves more complicated than the simple acoustic waves. Let us consider perturbing a uniform plasma with a uniform magnetic field $\mathbf{B}_0$. To simplify the treatment, we neglect the dissipative effects like viscosity, heat conduction and electrical resistivity. If these effects are included, then they are found to cause a damping of the wave.

In the spirit of usual perturbation analysis as discussed in §6.2, we write the density, the pressure, the velocity and the magnetic field as $\rho_0 + \rho_1(\mathbf{x}, t)$, $p_0 + p_1(\mathbf{x}, t)$, $\mathbf{v}_1(\mathbf{x}, t)$ and $\mathbf{B}_0 + \mathbf{B}_1(\mathbf{x}, t)$. The next step is to substitute these in the basic MHD equations and then linearize them. Since we are assuming adiabatic conditions by neglecting heat conduction, the pressure perturbations would be related to density perturbations as discussed in §6.2. The basic relations (6.11) and (6.13) are written down again:

$$p_1 = c_s^2 \rho_1, \tag{14.37}$$

with

$$c_s^2 = \frac{\gamma p_0}{\rho_0}. \tag{14.38}$$

Since c_s given by (14.38) is the speed with which acoustic waves would propagate in the absence of magnetic field, we would refer to c_s as the sound speed. On perturbing and linearizing the equation of continuity, we again get (6.15), which is

$$\frac{\partial \rho_1}{\partial t} + \rho_0 \nabla \cdot \mathbf{v}_1 = 0. \tag{14.39}$$

We now have to consider equation (14.3). There is no external force **F** in the present situation and the viscosity term is neglected. Then, perturbing and linearizing lead to

$$\rho_0 \frac{\partial \mathbf{v}_1}{\partial t} = -c_s^2 \nabla \rho_1 + \frac{(\nabla \times \mathbf{B}_1) \times \mathbf{B}_0}{4\pi}, \tag{14.40}$$

where we have substituted for p_1 from (14.37). Lastly, we have to perturb the induction equation (14.10). Neglecting the resistivity term, the linearized perturbation equation is readily found to be

$$\frac{\partial \mathbf{B}_1}{\partial t} = \nabla \times (\mathbf{v}_1 \times \mathbf{B}_0). \tag{14.41}$$

The evolutions of the basic perturbation variables ρ_1, $\mathbf{v}_1$ and $\mathbf{B}_1$ are governed by equations (14.39), (14.40) and (14.41). The energy equation has already been taken care of on using (14.37).

On differentiating (14.40) with respect to time and making use of (14.39), (14.41),

$$\frac{\partial^2 \mathbf{v}_1}{\partial t^2} = c_s^2 \nabla(\nabla \cdot \mathbf{v}_1) + [\nabla \times \{\nabla \times (\mathbf{v}_1 \times \mathbf{v}_A)\}] \times \mathbf{v}_A, \tag{14.42}$$

where

$$\mathbf{v}_A = \frac{\mathbf{B}_0}{\sqrt{4\pi\rho_0}} \tag{14.43}$$

has the dimension of velocity and is known as the *Alfvén velocity*—in honour of Alfvén (1942a,b) who pioneered the study of hydromagnetic waves. As in §6.2, the uniformity of the system prompts us to look for solutions in which all the perturbation quantities vary as $\exp[i(\mathbf{k} \cdot \mathbf{x} - \omega t)]$. Then we have to replace $\partial/\partial t$ by $-i\omega$ and ∇ by $i\mathbf{k}$. On doing this in (14.42), we end up with

$$\omega^2 \mathbf{v}_1 = (c_s^2 + v_A^2)(\mathbf{k} \cdot \mathbf{v}_1)\mathbf{k} + \mathbf{v}_A \cdot \mathbf{k}[(\mathbf{v}_A \cdot \mathbf{k})\mathbf{v}_1 - (\mathbf{v}_A \cdot \mathbf{v}_1)\mathbf{k} - (\mathbf{k} \cdot \mathbf{v}_1)\mathbf{v}_A]. \tag{14.44}$$

This is the basic dispersion relation for hydromagnetic waves relating **k** and ω when c_s and $\mathbf{v}_A$ are given.

Let us now pause for a moment and take stock of the situation. We wish to study the nature of the waves produced by the perturbations we are considering. Once we obtain a dispersion relation between **k** and ω, it is in principle possible to calculate the phase and group velocities. The dispersion relation (14.44), however, looks somewhat complicated. Although it contains all the important information we want to find out, it would require some amount of algebra to isolate the necessary results. Before proceeding to a general discussion, let us first show that (14.44) admits of one kind of relatively simple wave. The magnetic field and the propagation vector (i.e. the vectors $\mathbf{v}_A$ and $\mathbf{k}$) define a plane. We consider purely transverse disturbances in which the displacement and therefore the velocity $\mathbf{v}_1$ are perpendicular to this plane. Then (14.44) reduces to

$$\omega^2 = (\mathbf{v}_A \cdot \mathbf{k})^2 \qquad (14.45)$$

so that

$$\omega = \pm v_A k \cos\theta,$$

where θ is the angle between the magnetic field and the propagation vector. We note that the group velocity $\nabla_{\mathbf{k}}\omega$ is equal to $\mathbf{v}_A$. Thus we are considering a wave in which a disturbance would move along the magnetic field with the velocity $\mathbf{v}_A$, the displacement being always in the transverse direction. Figure 14.6 sketches the appearance of the magnetic field lines during the propagation of such a wave. This type of wave was first predicted theoretically by Alfvén (1942a,b) and is therefore known as the *Alfvén wave*. The existence of such a wave was afterwards demonstrated by Lundquist (1949) by carrying out experiments with mercury subjected to a strong magnetic field. We saw in §14.1 that a magnetic field $\mathbf{B}_0$ has a tension $B_0^2/4\pi$ associated with it along the field lines. A magnetic field in a plasma can, therefore, be thought to be like a stretched string. Whenever the magnetic field lines are distorted by a transverse perturbation, the magnetic tension tries to oppose the distortion. Just as a transverse wave can be started in a string by plucking it, similarly we have the transverse Alfvén wave moving along the field lines. The velocity of the transverse wave moving along a stretched string can be shown to be something like $\sqrt{\text{tension/density}}$ (see, for example, Joos 1958, Chapter VIII, §8). Keeping in mind that the magnetic tension is given by $B_0^2/4\pi$, the expression (14.43) for the Alfvén velocity is exactly like that.

Figure 14.6 A sketch showing the distortions of magnetic field lines during the propagation of an Alfvén wave along them.

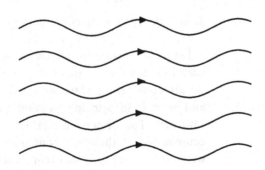

We now come to a general discussion of the wave modes permitted by (14.44). Since (14.44) is a vector equation linear in $\mathbf{v}_1$, it can be broken into three scalar components of the following form in any Cartesian coordinates:

$$T_x(\omega^2, \mathbf{k})v_{1x} + T_y(\omega^2, \mathbf{k})v_{1y} + T_z(\omega^2, \mathbf{k})v_{1z} = 0, \qquad (14.46)$$

where v_{1x}, v_{1y}, v_{1z} are the components of $\mathbf{v}_1$ and T_x, T_y, T_z are their coefficients. If we have three scalar equations of the form (14.46), the determinant has to be zero for the sake of consistency. The detailed calculations are given as Exercise 14.6. Even without doing the detailed calculations, one can at once see that setting a 3×3 determinant equal to zero would give three possible solutions of ω^2 for each value of $\mathbf{k}$. Other things being the same, $-\omega$ would be a solution if $+\omega$ is a solution. Since $+\omega$ and $-\omega$ merely correspond to waves propagating in opposite directions, the three solutions of ω^2 corresponding to any $\mathbf{k}$ essentially implies the existence of three kinds of wave modes. One of them is the Alfvén wave discussed above. The other two are of more complicated nature, as can be found by working out Exercise 14.6. The existence of these wave modes in a magnetized gas was demonstrated by Herlofson (1950).

Any wave in nature is driven by some restoring force which opposes displacements in the system. In the present context of the MHD model, two types of restoring forces are possible: one arising out of pressure gradients; and the other arising out of magnetic stresses. The simple Alfvén wave is purely transverse without involving any density or pressure variations, the magnetic tension being the only restoring force for it. The other two waves are more complicated due to the fact that they are mixtures of acoustic and magnetic waves, where both types of restoring forces are present. For one of these modes, the pressure and magnetic restoring forces are roughly in phase, making the mode propagate fast. It is called the *fast mode*. The other mode, for which these restoring forces are roughly out of phase, is known as the *slow* mode. One can also show that the displacements associated with the three modes make up a triad of orthogonal vectors (see Exercise 14.6). So any arbitrary disturbance in our system can be represented as a superposition of the Alfvén, fast and slow modes.

14.6 Magnetoconvection and sunspots

We saw in Chapter 7 that perturbations in neutral fluid configurations can give rise to waves in some circumstances and may lead to instabilities in other circumstances. The previous section provided an example of waves produced in a magnetofluid in a particular situation. We expect that there would be other types of MHD configurations in which a perturbation would lead to instabilities. In fact, all the calculations presented in §7.3–7.5 can be extended to the more complicated situation of a magnetofluid with a magnetic field embedded in it. We refer the reader to the classic monograph of Chandrasekhar (1961) for a systematic discussion of how various hydrodynamic instabilities are modified due to the presence of magnetic fields. To give the reader an idea of the subject, we briefly discuss how the theory of Rayleigh–Bénard convection has to be extended if we are dealing with a magnetofluid with a uniform vertical magnetic field instead of an ordinary fluid. This subject, known as *magnetoconvection*, is of considerable interest in astrophysics.

Instead of perturbing the hydrodynamic equations as done in §7.3, we now have to perturb the MHD equations, taking into account the presence of a vertical magnetic field B_0. Needless to say, this is a much more complicated analysis, which was carried out by Thomson (1951) and Chandrasekhar (1952). We merely present the main results without the full calculations. We saw in §7.3 that convective instability in neutral fluids is always of a non-oscillatory nature. In the case of a fluid with a vertical magnetic field, the detailed analysis shows that it is possible for the convective instability to begin in the form of growing oscillations, provided λ defined in (14.11) is less than the thermometric conductivity κ defined in (7.9). If λ is small compared to other transport coefficients, then the magnetic field is nearly frozen in the fluid and distortions in the field lines can give rise to Alfvén-type oscillations. It is no wonder that convective instability in such a situation is of an oscillatory nature. If, however, $\kappa < \lambda$, then such oscillatory onset of convection is not possible. This happens to be the case in most terrestrial situations, where magnetoconvection is of non-oscillatory nature. For example, mercury at room temperature has $\kappa = 4.5 \times 10^{-2}$ cm^2 s^{-1} and $\lambda = 7.6 \times 10^3$ cm^2 s^{-1}.

We consider a fluid with $\kappa < \lambda$ so that the magnetoconvection is non-oscillatory. In §7.3, we introduced a very important dimensionless number R in (7.23). It is a measure of the temperature gradient in the fluid and is known as the Rayleigh number. If a uniform vertical magnetic field B_0 is present, we can introduce another important dimensionless number in the problem. It is

$$Q = \frac{B_0^2 d^2}{4\pi \rho_b \nu \lambda}, \tag{14.47}$$

where the various symbols are as defined in §7.3. This dimensionless number is often referred to as the *Chandrasekhar number*. We saw in §7.3 that the Rayleigh number which makes a perturbation with horizontal wavenumber k' (measured with respect to the depth d as the unit of length) marginally unstable is given by (7.25) for the simplest boundary conditions. In the presence of a vertical magnetic field, the Rayleigh number necessary for making such a perturbation unstable is found to be

$$R = \frac{\pi^2 + k'^2}{k'^2}[(\pi^2 + k'^2)^2 + \pi^2 Q]. \tag{14.48}$$

If the Chandrasekhar number Q is set to zero, then we get back the case of ordinary hydrodynamic convection and (14.48) reduces to (7.25).

Since perturbations with different wavenumbers k' are in general present in the system, the system becomes unstable if any k' is unstable. To obtain the critical Rayleigh number R_c for convection to be possible, we then find out when R given by (14.48) is a minimum as a function of k'. The procedure is exactly the same as that followed in §7.3. Since (7.25) was simpler than (14.48), one could there find the analytical expression $(27/4)\pi^4$ for the critical Rayleigh number. Here, for a given value of the Chandrasekhar number Q, one can calculate a

Figure 14.7 A plot showing the experimentally determined critical Rayleigh number R_c for different values of the Chandrasekhar number Q. The smooth curve is from theoretical calculations. Adapted from Chandrasekhar (1961).

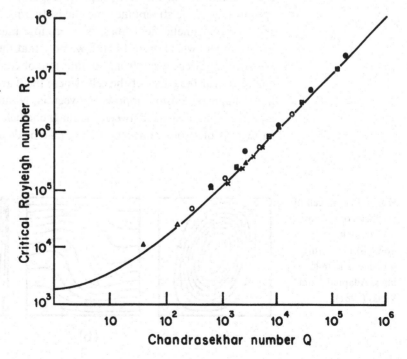

numerical value of the critical Rayleigh number R_c. Nakagawa (1955) carried out careful experiments with mercury to determine R_c for different values of Q. Figure 14.7 adapted from Chandrasekhar (1961, Chapter IV) shows Nakagawa's data plotted with a curve obtained from theoretical calculations with boundary conditions appropriate for comparing with experiments. For small Q, R_c asymptotically tends to the critical Rayleigh number for pure hydrodynamic convection in the absence of a magnetic field. On the other hand, when Q is large, R_c increases monotonically with Q. This means that, when the magnetic field is stronger, it is harder to start convection. Understanding this result physically is not difficult. Since convection distorts magnetic field lines, convective motions are opposed by magnetic tension. We therefore need a steeper temperature gradient to drive the convection when a strong magnetic field is present.

The linear stability theory only gives the condition for convection to be possible. In order to understand the nature of magnetoconvection, it is necessary to perform numerical simulations using the full nonlinear equations with a Rayleigh number above the critical value. One of the first impressive simulations was carried out by Weiss (1981). It was found that the fluid motions eventually settle into steady patterns of convection cells. Figure 14.8 taken from Weiss (1981) displays such a convection cell in the vertical plane. Figures 14.8(a) and 14.8(b) respectively give the streamlines and the isotherms. Figure 14.8(c), which shows the magnetic field lines, is of utmost interest. On comparing Figure 14.8(c) with Figure 14.8(a), we note that the magnetic field lines are virtually swept away from the interiors of convection cells and are confined near the edges of the cells where there are little fluid motions. Since magnetic tension opposes convection, Nature seems to arrive at a partitioning of space between the magnetic field and the convection. In regions of strong convection, magnetic fields are excluded. On the

Figure 14.8 A cell of magnetoconvection. (a) Streamlines of flow. (b) Isotherms. (c) Magnetic field lines. Adapted from Weiss (1981).

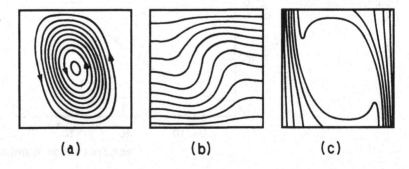

(a)   **(b)**   **(c)**

other hand, in the regions of strong magnetic field, there is hardly any convection.

The theory of magnetoconvection can directly be applied to explain many aspects of sunspots. Figure 14.9 shows a large sunspot. Sunspots appear darker because of their lower temperature compared to the surrounding solar surface. The typical size of a fully developed sunspot is about 10,000 km, although a very big sunspot would be large enough for the whole Earth to be immersed in it. An important clue to the physical nature of sunspots was found when Hale (1908) discovered the existence of strong magnetic fields in sunspots from the Zeeman splitting of sunspot spectra. The typical value of the magnetic field in sunspots is about 3000 G. This magnetic field is presumably sitting in the layers just below the solar surface where the heat is transported by convection. In fact, we see granules or convection cells around the sunspot in Figure 14.9. Biermann (1941) suggested that the lower temperature of sunspots is caused by the inhibition of convection by the magnetic stresses within sunspots. Detailed simulations of magnetoconvection indeed lend support to this idea. In fact, the top left corner of Figure 14.8(c) is very much like a sunspot. The magnetic field there has certainly become concentrated. We note in Figure 14.8(b) that

Figure 14.9 Photograph of a fully grown sunspot with granulation visible around. Courtesy: W. Schmidt.

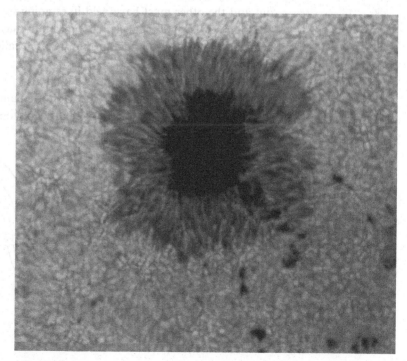

this corner is also a region of reduced temperature where the isotherms dip downward. To an observer looking from the top, this region would appear as a cooler and darker spot on the surface from where magnetic field lines are sticking out. In spite of this success in explaining some features of sunspots on the basis of magnetoconvection theory, it should be emphasized that sunspots are still not very well-understood objects. We still do not know what determines the typical magnetic field strengths or typical sizes of sunspots.

14.7 Bipolar magnetic regions and magnetic buoyancy

A few years after the discovery of the existence of magnetic fields in sunspots (Hale 1908), Hale made another important discovery. Often two large sunspots appear side by side. Hale *et al.* (1919) discovered that the two sunspots in such a pair almost always have opposite polarities. The most obvious explanation for this is that a strand of magnetic field has come out through the solar surface as shown in Figure 14.10(b). If the two sunspots are merely the two locations where this strand of magnetic field intersects the solar surface, then we readily see that one sunspot must have magnetic field lines coming out and the other must have field lines going in. We now address the question how such a magnetic configuration may come about. Such a configuration with two large sunspots is often called a *bipolar magnetic region.*

We discussed in the previous section that magnetic fields in the solar convection zone may be expected to remain concentrated within localized regions. Let us consider a nearly horizontal cylindrical region

Figure 14.10
Magnetic buoyancy
of a flux tube. (a) A
nearly horizontal flux
tube under the solar
surface. (b) The flux
tube after its upper
part has risen
through the solar
surface.

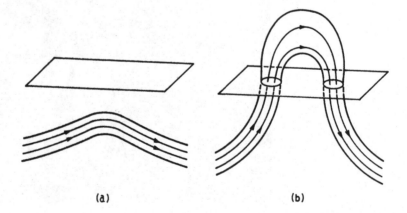

(a)   (b)

within which some magnetic field is concentrated, as sketched in Figure 14.10(a). Such a region of concentrated magnetic field with very little magnetic field outside is often called a *magnetic flux tube*. Parker (1955a) pointed out that a horizontal flux tube may be buoyant. The argument is quite straightforward. Let p_i be the gas pressure inside the magnetic flux tube and p_e be the external pressure. We have seen in (14.5) that a magnetic field causes a pressure $B^2/8\pi$ wherever it exists. In order to have a pressure balance across the bounding surface of the flux tube, we must have

$$p_e = p_i + \frac{B^2}{8\pi}. \tag{14.49}$$

It readily follows that

$$p_i < p_e. \tag{14.50}$$

This usually, though not always, implies that the internal density ρ_i is also less than the external density ρ_e. In the particular case when the temperature inside and outside are both T, (14.49) leads to

$$R\rho_e T = R\rho_i T + \frac{B^2}{8\pi}, \tag{14.51}$$

from which we obtain

$$\frac{\rho_e - \rho_i}{\rho_e} = \frac{B^2}{8\pi p_e}. \tag{14.52}$$

We thus see that the fluid in the interior of the flux tube is lighter and must be buoyant. In the limit of high $\mathscr{R}_M$, the magnetic field is frozen in this lighter fluid. As a result, the flux tube as an entity becomes buoyant and rises against a gravitational field. This very important effect, discovered by Parker (1955a), is known as the *magnetic buoyancy*. Since (14.50) does not *always* imply that the interior of a flux tube is lighter, it is possible that one part of a flux tube becomes buoyant and not the other parts. Here we shall not get into a discussion as to how this may come about. Suppose only the middle part of the flux tube shown in Figure 14.10(a) has become buoyant. Then this middle part is expected to rise, eventually piercing through the surface and creating the configuration of Figure 14.10(b). With the help of the idea of magnetic buoyancy, one can thus explain how a bipolar magnetic region arises. Figure 14.11 is the photograph of a freshly emerged bipolar magnetic region showing two large sunspots along with some smaller spots. The granules lying between the two large sunspots seem somewhat distorted and elongated. Looking at the photograph carefully, one almost gets the feeling that something has recently come up through the solar surface between the two large sunspots.

The global distribution of the bipolar magnetic regions on the whole solar disk presents an interesting pattern. Figure 14.12 is a magnetogram image of the whole solar disk, where regions of positive polarity are indicated by white and regions of negative polarity by black, the regions without appreciable magnetic field being represented in grey. One notes that most bipolar magnetic regions are roughly aligned parallel to the solar equator. In the magnetic bipolar regions in the northern hemisphere, one finds the positive polarity (white) to appear on the right side of the negative polarity (black). This is reversed in the southern hemisphere, where white appears to the left of black. How can we explain this distribution pattern of bipolar magnetic regions?

It is known that the Sun does not rotate like a solid body. The regions near the equator rotate with a higher angular velocity compared to regions near the poles. Let us consider a magnetic field line passing through the solar interior as shown in Figure 14.13(a). Since the magnetic field line must be nearly frozen in the plasma due to the high $\mathcal{R}_M$, we expect the differential rotation to stretch out this field line as shown in Figure 14.13(b). Thus the differential rotation has a tendency to produce strong magnetic fields in the toroidal direction,

Figure 14.11 A photograph showing a bipolar magnetic region produced by fresh flux that emerged from underneath the solar surface. From Zwaan (1985). (©Kluwer Academic Publishers. Reproduced with permission from *Solar Physics*.)

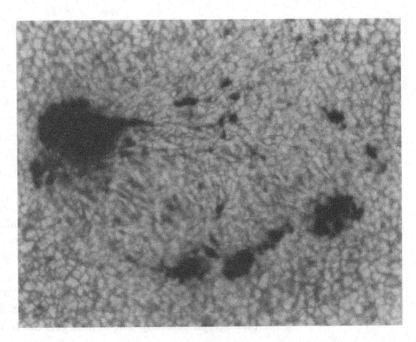

and the magnetic fields in the interior of the Sun are believed to be predominantly toroidal. Parts of the toroidal field may then become buoyant and produce bipolar magnetic regions by piercing through the solar surface. It is straightforward to see from Figure 14.13(b) that the bipolar regions in the two hemispheres would have opposite polarity alignments. We thus see that the global distribution of bipolar

Figure 14.12 A magnetogram picture of the full solar disk. The regions with positive and negative polarities are respectively shown in white and black, with grey indicating regions where the magnetic field is weak. Courtesy: K. Harvey.

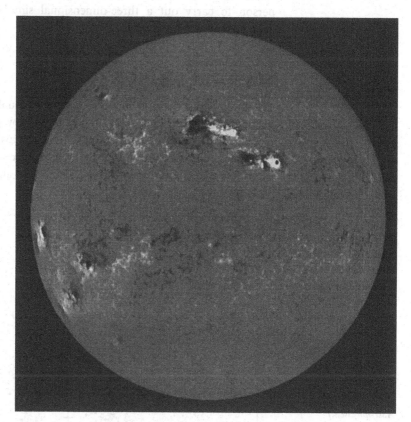

Figure 14.13 The production of a strong toroidal magnetic field underneath the solar surface. (a) An initial poloidal field line. (b) A sketch of the field line after it has been stretched by the faster rotation near the equatorial region.

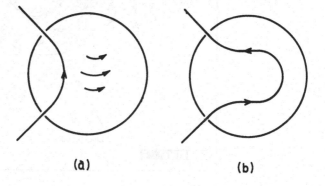

(a) (b)

magnetic regions can be rather beautifully explained by some simple ideas based on the basic principles of MHD. Our discussion here has been rather qualitative. Starting from MHD equations, Spruit (1981) obtained an equation for the dynamics of a thin flux tube embedded in an unmagnetized atmosphere. Moreno-Insertis (1986) used this equation to model the emergence of bipolar active regions through a two-dimensional simulation, whereas Choudhuri (1989) was the first person to carry out a three-dimensional simulation in a spherical geometry appropriate for the Sun.

14.8 Parker instability

The interstellar medium inside a galaxy is usually found to be distributed rather non-uniformly. Figure 14.14 shows how the interstellar medium is distributed in the galaxy M81. In parts of the spiral arms, the interstellar medium seems to form a succession of clumps like beads on a string. It was Parker (1966) who first pointed out that

Figure 14.14 The distribution of interstellar matter in the galaxy M81, as measured by radio emission from neutral atoms. From Rots (1975). (©Springer-Verlag. Reproduced with permission from *Astronomy and Astrophysics*.)

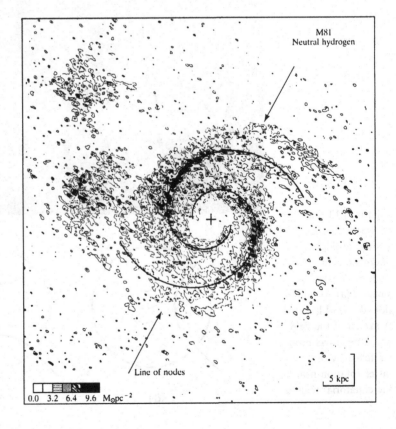

a uniform distribution of the interstellar medium would be unstable. This instability, known as the *Parker instability*, is related to magnetic buoyancy and is presumably the cause behind the interstellar medium fragmenting into clumps.

The magnetic field of the galaxy can be assumed to be frozen in the interstellar medium. Let us consider an initial configuration with the interstellar medium distributed uniformly in a layer having straight magnetic field lines passing through it. Now suppose the system has some small perturbations with parts of the magnetic field lines bulging upward, as sketched in Figure 14.15(a). From symmetry, the gravitational field is directed towards the central plane of the layer. So the gravitational field in the bulging region of magnetic field lines must be downward. If the magnetic field is frozen in the plasma, then the plasma can come down vertically in the bulge region only if the magnetic field lines are also brought down. It is, however, possible for the plasma to flow down along the magnetic field lines as indicated by the arrows in Figure 14.15(a). Alfvén's theorem of flux-freezing allows such flows without bringing down the field lines in the bulge, and hence we expect that the downward gravitational field in the bulge region would make the plasma flow in this fashion. As a result of the plasma draining down from the top region of the bulge, this region becomes lighter and more buoyant. We therefore expect this region to rise up further. In other words, the initial bulge keeps on getting enhanced, leading to an instability (Parker 1966). As the magnetic field lines

Figure 14.15 Sketch of Parker instability. (a) Perturbed magnetic field lines bulging out of the galactic plane. (b) The final configuration.

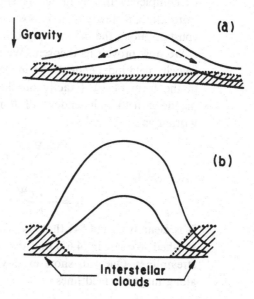

become more bent, the magnetic tension gets stronger. Eventually the magnetic tension halts the rise of the upper part of the bulge. This was clearly seen in the detailed numerical simulations of Parker instability by Mouschovias (1974).

Figure 14.15(b) sketches what the final configuration may look like. The magnetic field lines bulge out of the galactic plane, whereas the interstellar plasma collects in the valleys of the magnetic field lines. This is presumably the reason why the interstellar medium is intermittent and clumpy. We should point out that these qualitative ideas described in this section can be confirmed by detailed calculations.

14.9 Magnetic field as a transporter of angular momentum

If there are magnetic field lines coming out of a rotating astrophysical object, it is possible to take away angular momentum from the object with the help of magnetic stresses. To understand how this works, let us first discuss an important theorem known as *Ferraro's law of isorotation* (Ferraro 1937). We consider a rotating object which is completely axisymmetric around the rotation axis. Using cylindrical coordinates, this means that the rotation speed

$$\mathbf{v} = r\Omega(r, z)\hat{\mathbf{e}}_\theta \qquad (14.53)$$

is independent of the toroidal coordinate θ. Suppose the object has an axisymmetric poloidal magnetic field, which we can assume to be completely frozen in the plasma. According to Ferraro's law of isorotation, a steady state is possible only if the angular velocity Ω is constant along the magnetic field lines.

We now prove this theorem. We saw in §4.8 that an incompressible velocity field independent of z in Cartesian coordinates can be written in the form (4.60). Exactly on the same considerations, a poloidal magnetic field independent of θ in cylindrical coordinates can be written as

$$\mathbf{B} = \nabla \times \left[\frac{1}{r} \Psi(r, z)\,\hat{\mathbf{e}}_\theta \right], \qquad (14.54)$$

from which

$$B_r = -\frac{1}{r}\frac{\partial \Psi}{\partial z}, \qquad B_z = \frac{1}{r}\frac{\partial \Psi}{\partial r}. \qquad (14.55)$$

It is clear from (14.54) that the $\nabla \cdot \mathbf{B} = 0$ condition is automatically satisfied. We saw in (4.63) that the stream function was constant along streamlines. One can show exactly similarly that $\Psi(r, z)$ is constant along magnetic field lines.

We now consider the induction equation (14.10). Assuming that the magnetic field is completely frozen in the plasma, we neglect the last term. For steady state, the L.H.S. vanishes so that we must have

$$\nabla \times (\mathbf{v} \times \mathbf{B}) = 0.$$

Substituting (14.53) and (14.54) in this equation, we find the θ component to be the only non-zero component, giving us

$$\frac{\partial(\Omega, \Psi)}{\partial(r, z)} = 0. \tag{14.56}$$

This means that we must have $\Omega = f(\Psi)$. In other words, the angular velocity should be constant along magnetic field lines. The condition of flux freezing essentially demands that the field lines rotate like rigid objects to make the steady state possible.

If the angular velocity were to vary along field lines, then the poloidal field lines would be continuously stretched to produce toroidal field lines as shown in Figure 14.13 and a steady state would not be possible. When magnetic fields are stretched as in Figure 14.13, the magnetic energy increases, implying that some work must be done on the magnetic field. If the magnetic field is sufficiently strong, then magnetic stresses would resist the deformation indicated in Figure 14.13. As a result, a strong magnetic field would try to impose a rigid rotation on the system. We are now ready to understand how magnetic fields may help in transporting angular momentum. Let us look at a few systems of astrophysical importance.

14.9.1 Magnetic braking during star formation

Consider first the example of a star-forming region collapsing due to Jeans instability. In the very crude discussion of Jeans instability presented in §7.5, we did not consider the effect of rotation or magnetic field. We expect any collapsing gas cloud in the Galaxy to have some angular momentum from the galactic rotation and to be threaded by the general galactic magnetic field. If the angular momentum per unit mass Ωr^2 remains conserved during the collapse, then we expect the angular velocity Ω to keep on rising. This results in the angular velocity of the collapsing cloud being larger than that of the surrounding plasma. The magnetic field lines connecting the collapsing cloud with the surrounding plasma resist such a variation of angular velocity and provide a kind of braking mechanism to slow down the rapidly spinning collapsed cloud. This is known as *magnetic braking*. A rough estimate of the rate at which the angular momentum is removed can be done as follows. See Mestel (1968) for more details.

Suppose the surrounding plasma up to a distance of $r = a$ is rotating with the angular velocity Ω of the collapsing cloud and the angular velocity is much less beyond $r = a$. The magnetic stresses would try to spin up the plasma beyond $r = a$ to the angular velocity Ω. We have seen in §14.5 that magnetic disturbances propagate at the Alfvén speed v_A given by (14.43). Hence in time δt a shell of plasma between radii a and $a + v_A \delta t$ is spun to angular velocity Ω. The angular momentum associated per unit mass of this shell is $(2/3)\Omega a^2$ so that the total angular momentum added to this shell is $(8\pi/3)\Omega a^4 \rho v_A \delta t$. This must have come from the angular momentum of the central object, which is $(2/5)M\Omega a^2$. We then have

$$\frac{2}{5}Ma^2 \frac{d\Omega}{dt} = -\frac{8\pi}{3}a^4 \rho v_A \Omega. \qquad (14.57)$$

Substituting $v_A = B/\sqrt{4\pi\rho}$ in (14.57), one has an approximate equation for magnetic braking. A more realistic analysis is very complicated, since the Alfvén speed is in general different in different directions. The collapse in the presence of angular momentum is not spherically symmetric, and the assumption of spherical symmetry used above is a very gross over-simplification.

14.9.2 Magnetized winds

We sketched a hydrodynamic theory of the solar wind in §6.8. Since the Sun is rotating and there are magnetic fields stretching out in the solar atmosphere, these facts have to be incorporated in a more complete model of the solar wind. Do we expect Ferraro's law of isorotation to hold out in the solar wind? It may be noted that we took the fluid velocity in the form (14.53) to prove Ferraro's law. If there is an outward flow, as in the solar wind, then Ferraro's law does not in general hold. In the lower corona, however, the magnetic energy density $B^2/8\pi$ is much larger than the kinetic energy density $(1/2)\rho v^2$ associated with the outward flow. In the lower corona, therefore, we expect the magnetic stresses to establish near-rigid rotation in accordance with Ferraro's law. Far out in the solar wind, however, the angular velocity is certainly less than the angular velocity of the Sun and the magnetic field lines get curved as shown in Figure 14.16. It was realized by Parker (1958) in his original paper on the solar wind that the magnetic field would look like this. Hence these spirals of magnetic field are often called the *Parker spirals*.

The theory of the magnetized solar wind in the equatorial plane was worked out by Weber and Davis (1967), and then this theory was generalized by Sakurai (1985) to cover regions out of the equatorial plane. The detailed calculations are extremely complicated. The main result can be described without getting into the full calculations. The distance up to which the magnetic energy remains larger than the kinetic energy of the flow is known as the *Aflvén radius*. The plasma rotates approximately like a solid body out to the Aflvén radius. Now consider how much angular momentum is removed from the Sun by the solar wind. If there were no magnetic fields around the Sun, then the angular momentum per unit mass carried in the solar wind would have the value $\Omega_\odot R_\odot^2$ appropriate to the solar surface. But the magnetic field makes the plasma rotate at an angular velocity $\Omega_\odot$ out to the Alfvén radius r_A, where the angular momentum per unit mass is $\Omega_\odot r_A^2$. So it is this angular momentum per unit mass which is carried by the solar wind. It may be noted that the detailed analysis of Weber and Davis (1967) gave *exactly* this result (see Exercise 14.9). It is estimated that $r_A \approx 10R_\odot$. The angular momentum per unit mass carried by the solar wind is, therefore, about 100 times larger than the angular momentum per unit mass at the solar surface. Thus the solar wind is much more efficient in removing angular momentum than mass. The Sun has lost very little mass due to the solar wind since its creation. But the loss of angular momentum due to the solar wind has probably been more considerable.

Figure 14.16 A sketch of magnetic field lines in the equatorial plane of the solar wind.

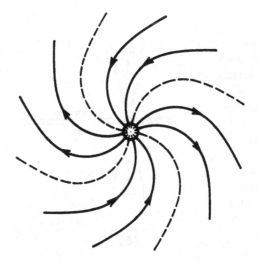

14.9.3 Jets from accretion disks

We mentioned in §6.7 that some galaxies have radio-emitting jets of vast size ejected from their central parts. The hydrodynamic model of Blandford and Rees (1974) discussed in that section had difficulties in accounting for some of the observed properties. We now describe the basic idea of a hydromagnetic jet model developed by Blandford and Payne (1982). The first question to be asked is about the energy source. Where does the energy for producing the jets come from? We pointed out in §5.7 that energy may be released at the galactic centre by the inflow of matter in an accretion disk around some massive central object (presumably a black hole). We therefore come up with the following scenario sketched in Figure 14.17(a). There is a black hole at the centre with an accretion disk in the equatorial plane with respect to the rotation axis. One manifestation of the energy release is the production of twin jets in the two polar directions. The basic problem is to understand how the energy released in the accretion disk ultimately produces such well-collimated jets. We saw in §5.7 that angular momentum has to be removed from the inner parts of the accretion disk to enable matter to fall inward. Blandford and Payne (1982) suggested the clever idea that the magnetic fields may be used to hurl a small amount of gas from the accretion disk carrying a large amount of angular momentum, the ejected gas ultimately becoming the jets. This simultaneously solves the twin problems of production of jets and removal of angular momentum from the accretion disk.

Figure 14.17(b) shows a magnetic field line PQ anchored at the point P in the accretion disk making an angle α with the equatorial plane. As discussed in §5.7, the material at P would move with the

Figure 14.17 (a) Accretion disk and jets around a central object. (b) Sketch of a magnetic field line PQ coming out of an accretion disk, indicating the various forces acting on an element of plasma at the point R on the line.

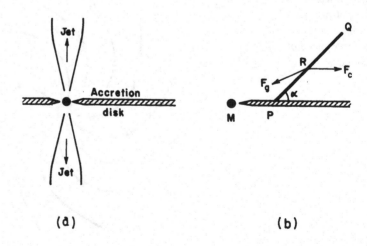

(a)   (b)

Keplerian angular velocity given by (5.40), which we write as

$$\Omega_0 = \left(\frac{GM}{r_0^3}\right)^{1/2}, \qquad (14.58)$$

where r_0 is the distance of P from the central mass M. We expect Ferraro's law to hold in regions near the accretion disk, as was the case in the inner corona of the Sun. The plasma lying on the field line PQ, therefore, moves with angular velocity Ω_0 appropriate for Keplerian motion at P. Let us use cylindrical coordinates and fix our attention on the point R on the field line. The forces acting there are the centrifugal force $\mathbf{F_c} = \Omega_0^2 r$ directed away from the rotation axis and the gravitational force $\mathbf{F_g}$ produced by the central mass M. There can also be a magnetic stress $\mathbf{j} \times \mathbf{B}/c$ which, however, plays no role for plasma motions along the magnetic field lines. Hence, if the resultant of $\mathbf{F_c}$ and $\mathbf{F_g}$ has a component along RQ in the outward direction, then the plasma at R would be hurled away from the accretion disk. To find the force arising out of $\mathbf{F_c}$ and $\mathbf{F_g}$, let us write down the corresponding potential, which is

$$\phi(r, z) = -\frac{GM}{r_0}\left[\frac{1}{2}\left(\frac{r}{r_0}\right)^2 + \frac{r_0}{(r^2 + z^2)^{1/2}}\right], \qquad (14.59)$$

where the potential for the centrifugal force has been taken as $-(1/2)\Omega_0^2 r^2$ with Ω_0 given by (14.58). We now write $r = r_0 + r'$ and expand all the terms in (14.59) assuming $|r'|, |z| < |r_0|$. Keeping terms to the order $1/r_0^2$, (14.59) gives

$$\phi(r, z) = -\frac{GM}{r_0}\left[\frac{3}{2} + \frac{3}{2}\frac{r'^2}{r_0^2} - \frac{1}{2}\frac{z^2}{r_0^2}\right]. \qquad (14.60)$$

Let s be the distance measured from P along the field line so that $r' = s\cos\alpha$, $z = s\sin\alpha$. To get the force along the field line, we have to substitute these in (14.60) and then take the negative derivative with respect to s. This leads to

$$-\frac{\partial\phi}{\partial s} = \frac{GMs}{r_0^3}(3\cos^2\alpha - \sin^2\alpha). \qquad (14.61)$$

This force is in the outward direction if $\alpha < 60°$. In other words, if a magnetic field line is anchored to an accretion disk making an angle less than 60°, then plasma is driven away along that field line.

It is not difficult to see that this mechanism would efficiently remove angular momentum like the solar wind. The magnetic stresses would force the outflowing plasma to rotate with the angular speed Ω_0 out to some Alfvén distance, beyond which magnetic energy falls below the flow kinetic energy. Hence, as in the solar wind, the angular momentum per unit mass carried by the outflow would be much larger than that at P. But it may not be obvious that the outflow ultimately gives rise to jets collimated along the polar axes. This was demonstrated by Blandford and Payne (1982) by detailed calculations. Beyond the Alfvén distance, the plasma rotates with angular velocity less than Ω_0 and the magnetic field lines get twisted. The stresses associated with these twisted field lines eventually force the outflow to be collimated along the polar axes. Blandford and Payne (1982) carried out their calculations by assuming the different magnetic field lines to be *self-similar*. If one makes no further assumptions beyond axisymmetry, then one is led to a generalized Grad–Shafranov equation (see Exercise 14.4) in the rz plane. This more general approach has been developed by Lovelace *et al.* (1986). It may be noted that a similar approach for the solar wind has been worked out by Low and Tsinganos (1986).

If the central object accreting matter has a strong magnetic field, then the accretion flow may get modified. Ghosh and Lamb (1978) developed the theory of accretion onto magnetized neutron stars.

14.10 MHD applied to weakly ionized plasmas

In §13.4 we derived the MHD model for fully ionized plasmas. We now wish to make a few comments on the applicability of the MHD model to weakly ionized plasmas. We saw in §13.1 that a weakly ionized plasma can be regarded as a mixture of three inter-penetrating fluids: the electron fluid; the ion fluid; and the neutral fluid. Let us now consider phenomena with length scales larger than the Debye length and time scales larger than the inverse of plasma frequency. Then we can neglect any charge separation so that the electron and ion fluids can together be regarded as a single ionized fluid. If the plasma is only weakly ionized, then this ionized fluid is mixed with a much denser neutral fluid. If the neutral fluid were not there, then (14.2) or (14.5) would have been the equation of motion of the ionized fluid. Since the ionized fluid must lose momentum due to collisions with the neutral atoms, we have to include a collision term as in (13.1). Let us assume the neutral fluid to be at rest. Noting that the momentum of the ionized fluid mainly resides with the ions, the momentum loss rate per unit volume for the ionized fluid can be written as $-m_i n v_c \mathbf{v} = -\rho v_c \mathbf{v}$, where v_c is the collision frequency between the ions and the neutral atoms. Adding such a term in (14.5), we have

$$\frac{d\mathbf{v}}{dt} = \mathbf{F} - \frac{1}{\rho}\nabla\left(p + \frac{B^2}{8\pi}\right) + \frac{(\mathbf{B}\cdot\nabla)\mathbf{B}}{4\pi\rho} - v_c\mathbf{v}, \qquad (14.62)$$

if the viscosity term can be neglected.

Since the interstellar medium is weakly ionized, (14.62) may be applied to study the relative motion between the ionized and the neutral components in the interstellar medium. This can be very important in the star-forming regions. A cloud collapsing due to the Jeans instability (see §7.5) usually has some angular momentum and magnetic field associated with it. We pointed out in §9.4 that the collapse makes the centrifugal force increase more rapidly than gravity. Hence, unless some angular momentum is removed from the collapsing cloud, eventually the collapse would come to a halt. We saw in §14.9.1 that the magnetic field may provide the braking necessary to slow down the rotation. But, as we saw in §14.2, the magnetic field also would become very strong if the magnetic flux is frozen in the collapsing cloud. The pressure of the very strong magnetic field also may halt the collapse. So the magnetic flux also needs to be removed. We note that the magnetic pressure force acts on the ionized fluid alone. Hence we expect the ionized fluid to spread out from the central region due to the magnetic pressure, while the neutral fluid remains unaffected. In a steady state, the outward flow of the ionized fluid given by (14.62) is

$$v_c\mathbf{v} = -\frac{1}{\rho}\nabla\left(\frac{B^2}{8\pi}\right),$$

where we have neglected the other force terms besides the magnetic pressure. This outflow of the ionized fluid through the neutral atoms is often called *ambipolar diffusion*, since this process is somewhat similar to the process of ambipolar diffusion introduced in §13.1. We now need to note that Alfvén's theorem of flux-freezing applies only to the ionized fluid and not to the neutral fluid. The ionized fluid, therefore, carries the magnetic flux with it while spreading out of the central region. Thus, ambipolar diffusion leads to a decrease of magnetic field in the star-forming region. It was Mestel and Spitzer (1956) who first pointed out this way of getting rid of the magnetic flux so that the gravitational collapse can proceed.

Exercises

14.1 Taking a dot product of the induction equation with $\mathbf{B}/4\pi$, show that

$$\frac{\partial}{\partial t}\left(\frac{B^2}{8\pi}\right) + \frac{\partial}{\partial x_i}\left(v_i\frac{B^2}{8\pi}\right) = -\mathcal{M}_{ij}\frac{\partial v_i}{\partial x_j} - \frac{j^2}{\sigma} - \frac{c}{4\pi\sigma}\nabla\cdot(\mathbf{j}\times\mathbf{B}),$$

where $\mathcal{M}_{ij}$ is as defined in (14.8). Given that $B^2/8\pi$ is the magnetic energy density per unit volume, give physical interpretations to the various terms in the equation.

14.2 Consider an ideal magnetofluid with zero viscosity, zero thermal conductivity and zero resistivity in the absence of any external body force. Show that the energy conservation equation (4.25) (with $K = 0$) is modified to

$$\frac{\partial}{\partial t}\left(\rho\epsilon + \frac{1}{2}\rho v^2\right) = -\nabla\cdot\left[\rho\mathbf{v}\left(w + \frac{1}{2}v^2\right)\right] - v_i\frac{\partial\mathcal{M}_{ij}}{\partial x_j}$$

on including the Lorentz force in the Euler equation. Now use the result of Exercise 14.1 to obtain

$$\frac{\partial}{\partial t}\left(\rho\epsilon + \frac{\rho v^2}{2} + \frac{B^2}{8\pi}\right) = -\frac{\partial}{\partial x_j}\left[v_j\left(\rho w + \frac{\rho v^2}{2} + \frac{B^2}{8\pi}\right) + v_i\mathcal{M}_{ij}\right].$$

14.3 Consider a constant initial magnetic field $\mathbf{B} = B_0\hat{\mathbf{e}}_y$ in a plasma of zero resistivity. Suppose a velocity field

$$\mathbf{v} = v_0 e^{-y^2}\hat{\mathbf{e}}_x$$

is switched on at time $t = 0$. Find out how the magnetic field evolves in time. Make a sketch of the magnetic field lines at some time after switching on the velocity field.

14.4 Consider a pressure-balanced magnetic field in which all variables are independent of z, i.e. $\partial/\partial z = 0$. As in the case of the two-dimensional velocity field in §4.8, we write the magnetic field as

$$\mathbf{B} = B_z(x,y)\hat{\mathbf{e}}_z + \nabla\times[A(x,y)\hat{\mathbf{e}}_z].$$

From (14.19), show that $p + B^2/8\pi$ has to be a function of A, which we write as

$$p + \frac{B^2}{8\pi} = F(A).$$

Show further that the function $F(A)$ has to satisfy the equation

$$\frac{\partial^2 A}{\partial x^2} + \frac{\partial^2 A}{\partial y^2} + 4\pi\frac{dF}{dA} = 0.$$

This is known as the *Grad–Shafranov equation*.

14.5 Consider a cylindrically symmetric nonlinear force-free field with the generating function $F(r)$ introduced in (14.34) given by

$$F(r) = \frac{B_0^2}{8\pi} \frac{1}{1 + (r/a)^2},$$

where B_0 and a are constants. Find out B_θ, B_z and calculate μ appearing in the force-free equation (14.30).

14.6 Use (14.44) to obtain three scalar equations of the form (14.46). Let θ be the angle between **k** and $\mathbf{B}_0$. Show that the phase velocities of the three wave modes are given by

$$v_{\text{ph}}^2 = \begin{cases} (v_A \cos \theta)^2, \\ \frac{1}{2}(c_s^2 + v_A^2) \pm \frac{1}{2}[(c_s^2 + v_A^2)^2 - 4c_s^2 v_A^2 \cos^2 \theta]^{1/2}. \end{cases}$$

Work out the eigenvectors for these three wave modes and show that the eigenvectors form a triad of orthogonal vectors.

14.7 Consider a shock wave with uniform magnetic fields B_1 and B_2 parallel to the shock front on the two sides, the velocity on both sides being perpendicular to the shock front. Figure out the conditions you have instead of (6.33–6.35) for this magnetized shock wave.

14.8 Consider a horizontal magnetic flux tube with magnetic field B and radius of cross-section a embedded in an isothermal atmosphere of perfect gas with constant gravity g. Let $\Lambda = p/\rho g$, which is a constant throughout an isothermal atmosphere. The flux tube rising due to magnetic buoyancy at speed U experiences a drag force per unit length given by

$$\tfrac{1}{2} C_D \rho U^2 a,$$

where C_D is a constant. Show that the flux tube eventually rises with an asymptotic speed

$$v_A \left(\frac{\pi a}{C_D \Lambda} \right)^{1/2},$$

where $v_A = B/\sqrt{4\pi\rho}$.

14.9 Consider the magnetized solar wind in the equatorial plane. Neglect viscosity and electrical resistivity, and assume steady state. From the induction equation, show that the components of magnetic and velocity fields at a distance r are related by

$$B_\phi = \frac{v_\phi - r\Omega_\odot}{v_r} B_r,$$

where $\Omega_\odot$ is the angular velocity of the Sun. From the ϕ component of the equation of motion, show further that

$$rv_\phi - \frac{B_r}{4\pi\rho v_r}rB_\phi = L$$

is a constant and can be interpreted as the angular momentum per unit mass carried jointly by plasma motion and magnetic stresses. Combine the above two equations to obtain

$$v_\phi = \Omega_\odot r\frac{(\mathcal{M}_A^2 L/r^2\Omega_\odot) - 1}{\mathcal{M}_A^2 - 1},$$

where $\mathcal{M}_A = \sqrt{4\pi\rho v_r}/B_r$ is the Alfvén Mach number. Argue on the basis of this equation that the angular momentum per unit mass carried away by the solar wind is

$$L = \Omega_\odot r_A^2,$$

where r_A is the Alfvén radius where $\mathcal{M}_A = 1$.

15 Theory of magnetic topologies

15.1 Introduction

In the previous chapter, we developed MHD following a pattern somewhat similar to the pattern followed earlier while developing hydrodynamics. After presenting the basic equations, we first considered the possibility of static equilibrium, and afterwards waves and instabilities were discussed. Although the mathematical analysis in the presence of a magnetic field becomes much more complicated than the corresponding analysis in the pure hydrodynamic case and consequently our discussions in Chapter 14 were often less complete than the earlier corresponding discussions in the pure hydrodynamic case, we have seen that the basic techniques and the methodology were the same.

We now wish to look at a class of MHD problems loosely called *topological problems*. Let us first consider a situation of ideal MHD, where we have a magnetofluid of zero resistivity. Then, according to Alfvén's theorem, the magnetic field is completely frozen in the plasma. We have pointed out one important consequence of Alfvén's theorem in §14.2. If two fluid elements lie on a magnetic field line, then they would always lie on one field line. We may have two far-away fluid elements in the ideal magnetofluid connected by a magnetic field line. No matter what happens to the magnetofluid or how it evolves in time, this connectivity between the two far-away fluid elements remains preserved if the resistivity is zero. The preservation of such connectivities may introduce some constraints on the dynamics of the system. Topology is a branch of mathematics in which one studies different transformations that preserve certain connectivities, and we use the word *topology* in the present context in the same sense.

Let us begin by clarifying the concept of *magnetic topologies*. Figure 15.1(a) shows two magnetic field lines M and N. The field line N can be wrapped around M as shown in Figure 15.1(b). It should be noted that no cutting or pasting of field lines was necessary in order to deform the configuration of Figure 15.1(a) to the configuration of Figure 15.1(b). We show another possible configuration of these two field lines in Figure 15.1(c). It must be clear that one has to cut and rejoin at least one field line in order to arrive at this configuration. We say that the configurations of Figures 15.1(a) and 15.1(b) are topologically equivalent, whereas the configuration of Figure 15.1(c) is topologically different from the other two. After giving the general idea, let us now give the mathematical definition. If two magnetic configurations $\mathbf{B}_1(\mathbf{x})$ and $\mathbf{B}_2(\mathbf{x})$ are such that one of them can be deformed into the other by continuous displacements *without cutting or pasting field lines anywhere*, then the two magnetic configurations are said to have the same *magnetic topology*. If this is not possible, then the magnetic topologies are different.

If a magnetofluid is ideal (i.e. has zero resistivity), then its magnetic topology can never change. So, as a result of any dynamics, it has to evolve through successive configurations which are all topologically equivalent. Suppose the plasma-β of the magnetofluid as defined in (14.20) is small. Then the system can be in equilibrium only if the force-free equation

$$\nabla \times \mathbf{B} = \mu \mathbf{B} \qquad (15.1)$$

is satisfied. Suppose the initial magnetic configuration $\mathbf{B}_i(\mathbf{x})$ is such that (15.1) is not satisfied. Then there must be initial unbalanced magnetic forces in the system, giving rise to motions which would try to bring the forces in balance. If the system has some viscosity (although no electrical resistivity), then we may expect that eventually the motions would damp out and the system would relax to an equilibrium configuration. But this can be possible only if $\mathbf{B}_i(\mathbf{x})$ has a topologically equivalent configuration for which (15.1) is satisfied and to which $\mathbf{B}_i(\mathbf{x})$ can relax. Is it always guaranteed that any arbitrary initial magnetic

Figure 15.1 Three configurations for two given magnetic field lines. The configurations (a) and (b) are topologically equivalent, whereas the configuration (c) is topologically different from them.

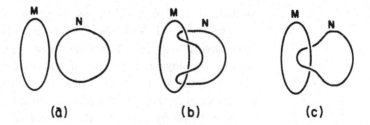

(a)   (b)   (c)

configuration $\mathbf{B}_i(\mathbf{x})$ would have a topologically equivalent configuration satisfying (15.1)? If not, then what happens? If we assume that the system will be unable to reach an equilibrium due to the topological constraints and the motions will continue forever, then also we get into difficulties. Since motions would always be associated with the viscous dissipation of energy, it would seem that the possibility of perpetual motions would be ruled out by the conservation of energy. What then happens to the system, if eventual equilibrium is ruled out by topological constraints and perpetual motions are ruled out by energy conservation? The answer to this question is still not completely known—at least not known in a way that appears convincing to everyone!

The above discussion should make it clear that topological problems are quite tricky to handle. Most of our calculations in different areas of physics are based on differential equations. Since differential equations relate various derivatives in local regions, they are the ideal tools for studying local properties of a system. On the other hand, topological connectivities of a system introduce non-local, global considerations. That is why differential equations, our most trusted tool in mathematical physics, prove inadequate in handling topological problems. Although the progress in this field has been very slow and there are few really solid results, we still decided to provide an introduction to this subject in this chapter. We believe that topological considerations are going to play increasingly important roles in plasma physics research in future—both in the laboratory context and in the astrophysical context. Magnetic topologies become particularly important in any *relaxation process*, where a plasma initially with unbalanced forces tries to find an equilibrium. The last two sections of this chapter will be devoted to two such relaxation problems, one in the laboratory and one in the solar corona.

15.2 Magnetic reconnection

We saw in the previous section that magnetic topologies are exactly preserved in a magnetofluid with zero electrical resistivity. What happens if the magnetofluid has a very small, but finite resistivity? Let us recall our discussion of ordinary hydrodynamics with very small viscosity presented in Chapter 5. In the Navier–Stokes equation (5.10), viscosity appears as a coefficient in front of the second derivative $\nabla^2 \mathbf{v}$. Even when the viscosity is small, this term may become important if there is a large velocity gradient. We saw that large velocity gradients are often present next to a solid boundary. We then have to include viscosity while treating a boundary layer, but otherwise the viscosity term can be neglected when the equations are applied to regions outside the boundary layer. Since electrical resistivity appears in the induction equation (14.10) as a coefficient in front of the second derivative $\nabla^2 \mathbf{B}$, somewhat similar considerations apply here also. Even if the electrical resistivity is small, its effect can become important in a layer where the magnetic field gradient is large. Since large gradients of magnetic field are associated with large current densities, such regions are often called *current sheets*. In a low-resistivity plasma, cutting and pasting of field lines can take place within current sheets, but the magnetic fields may be taken to be frozen in the plasma outside the current sheets and magnetic topologies are preserved everywhere except in the current sheets.

Figure 15.2 shows a typical current sheet with oppositely directed magnetic fields above and below. The large value of $\nabla^2 \mathbf{B}$ in the central region would make the electrical resistivity term important there and hence the magnetic field would decay away in the central region. Since magnetic fields have the pressure $B^2/8\pi$ associated with them, a decrease in the magnetic field would cause a pressure decrease in the central region. If the plasma-β defined in (14.20) turns out to be very large, then the magnetic pressure constitutes a negligible fraction of the total pressure and a decrease in the magnetic pressure is not expected to have any dramatic consequence. But if the plasma-β is of the order of 1 or smaller, then the decay of the magnetic field in the central region would cause an appreciable depletion of the total

Figure 15.2 Magnetic reconnection in a current sheet. See the text for explanations.

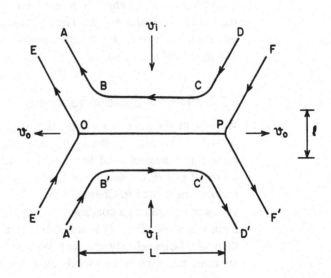

pressure there, and we expect that the plasma from above and below with fresh magnetic fields would be sucked into the central region. This fresh magnetic field would then decay and more plasma from above and below would be sucked in to compensate for the pressure decrease due to this decay. This process, known as *magnetic reconnection* or *neutral-point reconnection*, may go on as long as fresh magnetic fields are brought to the central region. Although Dungey (1953) hinted at the possibility of such a process, the first formulation was due to Parker (1957) and Sweet (1958).

Let us look at Figure 15.2 more carefully to understand the physics of magnetic reconnection. The field lines $ABCD$ and $A'B'C'D'$ are moving with inward velocity v_i towards the central region. Eventually the central parts BC and $B'C'$ of these field lines decay away. The part AB is moved to EO and the part $A'B'$ to $E'O$. These parts originally belonging to different field lines now make up one field line EOE'. Similarly the parts CD and $C'D'$ eventually make up the field line FPF'. We thus see that the cutting and pasting of field lines take place in the central region. Since plasmas from the top and the bottom in Figure 15.2 push against the central region, the plasma in the central region is eventually squeezed out sideways through the points O and P. Let v_o be the outward velocity with which reconnected field lines EOE' and FPF' move away from the reconnection region. Our aim now is to estimate the incoming velocity v_i, which essentially gives the rate at which the reconnection proceeds.

Carrying a full mathematical analysis of the problem is extremely difficult. Parker (1957) replaced the differential equations by approximate algebraic equations and showed that rough estimates of various quantities can be obtained in that fashion. We also follow that approach here. Let L be the width of the central region over which the magnetic field decays, as indicated in Figure 15.2. After the magnetic field has decayed, the field-free plasma is squeezed through the points O and P. If l be the thickness of the outflowing plasma, then the equation of continuity (14.1) can be replaced by the approximate mass conservation condition

$$v_i L \approx v_o l. \tag{15.2}$$

Since the magnetic stresses drive the outward flow, we expect that the kinetic energy of the outward flow should be comparable to the magnetic energy around the current sheet. We can therefore write

$$\frac{1}{2}\rho v_o^2 \approx \frac{B^2}{8\pi}, \tag{15.3}$$

which replaces the equation of motion (14.2). Lastly, we have to consider the induction equation (14.10), which in the steady state becomes

$$\nabla \times (\mathbf{v} \times \mathbf{B}) + \lambda \nabla^2 \mathbf{B} = 0.$$

The diffusion term, which is of order $\lambda B/l^2$, has to be balanced by the other term corresponding to the supply of fresh magnetic flux at velocity v_i. Since this term $\nabla \times (\mathbf{v} \times \mathbf{B})$ should be of order $v_i B/l$, we have

$$\frac{v_i B}{l} \approx \frac{\lambda B}{l^2}. \tag{15.4}$$

We now carry out an approximate analysis based on (15.2–15.4), which replace the full equations of MHD.

It follows from (15.3) that

$$v_o \approx v_A, \tag{15.5}$$

where

$$v_A = \frac{B}{\sqrt{4\pi\rho}} \tag{15.6}$$

is the Alfvén speed introduced in (14.43). We also have

$$l \approx \frac{\lambda}{v_i} \tag{15.7}$$

from (15.4). Substituting (15.5) and (15.7) in (15.2), we obtain

$$v_i \approx \frac{l v_o}{L} \approx \frac{\lambda v_A}{v_i L}$$

so that

$$v_i^2 \approx \frac{v_A^2}{v_A L/\lambda}. \tag{15.8}$$

It should be noted that $v_A L/\lambda$ here is the appropriate magnetic Reynolds number $\mathscr{R}_M$ as defined in (14.12). We can therefore write (15.8) as

$$v_i \approx \frac{v_A}{\sqrt{\mathscr{R}_M}}. \tag{15.9}$$

We have seen in §14.2 that $\mathscr{R}_M$ is typically a very large number in astrophysical situations. Therefore (15.9) implies that the reconnection proceeds at a rate which is a tiny fraction of the Alfvén speed—the speed at which magnetic disturbances propagate. The reconnection rate given by (15.9) is known as the *Sweet–Parker reconnection rate* (Sweet 1958; Parker 1957).

Often one observes gigantic explosions on the surface of the Sun, known as *solar flares*. In a large flare, an energy of the order of 10^{32} erg may be released in a few minutes. It is believed that the complicated magnetic fields above the Sun's surface occasionally lead to the sudden formation of current sheets, and the enormous amount of magnetic energy dissipated very rapidly in these current sheets is what gives rise to a flare. In order for such a large amount of energy to be released in such a short time, the reconnection has to proceed fairly fast—at rates comparable to the Alfvén speed. The Sweet–Parker rate appears to be quite inadequate for the purpose of explaining flares. Hence attempts have been made to suggest scenarios in which the reconnection proceeds at a faster rate. One of the famous models for faster reconnection is due to Petschek (1964). We shall not discuss this model here, since the mathematics is slightly involved. The scenario is somewhat different from that shown in Figure 15.2, and the final reconnection rate turns out to be

$$v_i \approx \frac{v_A}{\log \mathscr{R}_M}. \tag{15.10}$$

This is known as the *Petschek reconnection rate*. When $\mathscr{R}_M$ is sufficiently large, it is easy to see that the Petschek rate given by (15.10) would correspond to a much faster inflow compared to the Sweet–Parker rate given by (15.9). Priest and Forbes (1986) developed a unified theory in which the Sweet–Parker and Pestschek rates follow as special cases. Since the reconnection rate may depend quite sensitively on the boundary conditions far away from the current sheet, building detailed and realistic models of magnetic reconnection is an extremely challenging theoretical problem. We have given only a very brief introduction to this complex subject. Furth, Killeen and Rosenbluth (1963) showed in a complicated paper that large current sheets may become unstable under certain circumstances. This instability, known as the *tearing mode instability*, breaks up the current sheet into many islands as sketched in Figure 15.3. When this instability occurs, the reconnection process becomes even more complicated.

Figure 15.3 A current sheet broken up by the tearing mode instability.

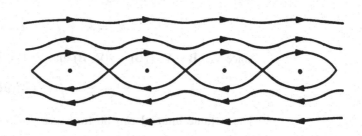

15.3 Magnetic helicity. Woltjer's theorems

We pointed out in §15.1 that differential equations are inadequate for handling problems involving magnetic topologies, because a magnetic topology is of the nature of a global property of the system and differential equations are local equations. We now discuss one integral quantity known as *magnetic helicity*, which is global in nature and has a close correspondence with topology.

Since the magnetic field is a solenoidal vector, it can be written as the curl of another vector field, i.e.

$$\mathbf{B} = \nabla \times \mathbf{A}, \tag{15.11}$$

where $\mathbf{A}$ is known as the *vector potential*. It is well known that the vector potential is not unique. Consider another vector field

$$\mathbf{A}' = \mathbf{A} + \nabla\chi, \tag{15.12}$$

where χ is any arbitrary scalar field. We note that $\mathbf{A}'$ would be as good a vector potential for $\mathbf{B}$ as $\mathbf{A}$. This arbitrariness in defining the vector potential is known as the *gauge freedom*.

We now consider a magnetic field in a region such that all field lines close within that region, with no field lines crossing the boundary. The condition for this is that the normal component $B_n = 0$ on the bounding surface of the region. The magnetic helicity of the magnetic field in this region is defined as

$$\mathcal{H} = \int \mathbf{A} \cdot \mathbf{B}\, dV, \tag{15.13}$$

where the integration has to be done over the whole region of which the boundary satisfies the condition $B_n = 0$. Let us now assume that we use the vector potential $\mathbf{A}'$ given by (15.12) instead of $\mathbf{A}$ to define the magnetic helicity, i.e.

$$\mathcal{H}' = \int \mathbf{A}' \cdot \mathbf{B}\, dV. \tag{15.14}$$

We now show that the magnetic helicity is gauge-invariant by proving that $\mathcal{H}' = \mathcal{H}$. Substituting for $\mathbf{A}'$ from (15.12) into (15.14), we get

$$\mathcal{H}' = \mathcal{H} + \int \nabla\chi \cdot \mathbf{B}\, dV. \tag{15.15}$$

Since $\nabla\chi \cdot \mathbf{B} = \nabla \cdot (\chi\mathbf{B}) - \chi(\nabla \cdot \mathbf{B})$ and $\nabla \cdot \mathbf{B} = 0$, we write

$$\mathcal{H}' = \mathcal{H} + \int \nabla \cdot (\chi\mathbf{B})\, dV.$$

The volume integral can be transformed by Gauss's theorem to the surface integral $\int \chi \mathbf{B} \cdot d\mathbf{S}$ over the surface bounding this volume. Since $B_n = 0$ everywhere on this surface, the surface integral vanishes so that

$$\mathscr{H}' = \mathscr{H}.$$

Thus magnetic helicity does not change on changing the gauge. It is of the nature of a real physical quantity, which is independent of the choice of gauge. Berger and Field (1984) proposed a scheme of extending the concept of magnetic helicity even to systems for which B_n is not equal to zero on the boundary. Here, however, we restrict our discussion only to systems within which all the magnetic field lines close.

To understand the connection between magnetic helicity and topology, let us consider an example discussed by Moffatt (1969). We calculate the magnetic helicity of the simple system shown in Figure 15.4, where we have two interconnected tubes C_1 and C_2 within which there are magnetic fields, but no magnetic field exists outside. Let the magnetic fluxes through these two tubes be Φ_1 and Φ_2 respectively. To find the magnetic helicity of the total system, let us consider a small volume element dV of tube C_1 of which the contribution to magnetic helicity is $\mathbf{A} \cdot \mathbf{B}\, dV$. Replacing $\mathbf{B}\, dV$ by $\Phi_1\, d\mathbf{x}$, the contribution from the whole of the tube C_1 is

$$\mathscr{H}_1 = \Phi_1 \oint_{C_1} \mathbf{A} \cdot d\mathbf{x}. \tag{15.16}$$

The line integral $\oint_{C_1} \mathbf{A} \cdot d\mathbf{x}$ can be transformed by the Stokes's theorem to the surface integral $\int (\nabla \times \mathbf{A}) \cdot d\mathbf{S} = \int \mathbf{B} \cdot d\mathbf{S}$, which is equal to the magnetic flux passing through the circuit C_1 and which turns out to be Φ_2 in the present case. From (15.16), we therefore have

$$\mathscr{H}_1 = \Phi_1 \Phi_2.$$

Figure 15.4 Two interconnected magnetic flux tubes of which the magnetic helicity is calculated in the text.

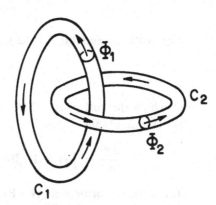

Since the other tube C_2 would make the same contribution to magnetic helicity, the magnetic helicity of the whole system is obviously given by

$$\mathcal{H} = 2\Phi_1\Phi_2. \qquad (15.17)$$

The significance of this result is that the magnetic helicity depends only on the fact that the two fluxes Φ_1 and Φ_2 are interlinked. The value of the magnetic helicity does not change if we deform the two tubes C_1 and C_2 as long as the linkage remains the same. If, however, one tube could be cut and removed so that the linkage between C_1 and C_2 were broken, then we readily see that the magnetic helicity would go to zero. We thus note that there is a direct correspondence between magnetic helicity and topology. As long as the topology does not change, the magnetic helicity is an invariant.

We know that the magnetic topology is preserved in a magnetofluid with zero resistivity. Since the magnetic helicity is directly related to the topology, we expect that the magnetic helicity also would be an invariant for an ideal magnetofluid. As it happens, this was established by Woltjer (1958) as a mathematical theorem a few years before the topological significance of magnetic helicity was generally appreciated. We now give that mathematical proof. The rate of change of magnetic helicity follows from differentiating (15.13), i.e.

$$\frac{d\mathcal{H}}{dt} = \int \frac{\partial}{\partial t}(\mathbf{A} \cdot \mathbf{B})\,dV = \int \frac{\partial \mathbf{A}}{\partial t} \cdot \mathbf{B}\,dV + \int \mathbf{A} \cdot \frac{\partial \mathbf{B}}{\partial t}\,dV. \qquad (15.18)$$

It may be noted that the Eulerian derivative of any local quantity integrated over the whole volume gives the Lagrangian variation of the total integrated quantity. For zero resistivity, the evolution of magnetic field is given by

$$\frac{\partial \mathbf{B}}{\partial t} = \nabla \times (\mathbf{v} \times \mathbf{B}). \qquad (15.19)$$

The evolution equation of the vector potential $\mathbf{A}$ may be taken to be

$$\frac{\partial \mathbf{A}}{\partial t} = \mathbf{v} \times \mathbf{B}, \qquad (15.20)$$

of which the curl gives (15.19). Substituting (15.19) and (15.20) into (15.18), we obtain

$$\frac{d\mathcal{H}}{dt} = \int (\mathbf{v} \times \mathbf{B}) \cdot \mathbf{B}\,dV + \int \mathbf{A} \cdot \nabla \times (\mathbf{v} \times \mathbf{B})\,dV.$$

The triple vector product $(\mathbf{v} \times \mathbf{B}) \cdot \mathbf{B}$ is equal to zero. Transforming the second term with the vector identity (A.7),

$$\frac{d\mathscr{H}}{dt} = \int \nabla \cdot [(\mathbf{v} \times \mathbf{B}) \times \mathbf{A}] \, dV + \int (\mathbf{v} \times \mathbf{B}) \cdot (\nabla \times \mathbf{A}) \, dV.$$

Again the second term is zero and the first term can be transformed to a surface integral by Gauss's theorem so that

$$\frac{d\mathscr{H}}{dt} = \oint [(\mathbf{v} \times \mathbf{B}) \times \mathbf{A}] \cdot d\mathbf{S}. \tag{15.21}$$

If there are no flows or no outgoing magnetic field lines through the bounding surface, then it is easily seen that the surface integral vanishes. Hence

$$\frac{d\mathscr{H}}{dt} = 0 \tag{15.22}$$

if the magnetofluid has zero resistivity. This completes our proof.

Woltjer (1958) proved another intriguing theorem of which the significance was appreciated much later. Let us try to find the minimum of the magnetic energy

$$W = \int \frac{B^2}{8\pi} \, dV \tag{15.23}$$

for a system subject to the constraint that its magnetic helicity is kept constant. Woltjer (1958) showed that the solution of the problem is a linear force-free field, which minimizes the energy when the magnetic helicity is constant. According to the principles of variational calculus, the solution would be given by

$$\delta W - \frac{\mu}{8\pi} \delta \mathscr{H} = 0, \tag{15.24}$$

where $\mu/8\pi$ is the Lagrange multiplier. From the expressions (15.13) and (15.23) for $\mathscr{H}$ and W, we have

$$\delta W - \frac{\mu}{8\pi} \delta \mathscr{H} = \int \left[\frac{\mathbf{B} \cdot \delta \mathbf{B}}{4\pi} - \mu \frac{\delta \mathbf{A} \cdot \mathbf{B}}{8\pi} - \mu \frac{\mathbf{A} \cdot \delta \mathbf{B}}{8\pi} \right] dV. \tag{15.25}$$

Since $\delta \mathbf{B} = \nabla \times \delta \mathbf{A}$, the last term on the R.H.S. is equal to

$$\int \frac{\mathbf{A} \cdot \delta \mathbf{B}}{8\pi} dV = \int \frac{\mathbf{A} \cdot (\nabla \times \delta \mathbf{A})}{8\pi} dV$$

$$= \int \left[\frac{\nabla \cdot (\delta \mathbf{A} \times \mathbf{A})}{8\pi} + \frac{\delta \mathbf{A} \cdot (\nabla \times \mathbf{A})}{8\pi} \right] dV.$$

on making use of (A.7). The first term on the R.H.S. can again be transformed by Gauss's theorem to a surface integral, which vanishes if the variations on the bounding surface are zero. Then

$$\int \frac{\mathbf{A} \cdot \delta \mathbf{B}}{8\pi} dV = \int \frac{\delta \mathbf{A} \cdot \mathbf{B}}{8\pi} dV.$$

On substituting this in (15.25),

$$\delta W - \frac{\mu}{8\pi} \delta \mathcal{H} = \int \frac{\mathbf{B}}{4\pi} \cdot [\delta \mathbf{B} - \mu \delta \mathbf{A}] \, dV. \qquad (15.26)$$

It follows from (15.26) that the condition (15.24) is satisfied if

$$\mathbf{B} = \mu \mathbf{A},$$

of which the curl is

$$\nabla \times \mathbf{B} = \mu \mathbf{B}, \qquad (15.27)$$

where μ, being the Lagrange multiplier, is a constant so that (15.27) is the equation of linear force-free field. A magnetic configuration satisfying this equation makes the magnetic energy minimum subject to the constancy of magnetic helicity. Thus the proof of Woltjer's second theorem is completed. The possible physical significance of this theorem is discussed in the next section.

15.4 Taylor's theory of plasma relaxation

In a paper of remarkable physical intuition and insight, Taylor (1974) pointed out that Woltjer's second theorem may actually be quite relevant in plasma relaxation processes taking place in Nature. Suppose a plasma with low resistivity is suddenly subjected to magnetic fields and currents. Initially the magnetic stresses in the plasma would be out of balance, resulting in internal motions until the plasma eventually relaxes to an equilibrium. If it is a low-β plasma, Taylor (1974) proposed that one can take the magnetic helicity to be a constant during the plasma relaxation, while magnetic energy becomes a minimum. Hence, in accordance with Woltjer's second theorem, the final state after plasma relaxation would be a linear force-free field satisfying (15.27), where μ is determined by the fact that magnetic helicity has to have the same value in the final state as it had initially.

Taylor's proposal was of the nature of a hypothesis or a conjecture. It is not something which can be proved mathematically. But let us give some arguments in support of the proposal. We can make a Fourier decomposition of the vector potential in the following way

$$\mathbf{A}(\mathbf{x}) = \frac{1}{(2\pi)^{3/2}} \int \tilde{\mathbf{A}}(\mathbf{k}) \exp(i\mathbf{k} \cdot \mathbf{x}) \, d^3 k \qquad (15.28)$$

so that the magnetic field is given by

$$\mathbf{B}(\mathbf{x}) = \frac{i}{(2\pi)^{3/2}} \int \mathbf{k} \times \tilde{\mathbf{A}}(\mathbf{k}) \exp(i\mathbf{k} \cdot \mathbf{x}) \, d^3 k. \qquad (15.29)$$

Then the magnetic energy can easily be shown to be given by

$$W = \frac{1}{8\pi} \int |\mathbf{k} \times \tilde{\mathbf{A}}(\mathbf{k})|^2 \, d^3k, \tag{15.30}$$

whereas the magnetic helicity is

$$\mathscr{H} = i \int \tilde{\mathbf{A}}^*(\mathbf{k}) \cdot [\mathbf{k} \times \tilde{\mathbf{A}}(\mathbf{k})] \, d^3k. \tag{15.31}$$

It follows from (15.30) that the spectrum of magnetic energy roughly goes as $\approx k^2 \tilde{A}^2(\mathbf{k})$, while the spectrum of magnetic helicity goes as $\approx k\tilde{A}^2(\mathbf{k})$ in accordance with (15.31). The higher wavenumbers therefore have a greater weight in the spectrum of magnetic energy than in the spectrum of magnetic helicity. Since the resistive diffusion of the magnetic field is due to the $\lambda \nabla^2 \mathbf{B}$ term in the induction equation (14.10), regions with sharper gradients such as the current sheets are the places where the resistive diffusion takes place. In the language of the Fourier space, it is the higher wavenumber modes which are more affected by the resistive diffusion. Since these modes are more important in magnetic energy rather than in magnetic helicity, we naturally conclude that the effect of resistivity on magnetic helicity must be small compared to its effect on magnetic energy. Hence, while magnetic energy decreases due to resistive decay, the effect on the magnetic helicity would be much less. During the plasma relaxation process, we may therefore expect the magnetic energy to become a minimum, while the magnetic helicity is not changed much.

In spite of the above arguments justifying Taylor's proposal, we do not have a sound proof of the proposal, and the best way of verifying the proposal is to look for experimental evidence. Taylor's theory had a spectacular success in explaining some puzzling experimental data, which had remained a mystery till that time. Suppose we have a cylindrical column of plasma. At some time, we drive a current I along the axis of the plasma. Such a current would heat the plasma and also try to pinch the plasma column by giving rise to a magnetic field in the θ direction. We saw in §14.3.2 that such a magnetic configuration would be unstable unless there is a magnetic field along the axis as well. So, when driving the current I, we also set up a magnetic field $\overline{B}$ along the axis by sending currents through external coils. This sudden introduction of the current and the magnetic field usually results in violent internal motions in the plasma, until the plasma eventually relaxes. Such experiments are being done for a long time. The important parameter in the problem turns out to be

$$\theta_{\mathrm{p}} = \frac{2I}{ca\overline{B}}, \tag{15.32}$$

where a is the radius of the plasma column. When θ_p exceeds the critical value of about 1.4, it is found experimentally that the magnetic field component B_z (where z is taken along the axis of the plasma column) in the regions away from the central axis has a reverse direction after relaxation compared to the direction in which the magnetic field $\overline{B}$ was put. This phenomenon, known as the *reversed field pinch*, was a great surprise when it was first discovered in the late 1950s (Honsaker *et al.* 1958). It may be noted that most of the plasma experiments use a torus geometry rather than a straight cylinder geometry. Since a part of the torus can be approximated as a cylinder, we shall discuss in the language of cylindrical geometry to keep things simpler.

According to Taylor's theory, the relaxed plasma configuration would be described by the linear force-free equation (15.27) in cylindrical geometry. We wrote down the appropriate solution in (14.33), which is

$$B_z = B_0 J_0(\mu r), \qquad B_\theta = B_0 J_1(\mu r). \qquad (15.33)$$

Figure 15.5 shows a plot of these field components. Since the Bessel function $J_0(\mu r)$ changes sign if the argument is larger than 2.404, we get a natural explanation for the reversed field pinch if $\mu a > 2.404$. Let us now consider what determines μ. Using (13.53), we can write down the force-free equation (15.27) as

$$\frac{4\pi}{c}\mathbf{j} = \mu \mathbf{B}.$$

On integrating this equation across a cross-section of the cylinder,

Figure 15.5 The Bessel function solution of the linear force-free field in a cylinder, as given by (15.33).

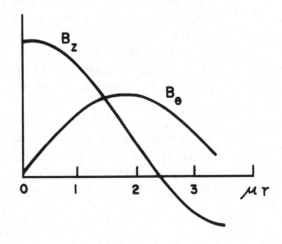

we obtain

$$\frac{4\pi}{c}I = \mu\pi a^2 \overline{B}, \tag{15.34}$$

where $\overline{B}$ is the average magnetic field put in the plasma. From (15.34), we can obtain μ in terms of I, $\overline{B}$ and a. From (15.32) and (15.34), we have

$$\theta_p = \tfrac{1}{2}\mu a. \tag{15.35}$$

The theoretical critical value of the parameter θ_p is then 1.202, beyond which field reversal is possible. This compares well with the experimental value of about 1.4. On substituting for B_z from (15.33), the average magnetic field $\overline{B}$ is found to be

$$\overline{B} = \frac{1}{\pi a^2}\int_0^a B_z \cdot 2\pi r \, dr = \frac{2B_0}{\mu a}J_1(\mu a). \tag{15.36}$$

One parameter which turns out to be very useful in the analysis of experimental data is the ratio of B_z at the outer edge to its average value

$$F = \frac{B_z(r = a)}{\overline{B}} = \frac{\mu a}{2}\frac{J_0(\mu a)}{J_1(\mu a)}$$

from (15.33) and (15.36). Then, using (15.35), we have

$$F = \theta_p \frac{J_0(2\theta_p)}{J_1(2\theta_p)}. \tag{15.37}$$

This equation gives the relation between the parameters F and θ_p

Figure 15.6 The plot of F against θ_p according to Taylor's hypothesis (the solid curve) along with some experimental data points. Adapted from Bodin and Newton (1980).

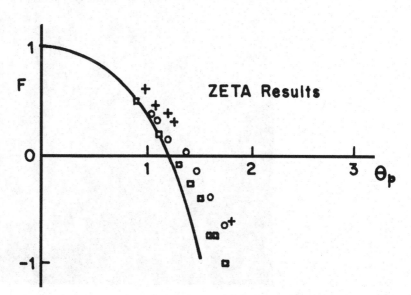

according to Taylor's theory. Figure 15.6 shows this theoretical relation between F and θ along with experimental data points obtained by measuring F for experimental setups with different θ_p.

The constancy of the magnetic helicity implies some constraints on magnetic topologies. The field reversal in pinch experiments is one of the most dramatic consequences of these topological constraints. Although the detailed dynamics of the plasma during the relaxation phase is extremely complicated, the beauty of Taylor's theory is that it predicts the final state after relaxation in terms of the initial quantities such as the current I and the magnetic field $\overline{B}$ put in the plasma.

15.5 Parker's theory of coronal heating

We now look at one astrophysical problem in which topological considerations play an important role. We have already pointed out in §4.4 that the solar corona is much hotter than the surface—the temperatures of the hottest regions being of the order 2×10^6 K. The

Figure 15.7 A soft X-ray image of the Sun obtained from the Japanese spacecraft *Yohkoh*. Courtesy: T. Sakurai.

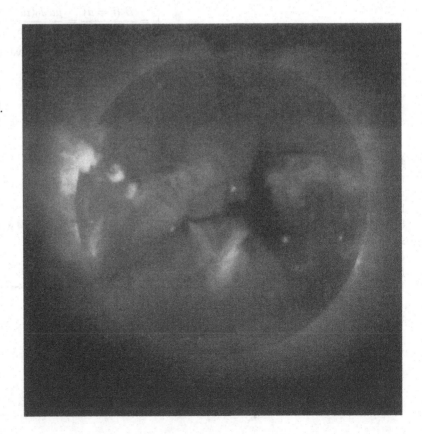

temperature distribution happens to be quite non-uniform. This is seen very dramatically in the soft X-ray images of the Sun taken from spacecrafts. One such image is shown in Figure 15.7. The solar surface can roughly be modelled as a black body with a temperature of about 6000 K. The surface looks dark in the X-ray image, since a black body of 6000 K emits very little X-radiation. The X-rays come mainly from those parts of the corona where the temperatures are of the order of million degrees and higher. On comparing the X-ray images with ordinary photographs of the Sun, it was found that the corona was hottest above those regions of the surface where lots of sunspots were seen. From such connections, it becomes evident that the magnetic fields spreading into the corona from the sunspots must be playing an important role in heating the corona. We indicated in Figure 14.10(b) that the magnetic field above a bipolar magnetic region on the solar surface exists in the form of a loop. Some loops can be seen in the X-ray images of the Sun like Figure 15.7. These are identified as the magnetic loops. A theory of coronal heating should therefore explain why magnetic loops seem to be the hottest regions in the corona.

The corona is made up of very tenuous gas. The particle number density at a distance $R_\odot$ above the solar surface is 10^6 cm^{-3} compared to about 10^{17} cm^{-3} in the photosphere. The high temperature essentially means that the kinetic energies of the gas particles are very high, although the actual heat capacity may not be very great due to the low density. From the radiation loss in the corona, one can estimate the rate at which heat should be supplied to the corona in order to compensate for this loss. This turns out to be of the order 10^7 erg cm^{-2} s^{-1} for the hottest regions of the corona. For comparison, the radiation emitted at the solar surface is 6.27×10^{10} erg cm^{-2} s^{-1}. The energy needed to heat the corona is, therefore, a very tiny fraction of the energy coming out of the Sun.

The initial theories of coronal heating were based on the idea that acoustic waves are generated by convective turbulence just below the photosphere and then they propagate to the corona, where they undergo viscous dissipation, producing the heat. We have shown in §6.4 that acoustic waves can steepen into shocks. Detailed calculations indeed showed that acoustic waves in the solar atmosphere steepen into shocks and thereafter dissipate too quickly (due to the large value of $\nu \nabla^2 \mathbf{v}$ at the shock waves) in the lower corona, before reaching the heights where the heat is to be deposited. It also became increasingly clear that magnetic fields have an important role in heating the corona. Osterbrock (1961) started the investigation whether MHD waves propagating along coronal magnetic fields could carry the necessary energy for heating the corona. We discussed the Alfvén mode in §14.5, which is non-compressive and does not steepen into shocks, unlike the compressive modes in which such steepenings take place quickly. It seems likely that those regions of the corona which have open magnetic field lines are heated by such non-compressive Alfvénic waves. But the closed magnetic loops are much hotter and it seems likely that they are heated by a different mechanism. In a famous and still controversial paper, Parker (1972) suggested a scenario as to how the closed magnetic loops may be heated.

Above the photosphere, the plasma-β defined by (14.20) turns out to be small so that the magnetic field controls the dynamics. On the other hand, the magnetic field is not dynamically important below the photosphere and is passively carried by the fluid motions. Any magnetic field line in the loop has its ends going underneath the photospheric surface at the two ends. The footpoints of the field lines in the photosphere are moved around by convective motions taking place below the photosphere, and the magnetic fields above the photosphere try to relax to new equilibrium configurations continuously as a result of changing footpoint positions. In order to study the dynamics of the system, it is *not* necessary to consider what is happening below the photospheric surface. We study the relaxation of the magnetic fields above the photosphere, taking the photosphere merely as the bounding surface on which the footpoints are moving. Parker (1972) argued that the curvature of the magnetic loop should not play any important role in the process of relaxation. Hence, just as torus geometries in laboratory plasma experiments are often modelled by cylindrical geometry, we take a straight magnetic loop with two surfaces at the two edges corresponding to the photospheric surfaces, as shown in Figure 15.8.

Figure 15.8 The twisting of magnetic field lines by footpoint motions. (a) An initial configuration with straight field lines. (b) The subsequent configuration after the bottom footpoint of one field line has been moved.

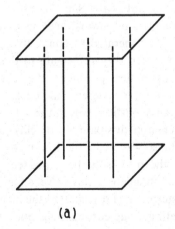

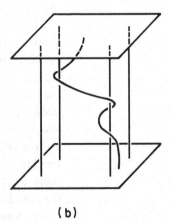

(a)   (b)

Let us consider the simplest problem in which initially all the magnetic field lines are assumed straight as in Figure 15.8(a). We expect the footpoints at both the bounding surfaces to be moved continuously. To have an idea of what happens, let us consider the hypothetical situation that the bottom footpoint of only one field line is moved in a way that gives rise to the configuration sketched in Figure 15.8(b). If we consider the topologies of the magnetic fields between the end surfaces, it is clear that the topologies of Figure 15.8(a) and Figure 15.8(b) are *not* equivalent. If we do not move the footpoints, the configuration of Figure 15.8(a) cannot be deformed to the configuration of Figure 15.8(b) without cutting or pasting field lines. If the plasma were perfectly conducting, then the configuration of Figure 15.8(b) would try to relax to a topologically equivalent configuration satisfying the equilibrium equation (14.19), which can be written as

$$-\nabla\left(p + \frac{B^2}{8\pi}\right) + \frac{(\mathbf{B}\cdot\nabla)\mathbf{B}}{4\pi} = 0 \qquad (15.38)$$

on using (14.4).

Since a magnetic loop in the coronal plasma is a system with very high $\mathscr{R}_M$, let us first assume that the magnetic fields are perfectly frozen and see what conclusions follow from that. Choosing the z axis along the direction of the initial magnetic field, we write the magnetic field as

$$\mathbf{B} = B_0\,\hat{\mathbf{e}}_z + \mathbf{b}, \qquad (15.39)$$

where B_0 is the initial uniform field and $\mathbf{b}$ is the departure from it caused by the footpoint motions. Let us now substitute (15.39) into (15.38) and throw away terms quadratic in $\mathbf{b}$, which is assumed to be small compared to $B_0\hat{\mathbf{e}}_z$. Remembering that the spatial derivatives of B_0 are zero, we get

$$-\nabla\left(p + \frac{B_0 b_z}{4\pi}\right) + \frac{B_0}{4\pi}\frac{\partial \mathbf{b}}{\partial z} = 0. \qquad (15.40)$$

We take the divergence of this equation, keeping in mind that $\nabla.\mathbf{b} = 0$. This gives the Laplace equation

$$\nabla^2\left(p + \frac{B_0 b_z}{4\pi}\right) = 0. \qquad (15.41)$$

Now let us consider an infinite plane passing through the middle of our system and parallel to the end surfaces. We must have (15.41) holding everywhere on this plane. There are some well-known analytic properties of the Laplace equation, which we can now invoke. For example, the solution of the Laplace equation has to be a constant if we do not allow the solution to blow up anywhere including infinity. We therefore must have

$$p + \frac{B_0 b_z}{4\pi} = \text{constant}, \qquad (15.42)$$

since we do not expect this quantity to blow up anywhere within the infinite plane we are considering. From (15.40) and (15.42), it readily follows that

$$\frac{\partial \mathbf{b}}{\partial z} = 0. \qquad (15.43)$$

Let us now try to interpret the mathematics we have just gone through. We have concluded that the relaxed equilibrium configuration should satisfy (15.43), i.e. it must be invariant in the z direction. Now let us look carefully at Figure 15.8(b). We see one field line wrapped around two field lines in such a way as to make it clear that no topologically equivalent configuration is possible which is invariant in the z direction. We are therefore in a very peculiar situation. The equilibrium has to satisfy a condition which apparently cannot be met by any topologically equivalent configuration. What do we make out of this? The above mathematical analysis tacitly assumes that all the quantities appearing in the equations are continuous, differentiable and generally analytically well behaved. Parker (1972) argued that the only way out of the difficulties is to suppose that some physical quantities have discontinuities so that the above analysis does not hold. If this argument is correct, then it may be possible to have relaxed equilibrium configurations with discontinuities in them such that (15.43) is circumvented. Now, in this ideal model with zero resistivity, we are led to the possibility of discontinuities. In reality, since the plasma has a small resistivity, the discontinuities in the magnetic field would be current sheets, in which magnetic reconnection would take place. In a magnetic reconnection, some magnetic field is dissipated, and presumably the magnetic energy goes into heat. If our ideal theory predicts the occurrence of discontinuities, then in reality those would be the regions where heat would be produced due to magnetic reconnection.

But are the above arguments of Parker (1972) correct? From the time of Parker's first paper, a fierce controversy has raged on this subject, and still there is no common view which is accepted by everybody. In this elementary textbook, we merely reproduced the arguments of the original paper (Parker 1972) without getting into the subtleties of the subsequent arguments. The present author's view is that Parker's ideas are basically correct, although the argument given above is not fully satisfactory. As the footpoints of a magnetic loop in the solar corona disturb the magnetic configurations above, current sheets form as a result of the plasma trying to reach equilibrium configurations. The magnetic energy dissipated in these current sheets heat the coronal loops. Readers desirous of learning more about the subject should consult the recent monograph by Parker (1994).

Exercises

15.1 If the magnetic field inside a volume satisfies the force-free equation (15.27), show that the magnetic helicity $\mathscr{H}$ and the magnetic energy W as defined by (15.13) and (15.23) respectively satisfy the equation

$$8\pi W = \mu\mathscr{H} + \oint_S (\mathbf{A} \times \mathbf{B}) \cdot d\mathbf{S},$$

where the last term is a surface integral over the surface bounding the volume under consideration.

15.2 Consider a cylindrical plasma column of radius a within which the magnetic field has relaxed to the configuration (15.33) according to Taylor's hypothesis. If Φ is the magnetic flux through the column and $\mathscr{H}$ the helicity per unit length, show that the force-free constant μ for the relaxed field is given by

$$2\pi a \frac{\mathscr{H}}{\Phi^2} = \frac{\mu a [J_0^2(\mu a) + J_1^2(\mu a)] - 2J_0(\mu a)J_1(\mu a)}{J_1^2(\mu a)}.$$

16 Dynamo theory

16.1 Magnetic fields in the Cosmos

Around 1600 William Gilbert, physician to Queen Elizabeth I of England, proposed a bold hypothesis to explain why a suspended compass needle points in the north–south direction. He suggested that the whole Earth is a huge magnet and attracts the compass needle. This is probably the first time that somebody proposed an astronomical object—the planet Earth—to have a large-scale magnetic field. Initially it was thought that the Earth's magnetism was of ferromagnetic origin. By the end of the nineteenth century, it became clear that a ferromagnetic substance does not retain the magnetism when heated beyond a certain temperature (the Curie point). Since the interior of the Earth is believed to be hotter than the Curie temperature of any known ferromagnetic substance, it was apparent that one has to look for alternative explanations for the Earth's magnetic field.

Until the beginning of the twentieth century, it was not known whether other astronomical objects have magnetic fields as well. When Hale (1908) made the momentous discovery of magnetic fields in sunspots on the basis of the Zeeman splittings of sunspot spectra, the existence of magnetic fields outside the Earth's environment was conclusively established for the first time. Large sunspots can have magnetic fields of the order of 3000 G, which is much stronger than the Earth's field (the maximum value on the Earth's surface is about 0.6 G). One of the major achievements of twentieth century astronomy is to establish that magnetic fields are ubiquitous in the Universe. Many stars are much more strongly magnetic than the Sun. Some pulsars are believed to have magnetic fields as strong as 10^{12} G. The Galaxy has a vast magnetic field with field lines running roughly along the spiral arms and having a typical strength of 10^{-6} G.

The magnetic field of the Sun is periodic and evolves in time in a complex wave-like fashion. We have exhaustive observational data on the statistics of sunspots collected since the middle of the nineteenth century (i.e. before it was realized that sunspots are markers for solar magnetism). Whenever somebody proposes a model to explain the origin of astronomical magnetic fields, the first crucial test the model has to pass is to explain the behaviour of the solar magnetic fields as inferred from the statistics of sunspots. We, of course, have exhaustive data for the Earth's magnetic field also. But it is an *approximately* static field and hence is not as fascinating as the solar magnetic fields. We now know from geological records that the Earth's magnetic field has occasionally flipped directions in the past in a random fashion. The reason for this is still not properly understood. Between two such flippings, the Earth's magnetic field does not vary much. We now have a considerable amount of observational data on the magnetic fields of galaxies, and developing theoretical models to explain these data

Figure 16.1 Line segments indicating the direction of the magnetic field in different regions of the spiral galaxy M51. From Neininger (1992). (©Springer-Verlag. Reproduced with permission from *Astronomy and Astrophysics*.)

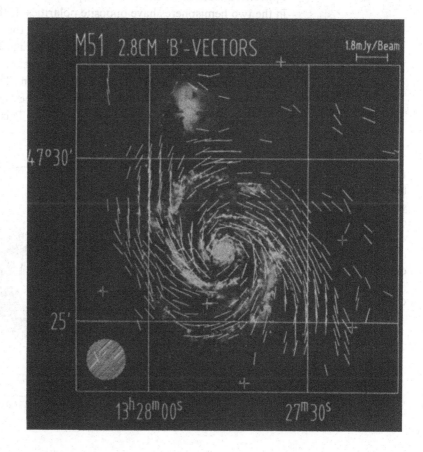

has become an important research activity in recent years. Figure 16.1 shows the distribution of magnetic field in the spiral galaxy M51 as inferred from the polarization of the radio emission (Neininger 1992). It is clear that the field lines run roughly along the spiral arms. The data for no astronomical magnetic field, however, can still match the data for the solar magnetic fields in detail, complexity and richness. We now summarize some of the salient features of the solar magnetic fields. We shall see towards the end of this chapter how most of these features can be explained.

We pointed out in §14.7 that often two large sunspots with opposite polarities appear side by side, implying the existence of a strong subsurface magnetic field in the azimuthal direction, i.e. in the ϕ direction if we introduce spherical coordinates with respect to the Sun's rotation axis. We also discussed in §14.7 that the origin of this strong azimuthal field B_ϕ can be attributed to stretching due to the internal differential rotation of the Sun. This azimuthal field B_ϕ has opposite directions in the two hemispheres, because bipolar sunspots in the two hemispheres have opposite polarities. If the right sunspot in the northern hemisphere has positive polarity, then the right sunspot in the southern hemisphere is negative (see Figure 14.12). The Sun has a magnetic cycle with a period of 22 years. There is a phase in the cycle when not many sunspots are seen. Then sunspots start appearing at around 40° latitude. As time goes on, newer sunspots tend to appear at lower and lower latitudes. This is clearly seen in

Figure 16.2 The butterfly diagram showing the distribution of sunspots in latitude (vertical axis) at different times (horizontal axis). Courtesy: K. Harvey.

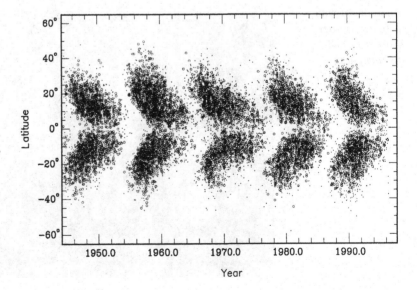

the so-called *butterfly diagram* first introduced by Maunder (1904). Figure 16.2 shows a butterfly diagram in which the horizontal axis is time. At any particular time, those ranges of latitude (vertical axis) are marked where sunspots appear. The butterfly pattern results from the equatorward shift of the latitude zones where sunspots are seen. Eventually one finds only very few sunspots near the equator. Then the next half-cycle begins with sunspots appearing again around 40° latitude. The polarity of bipolar sunspot pairs changes from one half-cycle (of 11 years' duration) to the next. If the right sunspots on a particular hemisphere are positive during a half-cycle, then they will be negative in the next half-cycle. These observations together imply that the subsurface azimuthal magnetic field B_ϕ in both the hemispheres must be propagating towards the equator like waves and must reverse its direction from half-cycle to half-cycle. Any theoretical model has to account for these observations.

16.2 Origin of astronomical magnetic fields as an MHD problem

One has to search in the MHD equations for an explanation for the astronomical magnetic fields. It is easy to see that the basic MHD equations discussed in §14.1 reduce to ordinary hydrodynamic equations if we set $\mathbf{B} = 0$. We know that the hydrodynamic equations constitute a self-consistent and well-behaved dynamical theory. Hence we can think of an Universe in which the magnetic field happens to be zero everywhere, even though the material of the Universe is in the plasma state. Such an Universe would evolve according to the equations of hydrodynamics and would be a completely well-behaved Universe. Why is our *real* Universe not like this *hypothetical* Universe? Why do we see magnetic fields all around? The main reason is that the $\mathbf{B} = 0$ state of a plasma can become unstable *under certain circumstances*. If this happens, then tiny magnetic fields (often present in plasmas due to statistical fluctuations) can grow to become stronger. The process by which the weak seed fields give rise to large-scale magnetic structures is called the *hydromagnetic dynamo process*.

Once a magnetic field is created in an astronomical object, it lasts for a very long time, even if there is no mechanism to sustain it. Let us first look at the question whether something like the dynamo process has to take place at the present time to explain the magnetic fields of the Earth, the Sun or the Galaxy. If there are no plasma motions in the interior of an astronomical body, it follows from the induction

Table 16.1 *The magnetic decay times of various astronomical objects*

Object	L (in cm)	σ (in e.s.u.)	τ (in years)
Earth	3×10^8	10^{16}	3×10^5
Sun	5×10^{10}	10^{17}	10^{11}
Galaxy	3×10^{20}	10^{10}	3×10^{23}

equation (14.10) that the magnetic field would decay away according to the equation

$$\frac{\partial \mathbf{B}}{\partial t} = \lambda \nabla^2 \mathbf{B}.$$

Taking τ to be the typical decay time and L to be the typical length scale, the L.H.S. is of the order $|B|/\tau$ and the R.H.S. of the order $\lambda |B|/L^2$. On equating them,

$$\tau \approx \frac{L^2}{\lambda} \approx \frac{4\pi\sigma L^2}{c^2}, \tag{16.1}$$

where we have made use of (14.11). The estimated decay times of the Earth, the Sun and the Galaxy are given in Table 16.1. The radii of the Earth's core and the Sun's core are taken as the typical length scales for them, whereas the thickness of the galactic disk (≈ 100 pc) is taken as the length scale for the Galaxy. The electrical conductivity of the molten metallic core of the Earth is about $\sigma \approx 10^{16}$ e.s.u. To estimate the electrical conductivity of the solar core or the interstellar medium in the Galaxy, we use (13.27). The electrical conductivity, which is the inverse of η, can be taken to be $\sigma \approx 10^7 T^{3/2}$ e.s.u. The values in Table 16.1 are obtained by taking the temperatures of the solar core and the interstellar medium to be about 10^7 and 10^2 K respectively.

The Earth's magnetic field has certainly lasted for a much longer time than the decay time shown in Table 16.1. Hence we need some mechanism to sustain the Earth's magnetic field. In the case of the Sun, we have the opposite problem. Although the decay time is fairly large, we know that the Sun's magnetic fields reverse every 11 years. So we need some mechanism to make the solar magnetic field oscillatory. Although the decay time of the galactic magnetic field is much larger than the age of the Universe, this magnetic field would have been wound up by the differential rotation of the Galaxy unless there is something which prevents that. Hence all these magnetic fields have to be continuously generated by the dynamo process. Since magnetic fields have energy $B^2/8\pi$ associated with them and energy cannot come from nowhere, some other form of energy has to be continuously converted into magnetic energy to compensate for the decay. We shall see later in this chapter that the convective motions inside the astronomical bodies can feed the magnetic field if certain conditions are satisfied. In an electromagnetic dynamo, a coil moves cutting flux lines and hence an e.m.f. is produced due to electromagnetic induction. We do not have electrical coils inside astronomical bodies. But we shall see that blobs of plasma in a region of convection can drag flux lines to produce an effective e.m.f. in a similar fashion as a result of electromagnetic induction (of which the effect is encapsulated in the induction equation).

To make life simpler and to concentrate on the basic physics, we consider the dynamo process in an incompressible fluid. It was pointed out in §4.2 that the continuity equation and the energy equation become redundant for an incompressible fluid. Hence we have to demonstrate the dynamo process on the basis of the equation of motion (14.3) and the induction equation (14.10), which describe how the velocity and magnetic fields interact with each other. This is still a very difficult problem, and only since the late 1970s have some calculations been carried out with big computers. The subject of dynamo theory developed historically by following a simpler approach. One can consider different kinds of velocity fields assumed as *given* and address the question whether they would sustain magnetic fields. This problem is called the *kinematic dynamo problem*. If the velocity field is given, then it is not necessary to bother about (14.3) and only the induction equation (14.10) has to be solved to find out how the magnetic field evolves under this given velocity field. It may be noted that (14.10) is linear in magnetic field **B** if the velocity field **v** is given. On the other hand, if we were to solve (14.3) and (14.10) simultaneously, treating both **B** and **v** as unknowns, then we would have to handle coupled nonlinear equations. Anybody with some experience in mathematical physics would know that solving a linear equation is much easier than solving coupled nonlinear equations. In this book, we confine ourselves only to the *kinematic dynamo problem* in order to demonstrate the existence of the dynamo process. But it should be kept in mind that the dynamo process is intrinsically a complex nonlinear process, because the Lorentz force of the magnetic field acts back on the velocity field so that assuming the velocity field as given is not realistic. The kinematic dynamo problem is an artificial way of making the problem linear and amenable to analytical treatments. We have mentioned that the dynamo process is essentially of the nature of an instability of the $\mathbf{B} = 0$ state of the plasma. We saw in Chapter 7 that instabilities arise out of nonlinear equations, although one can carry out a linear stability analysis to find out the conditions for the onset of instabilities. The kinematic dynamo problem is to be taken in a similar spirit. Although the actual problem is nonlinear, the linear problem of kinematic dynamo throws some light on the conditions under which the dynamo process is possible.

16.3 Cowling's theorem

When approaching a new equation, one first tries to understand it by solving it for sufficiently simple situations. Perhaps the simplest kinematic dynamo problem we can think of is to find whether a time-independent magnetic field, symmetric around an axis, can be maintained by some time-independent velocity field which is also symmetric around the same axis. Cowling (1934) showed that this is *not possible*. It has been realized in recent years that Cowling's original proof was not fully satisfactory, and attempts have been made to develop a more rigorous proof. In this elementary textbook, however, we present only Cowling's original proof of this important theorem.

We begin by assuming that it is *possible* to maintain a steady axisymmetric magnetic field $\mathbf{B}$ by a steady axisymmetric velocity field $\mathbf{v}$. We shall show that this assumption leads to an absurd conclusion and hence cannot be correct. If we introduce spherical coordinates

Figure 16.3 Field lines of an axisymmetric magnetic configuration projected on a meridional plane. The path of integration through the neutral point *N* is indicated by the dashed curve.

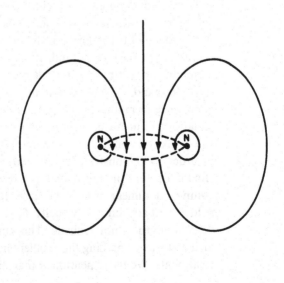

around the axis of symmetry, then axisymmetry means that $\partial/\partial\phi$ of any quantity is zero. Further $\partial/\partial t$ of all quantities are also taken to be zero. Consider a plane passing through the axis of symmetry. The projections of the magnetic field lines on this meridional plane must be closed curves. Even for the simplest dipolar configuration shown in Figure 16.3, there has to be one point N on each side of the symmetry axis which is encircled by magnetic field lines lying in the meridional plane. Such a point encircled by field lines is called a *neutral point*. For more complicated magnetic configurations, there can be more than one neutral point on each side. From (13.53), it is easy to see that j_ϕ has to be non-zero at a neutral point. This is the reason why it is not realistic to make the field lines of the dipolar geometry open by pushing the neutral point at an infinite distance from the symmetry axis, because that will imply a current ring at infinity.

We now take a line integral of Ohm's law (13.36) along the closed circle through the neutral point N going around the symmetry axis. This gives

$$\oint j_\phi \, dl = \sigma \left(\oint \mathbf{E} \cdot d\mathbf{l} + \frac{1}{c} \oint \mathbf{v} \times \mathbf{B} \cdot d\mathbf{l} \right). \qquad (16.2)$$

Since $\mathbf{B}$ can have only the ϕ component at the neutral point N, we have $\mathbf{B}$ and $d\mathbf{l}$ parallel there so that the triple scalar product $\mathbf{v} \times \mathbf{B} \cdot d\mathbf{l}$ on the R.H.S. of (16.2) vanishes. The other term on the R.H.S. can be transformed by Stokes's theorem (A.15):

$$\oint \mathbf{E} \cdot d\mathbf{l} = \int \nabla \times \mathbf{E} \cdot d\mathbf{S} = -\frac{1}{c} \int \frac{\partial \mathbf{B}}{\partial t} \cdot d\mathbf{S},$$

which is zero because of the assumed time-independence. Thus the R.H.S. of (16.2) vanishes. On the other hand, the L.H.S. of (16.2), which is equal to $j_\phi \times$ (circumference of the path of integration), cannot be zero, because j_ϕ is non-zero. Thus we have drawn the absurd conclusion that a non-zero quantity is equal to zero. So our starting assumption must be wrong. Hence it is not possible to sustain steady axisymmetric magnetic fields by steady axisymmetric flows. Thus Cowling's anti-dynamo theorem is established.

Cowling's theorem rules out sufficiently simple solutions of the kinematic dynamo problem. It was later shown that more complex (i.e. non-axisymmetric) steady flows can sustain steady magnetic fields. A discussion of such solutions, however, is beyond the scope of this book. We shall now discuss how Parker (1955b) invoked turbulent flows to solve the dynamo problem, because this topic has great astrophysical relevance.

16.4 Parker's turbulent dynamo. Qualitative idea

In this section, we present the qualitative ideas behind the turbulent dynamo theory of Parker (1955b). The mathematical formulation of these qualitative ideas will be taken up in §16.5.

It is useful to introduce spherical coordinates with respect to the rotation axis of the astrophysical object we are considering. The $B_\phi \hat{\mathbf{e}}_\phi$ component of the magnetic field is referred to as the azimuthal or (more usually) the *toroidal* magnetic field. The combination of the other components is called the *poloidal* magnetic field

$$\mathbf{B}_p = B_r \hat{\mathbf{e}}_r + B_\theta \hat{\mathbf{e}}_\theta.$$

The poloidal field lines can be represented by curves in the meridional plane (i.e. a plane passing through the axis of the coordinate system). We have already discussed in §14.7 that it is possible to generate the toroidal field by the stretching of poloidal field lines due to differential rotation. Figure 16.4(a) shows a body like the Sun of which the equator is rotating faster than the poles. A poloidal field line is shown to have been stretched by the differential rotation in the toroidal direction. It is easy to see that the toroidal fields produced in the two hemispheres have opposite signs, in accordance with the observations discussed in §14.7 and §16.1. Hence it is no problem to produce a toroidal field

Figure 16.4 Different stages of the dynamo process.

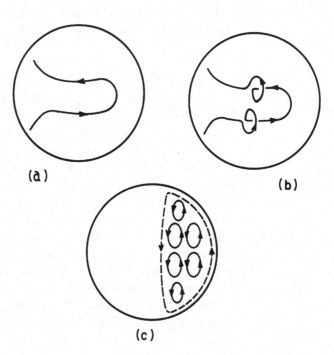

(a)

(b)

(c)

of the appropriate nature, if an astronomical body has some internal differential rotation and a poloidal field that can be stretched. But, if the poloidal field cannot be sustained, then it will eventually decay away and consequently the production of the toroidal field will also stop.

In a famous paper, Parker (1955b) gave the crucial idea of how the poloidal field can be generated. If there are turbulent convective motions inside the astronomical body, then the upward (or downward) moving plasma blobs stretch out the toroidal field in the upward (or downward) direction due to flux-freezing. The blobs of plasma moving in the upward direction generally spread out as they rise. If the convection takes place in a rotating frame of reference, then such spreading gives rise to vorticity in the blob as discussed in §9.2.2. The blobs therefore start rotating as they rise so that the fluid motions become helical in nature. We see the evidence of such helical motions in cyclones in the Earth's atmosphere. Figure 16.4(b) shows that a toroidal field line has been twisted by such helical turbulent motions in such a way that its projections in the meridional plane are magnetic loops. Several such magnetic loops produced by the helical turbulent motions are shown projected in the meridional plane in Figure 16.4(c). The arguments of §9.2.2 can easily be extended to show that the helical motions in the two hemispheres have opposite sense. If we keep this in mind and also note that B_ϕ has opposite directions in the two hemispheres, it then follows that the magnetic loops produced in the two hemispheres have the *same* sense. This is indicated in Figure 16.4(b). Because of the presence of turbulent diffusion in a region of convective turbulence, magnetic flux is only partially frozen in the plasma. As a result of the partial freezing, stretching of the poloidal field lines to produce the toroidal field

Figure 16.5
Schematic
representation of
Parker's idea of the
turbulent dynamo.

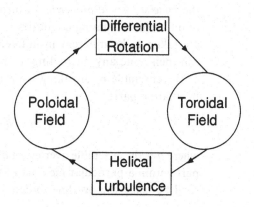

and twisting of the toroidal field lines to produce the magnetic loops are possible. Turbulent diffusion, however, is eventually expected to smoothen out the magnetic fields of the loops in Figure 16.4(c) to give rise to a large-scale magnetic field. Since all the loops in Figure 16.4(c) have the same sense, their diffusion gives rise to a global field with the same sense as indicated by the broken field line. Thus we ultimately end up with a poloidal field in the meridional plane starting from a toroidal field.

Figure 16.5 summarizes the main points of the argument. The poloidal and toroidal fields can sustain each other through a cyclic feedback process. The poloidal field can be stretched by the differential rotation to generate the toroidal field. The toroidal field, in its turn, can be twisted by the helical turbulence (associated with convection in a rotating frame) to give back a field in the poloidal plane. In the next two sections, we take up the challenge of putting these qualitative ideas into a mathematical formalism and quantitatively showing that they really work.

16.5 Mean field magnetohydrodynamics

The original mathematical treatment of Parker (1955b) was very much based on intuitive arguments. A more formal and systematic approach was developed later by Steenbeck, Krause and Rädler (1966). This more formal theory is known as *mean field magnetohydrodynamics*. Here we briefly summarize some of the main ideas.

Since turbulence plays an important role in this theory, it is necessary to devise a scheme to handle turbulence. This is done by invoking the concept of an *ensemble* discussed in §1.4 and also in §8.1 in the context of turbulence. In other words, we consider many replicas of the same astrophysical system, which are identical in all respects and hence have the *same statistical properties* of turbulence. However, the *actual values* of different fluctuating quantities (such as velocity and magnetic field) are taken to be different in different members of the ensemble. One can then write any fluctuating quantity q within a particular member of the ensemble by breaking it up into a mean part and a randomly fluctuating part:

$$q = \overline{q} + q', \tag{16.3}$$

where $\overline{q}$ is the ensemble average of q, and q' is its remaining fluctuating part within a particular member of the ensemble. As in Chapter 8, we shall be using the overline to denote the ensemble average. It is easy to see that

$$\overline{\overline{q}} = \overline{q}, \qquad \overline{q'} = 0. \tag{16.4}$$

If we use expansions of the form (16.3), then Ohm's law (13.36) becomes

$$\overline{\mathbf{j}} + \mathbf{j}' = \sigma \left[\overline{\mathbf{E}} + \mathbf{E}' + \frac{1}{c}(\overline{\mathbf{v}} + \mathbf{v}') \times (\overline{\mathbf{B}} + \mathbf{B}') \right]. \tag{16.5}$$

Let us now take the ensemble average of the whole equation. The ensemble average of the last term gives

$$\overline{(\overline{\mathbf{v}} + \mathbf{v}') \times (\overline{\mathbf{B}} + \mathbf{B}')} = \overline{\overline{\mathbf{v}} \times \overline{\mathbf{B}}} + \overline{\mathbf{v}' \times \overline{\mathbf{B}}} + \overline{\overline{\mathbf{v}} \times \mathbf{B}'} + \overline{\mathbf{v}' \times \mathbf{B}'}. \tag{16.6}$$

It is easy to see that $\overline{\overline{\mathbf{v}} \times \overline{\mathbf{B}}} = \overline{\mathbf{v}} \times \overline{\mathbf{B}}$. Since $\overline{\mathbf{B}}$ is a constant while averaging over the ensemble, it can be taken outside the averaging sign so that

$$\overline{\mathbf{v}' \times \overline{\mathbf{B}}} = \overline{\mathbf{v}'} \times \overline{\mathbf{B}} = 0,$$

since $\overline{\mathbf{v}'} = 0$. In the same way, $\overline{\overline{\mathbf{v}} \times \mathbf{B}'} = 0$. Hence (16.6) becomes

$$\overline{(\overline{\mathbf{v}} + \mathbf{v}') \times (\overline{\mathbf{B}} + \mathbf{B}')} = \overline{\mathbf{v}} \times \overline{\mathbf{B}} + \overline{\mathbf{v}' \times \mathbf{B}'}. \tag{16.7}$$

It is now easy to see that the ensemble average of (16.5) gives

$$\overline{\mathbf{j}} = \sigma \left(\overline{\mathbf{E}} + \frac{1}{c}\overline{\mathbf{v}} \times \overline{\mathbf{B}} + \frac{1}{c}\mathscr{E} \right), \tag{16.8}$$

where

$$\mathscr{E} = \overline{\mathbf{v}' \times \mathbf{B}'} \tag{16.9}$$

is known as the *mean e.m.f.* We see from (16.8) that the average quantities satisfy an equation similar to Ohm's law (13.36), the only difference being the additional term $\mathscr{E}$ which can act as an additional source of the mean current. Thus the turbulent fluctuations can give rise to an extra e.m.f. It should be clear to the reader that the origin of this e.m.f. is similar to the origin of the Reynolds stress discussed in §8.5.

If we substitute expansions of the form (16.3) in the induction equation (14.10), then we get

$$\frac{\partial \overline{\mathbf{B}}}{\partial t} + \frac{\partial \mathbf{B}'}{\partial t} = \nabla \times (\overline{\mathbf{v}} \times \overline{\mathbf{B}} + \mathbf{v}' \times \overline{\mathbf{B}} + \overline{\mathbf{v}} \times \mathbf{B}' + \mathbf{v}' \times \mathbf{B}') + \lambda \nabla^2 (\overline{\mathbf{B}} + \mathbf{B}'). \tag{16.10}$$

Again averaging this equation term by term gives

$$\frac{\partial \overline{\mathbf{B}}}{\partial t} = \nabla \times (\overline{\mathbf{v}} \times \overline{\mathbf{B}}) + \nabla \times \mathscr{E} + \lambda \nabla^2 \overline{\mathbf{B}}. \tag{16.11}$$

One could also have obtained this equation by substituting for $\overline{\mathbf{E}}$ from (16.8) into the averaged Maxwell equation

$$\frac{\partial \overline{\mathbf{B}}}{\partial t} = -c\nabla \times \overline{\mathbf{E}}.$$

We again see from (16.11) that the averaged quantities satisfy an equation very similar to the induction equation, the only difference arising out of the term involving the mean e.m.f. $\mathscr{E}$. This is the term which causes the mean fields to evolve in a way different from the way laminar fields satisfying MHD equations evolve. The secret of the turbulent dynamo lies in this term.

We have to evaluate the mean e.m.f. $\mathscr{E}$ to understand how the turbulent dynamo operates. Let us begin with a few comments on the nature of this term. First of all, if the velocity fluctuation $\mathbf{v}'$ and the magnetic fluctuation $\mathbf{B}'$ were completely *uncorrelated*, then we would have

$$\overline{\mathbf{v}' \times \mathbf{B}'} = \overline{\mathbf{v}'} \times \overline{\mathbf{B}'},$$

which clearly vanishes. Hence $\mathscr{E}$ can be non-zero only if there is a correlation between $\mathbf{v}'$ and $\mathbf{B}'$. Do we expect such a correlation to exist? Let us think how the fluctuating magnetic field $\mathbf{B}'$ arises. In the absence of turbulence, we would just have the mean magnetic field $\overline{\mathbf{B}}$. The turbulent velocity field $\mathbf{v}'$ in a member of the ensemble distorts the field lines of this mean field $\overline{\mathbf{B}}$ to produce the fluctuating component $\mathbf{B}'$. Since $\mathbf{B}'$ results from $\mathbf{v}'$, we certainly expect $\mathbf{B}'$ and $\mathbf{v}'$ to be correlated.

Before evaluating $\mathscr{E}$, let us address the question: what exactly do we mean by an *evaluation* of $\mathscr{E}$? The philosophy of kinematic dynamo theory is that we want to study the evolution of the mean magnetic field $\overline{\mathbf{B}}$, provided the mean velocity field $\overline{\mathbf{v}}$ and the statistical properties of $\mathbf{v}'$ are given. One sees from (16.11) that this objective will be achieved if we are able to express $\mathscr{E}$ in terms of $\overline{\mathbf{B}}$, $\overline{\mathbf{v}}$ and the statistical properties of $\mathbf{v}'$. The first step is to evaluate $\mathbf{B}'$ in terms of these quantities. By subtracting (16.11) from (16.10), we get the equation for the evolution of $\mathbf{B}'$ in a member of the ensemble:

$$\frac{\partial \mathbf{B}'}{\partial t} = \nabla \times (\mathbf{v}' \times \overline{\mathbf{B}} + \overline{\mathbf{v}} \times \mathbf{B}' + \mathbf{v}' \times \mathbf{B}' - \mathscr{E}) + \eta \nabla^2 \mathbf{B}'. \qquad (16.12)$$

Let us assume the turbulence to have a correlation time τ. In other words, if the velocity at a point inside a system is found to be $\mathbf{v}'$ at a certain time, then we expect the velocity to be not very different from $\mathbf{v}'$ for the time τ. After time τ, however, the velocity would have changed appreciably. We now want to find out how the field lines of the mean field $\overline{\mathbf{B}}$ would be distorted by a velocity field $\mathbf{v}'$ operating for a correlation time τ. A departure $\mathbf{B}'$ from the mean magnetic field will result from this operation. It is to be noted that all the terms on the R.H.S. of (16.12) except $\nabla \times (\mathbf{v}' \times \overline{\mathbf{B}})$ are linear in $\mathbf{B}'$ (keeping the definition of $\mathscr{E}$ given by (16.9) in mind). If we assume $\mathbf{B}'$ to remain sufficiently small, then we can neglect these other terms on the R.H.S. of (16.12) compared to $\nabla \times (\mathbf{v}' \times \overline{\mathbf{B}})$. This is called the *first-order smoothing approximation*, which gives us

$$\frac{\partial \mathbf{B}'}{\partial t} \approx \nabla \times (\mathbf{v}' \times \overline{\mathbf{B}}). \tag{16.13}$$

Hence the departure $\mathbf{B}'$ from the mean magnetic field resulting from $\mathbf{v}'$ operating for the correlation time τ is

$$\mathbf{B}' \approx \tau \nabla \times (\mathbf{v}' \times \overline{\mathbf{B}}),$$

i.e.

$$\mathbf{B}' \approx \tau(\overline{\mathbf{B}} \cdot \nabla)\mathbf{v}' - \tau(\mathbf{v}' \cdot \nabla)\overline{\mathbf{B}} \tag{16.14}$$

assuming $\nabla \cdot \mathbf{v}' = 0$ because of incompressibility. Since (16.14) gives the typical value of $\mathbf{B}'$ within a member of the ensemble, we have to substitute it in (16.9) to evaluate the mean e.m.f. $\mathscr{E}$.

To proceed further, it is convenient to use the Levi-Civita antisymmetric tensor ϵ_{ijk} introduced in Appendix A (see (A.22–A.24)). The i-th component of $\mathbf{v}' \times \mathbf{B}'$ is given by

$$(\mathbf{v}' \times \mathbf{B}')_i = \epsilon_{ijk} v'_j B'_k$$
$$= \epsilon_{ijk} v'_j \overline{B}_l \frac{\partial v'_k}{\partial x_l} \tau - \epsilon_{ijk} v'_j v'_l \frac{\partial \overline{B}_k}{\partial x_l} \tau$$

where we have substituted for B'_k from (16.14). On taking the ensemble average now, we get the i-th component of $\mathscr{E}$ (noting that the *dummy* indices j and l can be interchanged) :

$$\mathscr{E}_i = \alpha_{ij} \overline{B}_j + \beta_{ijk} \frac{\partial \overline{B}_k}{\partial x_j} \tag{16.15}$$

where

$$\alpha_{ij} = \epsilon_{ilk} \overline{v'_l \frac{\partial v'_k}{\partial x_j}} \tau, \qquad \beta_{ijk} = -\epsilon_{ilk} \overline{v'_l v'_j} \tau. \tag{16.16}$$

We have thus solved the problem of expressing the mean e.m.f. $\mathscr{E}$ in terms of the mean magnetic field $\overline{\mathbf{B}}$ and the coefficients α_{ij}, β_{ijk}, which depend on the statistical properties of the turbulent velocity field $\mathbf{v}'$ as seen from (16.16).

The general expression for the mean e.m.f. under the first-order smoothing approximation is given by (16.15) and (16.16). The expressions for α_{ij} and β_{ijk} can be simplified considerably if the turbulence is assumed to be *isotropic*. We would then expect them to have the following forms from considerations of symmetry:

$$\alpha_{ij} = \alpha\delta_{ij}, \qquad \beta_{ijk} = -\lambda_T\epsilon_{ijk}. \tag{16.17}$$

It is easy to see that

$$\alpha = \frac{1}{3}\alpha_{ii}$$

$$= \frac{1}{3}\epsilon_{ilk}\overline{v'_l\frac{\partial v'_k}{\partial x_i}}\tau$$

$$= -\frac{1}{3}\overline{v'_l\epsilon_{lik}\frac{\partial v'_k}{\partial x_i}}\tau,$$

which can again be put in the vectorial notation

$$\alpha = -\frac{1}{3}\overline{\mathbf{v}'\cdot(\nabla\times\mathbf{v}')}\tau. \tag{16.18}$$

We leave it as Exercise 16.1 for the reader to show that λ_T appearing in (16.17) is given by

$$\lambda_T = \frac{1}{3}\overline{\mathbf{v}'\cdot\mathbf{v}'}\tau. \tag{16.19}$$

Putting the form (16.17) in (16.15), we conclude that the mean e.m.f. for isotropic turbulence is given by

$$\mathscr{E} = \alpha\overline{\mathbf{B}} - \lambda_T\nabla\times\overline{\mathbf{B}}. \tag{16.20}$$

Substituting (16.20) in (16.11) finally gives

$$\frac{\partial\overline{\mathbf{B}}}{\partial t} = \nabla\times(\overline{\mathbf{v}}\times\overline{\mathbf{B}}) + \nabla\times(\alpha\overline{\mathbf{B}}) + (\lambda+\lambda_T)\nabla^2\overline{\mathbf{B}} \tag{16.21}$$

if we neglect the spatial variation of λ_T.

It is clear from (16.21) that λ_T is of the nature of a diffusion coefficient. The expression (16.19), which can be compared with (8.39), shows that λ_T is actually the *turbulent diffusion* which we introduced in §8.4. In the regions of turbulence inside astrophysical bodies, usually λ_T turns out to be much larger than molecular resistivity λ. Hence, in most problems of turbulent dynamo action, λ can be neglected compared to λ_T. We shall show in the next section that the coefficient α is crucial for the dynamo generation of magnetic fields. We see from (16.18) that α is directly proportional to $\overline{\mathbf{v}'\cdot(\nabla\times\mathbf{v}')}$, which is a measure of the helical motion in the turbulent fluid. If rising blobs of fluid tend to rotate preferentially in a certain sense, then $\overline{\mathbf{v}'\cdot(\nabla\times\mathbf{v}')}$ is non-zero, because the curl of the rotating motion is then in the same direction as the velocity of rise so that there are non-zero contributions to $\mathbf{v}' \cdot (\nabla \times \mathbf{v}')$. For the rising blobs to rotate preferentially in one direction, the convective turbulence has to be reflectionally *asymmetric* as expected in a rotating frame (see §9.2.2). If the turbulence were reflectionally symmetric, then the helicity of turbulence and therefore the α-coefficient would have been zero. We have already discussed in the previous section that helical turbulent motions can twist toroidal field lines to produce the poloidal field. It is the α-coefficient which encapsulates this effect of helical motions in the mathematical theory. This will become clearer in the next section. Although we are assuming the turbulence to be isotropic to simplify our mathematical equations (the turbulence in the solar convection zone is far from isotropic and this was already considered by Parker 1955b), the turbulence has to have the special character of non-zero net helicity. Then only is the dynamo action possible. Since turbulence in rotating frames usually has net helicity, we have to consider the possibility of dynamo action in turbulent plasmas having rotational motion.

We saw in §16.3 that Cowling's theorem rules out axisymmetric solutions of the dynamo problem by *laminar* flows (steady axisymmetric flows have got to be laminar). We took a line integral of Ohm's law (13.36) through the neutral point to prove this theorem. The mean fields satisfy (16.8), which differs from Ohm's law due to the extra term $\mathscr{E}$. When $\mathscr{E}$ is non-zero, the neutral point argument no longer goes through and hence Cowling's theorem does not hold for mean fields. So we may expect steady axisymmetric solutions for the mean magnetic field on the basis of mean field equations. Such solutions *actually exist*. Hence we have the paradoxical situation that the dynamo problem is easier to solve with turbulent flows rather than with laminar flows. Although the laminar equations are simpler than the mean field equations, Cowling's theorem forbids sufficiently simple solutions of the dynamo problem with laminar flows. On the other hand, it is fairly straightforward to find axisymmetric dynamo solutions of the mean field equations. We discuss some simple solutions of the mean field equations in the next section.

16.6 A simple dynamo solution

We saw in §16.4 that the main idea of Parker's turbulent dynamo (as summarized in Figure 16.5) is that the poloidal and toroidal magnetic fields can sustain each other—the toroidal field arises due to the action of differential rotation on the poloidal field and the poloidal field is then generated back from the toroidal field by helical turbulence. We

now want to discuss the mathematical representations of these ideas
on the basis of the mean field equations. Since we shall be concerned
only with mean fields in this section, let us drop the overline sign to
denote averages. So **v** and **B** will stand for the mean velocity field and
mean magnetic field respectively. Neglecting λ compared to λ_T, we
write (16.21) as

$$\frac{\partial \mathbf{B}}{\partial t} = \nabla \times (\mathbf{v} \times \mathbf{B}) + \nabla \times (\alpha \mathbf{B}) + \lambda_T \nabla^2 \mathbf{B}. \qquad (16.22)$$

This is the basic dynamo equation from which we have to extract the
appropriate solution to explain the features of solar magnetic fields
discussed in §16.1.

For applications to the Sun or to the Earth, one has to write down
(16.22) in spherical coordinates. Some of the basic features, however,
can be understood quite well by solving (16.22) in a local Cartesian coordinate system at a point in the spherical object. Since such
calculations are much simpler than the calculations in full spherical
geometry, we shall discuss here only the calculations in local Cartesian
coordinates. Figure 16.6 shows a local Cartesian system at a point in
the northern hemisphere of a spherical body like the Sun. The x axis
corresponds to the radially outward direction, the y axis to the toroidal
(i.e. ϕ) direction and the z axis to the direction of increasing latitude.
Since the statistics of sunspots show a tendency for equatorward mi-
gration, we want to find a wave-like solution of (16.22) propagating
in the negative z direction. As Cowling's theorem is not applicable for
mean fields, we expect to find solutions symmetric around the rotation

Figure 16.6 Local
Cartesian coordinates
at a point in the
northern hemisphere
of a spherical body.

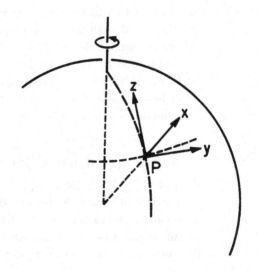

axis. With respect to the local Cartesian system, axisymmetry means symmetry in y, i.e. we look for solutions for which $\partial/\partial y = 0$.

The toroidal magnetic field in our representation is simply the component $B_y \hat{\mathbf{e}}_y$. The poloidal magnetic field is a two-dimensional solenoidal (i.e. zero divergence) vector field with field lines lying in the xz plane. We discussed in §4.8 how two-dimensional solenoidal velocity fields can be handled with stream functions. In exactly the same fashion, the poloidal magnetic field can be written as $\nabla \times [A(x, z)\hat{\mathbf{e}}_y]$, where $A(x, z)$ is a function which will be constant on the poloidal field lines in the xz plane. Hence the full magnetic field can be written as

$$\mathbf{B} = B_y(x, z)\hat{\mathbf{e}}_y + \nabla \times [A(x, z)\hat{\mathbf{e}}_y], \tag{16.23}$$

where B_y and A respectively represent the toroidal and the poloidal components. The mean velocity field is due to the differential rotation and hence has a component in the y direction alone. Although this velocity field v_y can be a function of both x and z, let us take it to be a function of x alone, since this results in simpler solutions. So we write

$$\mathbf{v} = v_y(x)\hat{\mathbf{e}}_y \tag{16.24}$$

such that the velocity shear is given by

$$G = \frac{\partial v_y}{\partial x}.$$

We shall now show that it is possible to have plane wave solutions propagating in the z direction, if the velocity shear G and the coefficients α, λ_T are taken as constants.

Substituting (16.23) and (16.24) in (16.22), it is straightforward to show that the y component of (16.22) becomes

$$\frac{\partial B_y}{\partial t} = GB_x - \alpha\nabla^2 A + \lambda_T\nabla^2 B_y, \tag{16.25}$$

since α is assumed constant. Apart from the y component, the other components of (16.22) can be put in the form

$$\nabla \times \left(\frac{\partial A}{\partial t}\hat{\mathbf{e}}_y - \alpha B_y\hat{\mathbf{e}}_y - \lambda_T\nabla^2 A\hat{\mathbf{e}}_y \right) = 0.$$

The easiest way to satisfy this equation is to take

$$\frac{\partial A}{\partial t} = \alpha B_y + \lambda_T\nabla^2 A. \tag{16.26}$$

It is straightforward to see that if the magnetic field is taken to be of the form (16.23) with B_y and A satisfying (16.25) and (16.26), then the basic dynamo equation (16.22) is satisfied. Hence we now have to solve the two simultaneous equations (16.25) and (16.26).

Before solving these two equations (16.25) and (16.26), let us try to understand their physical significance. The simpler equation (16.26) gives the evolution of the poloidal field. If the term αB_y were not there, then it would have been a simple diffusion equation and would imply that any poloidal field diffuses away. The additional term αB_y acts as the source term which generates the poloidal field. We have seen that α is a measure of helical motion in the turbulence. Hence this source term corresponds to the production of the poloidal field as a result of helical motions twisting the toroidal field B_y. The other equation (16.25) gives the evolution of the toroidal field and has two source terms (the first two terms on the R.H.S.). The term GB_x corresponds to the velocity shear of differential rotation stretching out the B_x component of the poloidal field to produce the toroidal field. Just as the helical motion can twist the toroidal field to produce the poloidal field, it can also twist the poloidal field to produce the toroidal field as well. This is the origin of the other source term $-\alpha \nabla^2 A$ in (16.25). If the astrophysical system has strong differential rotation, then the differential rotation term GB_x is much larger than the other source term $-\alpha \nabla^2 A$, and this other term can be neglected. Then (16.25) becomes

$$\frac{\partial B_y}{\partial t} = -G\frac{\partial A}{\partial z} + \lambda_T \nabla^2 B_y. \qquad (16.27)$$

Equations (16.26) and (16.27) together constitute the mathematical representation of Figure 16.5.

If the term $-\alpha \nabla^2 A$ in (16.25) is neglected (as we are doing here), then the dynamo is known as an $\alpha\omega$ dynamo. When one uses spherical coordinates, the rotation is usually denoted by ω (see Exercise 16.2). Hence, for the $\alpha\omega$ dynamo equations written in spherical geometry, the source term for the poloidal field involves α and the source term for the toroidal field involves ω, giving rise to the name $\alpha\omega$ dynamo. On the other hand, if the astrophysical system does not have much differential rotation, then the term GB_x in (16.25) can be neglected and the other source term $-\alpha \nabla^2 A$ retained. Such a dynamo is called an α^2 dynamo, because the source terms for both the components involve the α-coefficient. We do not discuss the α^2 dynamo in the text, but it is given as Exercise 16.3.

We now wish to show that (16.26) and (16.27) admit of propagating wave-like solutions of the type $\exp[i(\omega t + kz)]$ for constant α, λ_T and G. Let us begin by trying solutions of the form

$$A = \hat{A}\exp(\sigma t + ikz), \quad B_y = \hat{B}\exp(\sigma t + ikz). \qquad (16.28)$$

Substituting in (16.26) and (16.27), we have

$$(\sigma + \lambda_T k^2)\hat{A} = \alpha\hat{B}$$

and

$$(\sigma + \lambda_T k^2)\hat{B} = -ikG\hat{A}.$$

Combining these two equations, we get

$$(\sigma + \lambda_T k^2)^2 = -ik\alpha G, \qquad (16.29)$$

from which

$$\sigma = -\lambda_T k^2 \pm \left(\frac{i-1}{\sqrt{2}}\right)\sqrt{k\alpha G}. \qquad (16.30)$$

Let us take k to be positive. We now separately discuss the two cases of $\alpha G > 0$ and $\alpha G < 0$.

First consider $\alpha G > 0$. For the dynamo maintenance of magnetic fields, we must have Re $\sigma \geq 0$. This is possible only if we choose the negative sign in (16.30) so that

$$\sigma = -\lambda_T k^2 + \left(\frac{k\alpha G}{2}\right)^{1/2} - i\left(\frac{k\alpha G}{2}\right)^{1/2}. \qquad (16.31)$$

This expression makes it clear that the dynamo problem has the character of a stability calculation. If αG, which gives the combined effect of helical turbulence and differential rotation, is larger than a critical value, then only is it possible for magnetic fields to grow. Otherwise magnetic fields decay away. We introduce a dimensionless parameter called the dynamo number defined as

$$N_d = \frac{|\alpha G|}{\lambda_T^2 k^3}. \qquad (16.32)$$

It is easy to see from (16.31) that the condition for the dynamo growth of the magnetic field (i.e. for Re $\sigma \geq 0$) is

$$N_d \geq 2. \qquad (16.33)$$

For the marginally sustained magnetic field (i.e. for $N_d = 2$), the eigenmodes are of the form

$$A, B_y \sim \exp\left[-i\left(\frac{k\alpha G}{2}\right)^{1/2}t + ikz\right]. \qquad (16.34)$$

This corresponds to a wave propagating in the positive z direction, i.e. in the poleward direction (see Figure 16.6).

We now consider the case $\alpha G < 0$ for which again the negative sign in (16.30) has to be chosen such that

$$\sigma = -\lambda_T k^2 + \left(\frac{k|\alpha G|}{2}\right)^{1/2} + i\left(\frac{k|\alpha G|}{2}\right)^{1/2}. \qquad (16.35)$$

It is straightforward to see that the condition for dynamo growth (16.33) remains the same, with N_d still defined by (16.32). The marginally sustained eigenmodes, however, now become

$$A, B_y \sim \exp\left[i\left(\frac{k|\alpha G|}{2}\right)^{1/2} t + ikz\right]. \qquad (16.36)$$

This gives an equatorward propagating wave as desired in the solar context. On this ground, one may be tempted to conclude that αG in the northern hemisphere of the Sun must be negative.

We thus see that the dynamo equation admits of solutions which can account for the periodicity and equatorward propagation of solar magnetic fields. Let us recapitulate what determines the direction of the propagation vector. We took **v** along the y direction with its variation in the x direction (i.e. $\partial v_y / \partial x \neq 0$). Then the dynamo wave propagates in the third z direction. It is to be noted that the wavefronts for our solution correspond to infinite planes perpendicular to the z axis. This is because we took α, λ_T and G as constants. In a realistic situation, one has to solve the dynamo equation in a finite region with suitable boundary conditions on the boundaries. We shall not get into a discussion of boundary conditions here, because they make the problem considerably more complicated. We hope that the plane wave solution for constant coefficients gives some idea on how to approach dynamo problems. The equatorward propagating plane wave solution of the dynamo discussed above was first obtained by Parker (1955b). Steenbeck and Krause (1969) solved the dynamo equation in spherical geometry with realistic boundary conditions and reproduced the butterfly diagram theoretically. It may be noted that the condition $\alpha G < 0$ for equatorward propagation was obtained by assuming the mean velocity **v** to be entirely in the toroidal direction. It has recently been shown by Choudhuri, Schüssler and Dikpati (1995) that it may be possible to have equatorward propagation even with $\alpha G > 0$ if there is a suitable velocity field in the poloidal plane. This new result may be of considerable importance for the solar dynamo, since there are indications that both α and G may be positive in appropriate locations in the northern hemisphere of the Sun.

16.7 Concluding remarks

We saw in Chapter 8 that turbulence has remained one of the unsolved grand problems of physics. It is perhaps fitting that we close our discussion on plasmas with a process in which turbulence plays a key role. One often thinks of turbulence as something which destroys all order. However, the dynamo process is an example of a situation in which large-scale ordered structures actually emerge out of turbulence. Understanding how large-scale order may sometimes arise out of turbulence has become an important research topic today. In the astrophysical context, another example of such a situation is the differential rotation in the Sun which plays such a crucial role in the dynamo theory. Most probably, the differential rotation is itself sustained by stresses arising out of convective turbulence. That is why the Sun has not been able to settle into a pattern of rigid rotation. A mean field approach similar to mean field MHD discussed above has been developed to model the differential rotation of the Sun.

In spite of our lack of understanding of turbulence, it is remarkable that we have been able to build up models to explain many features of solar magnetism. But one should not be overjoyed by this success. Although the kinematic dynamo approach has yielded some suggestive results, we are still very far from having a *truly quantitative* model of any dynamo process in the astrophysical Universe. We still do not know how to calculate the coefficients α or λ_T from fundamental principles. In kinematic calculations, one typically uses values of these quantities obtained on the basis of some very rough hand-waving order-of-magnitude arguments. It follows from (16.19) that λ_T has to be positive. In the case of α, it is not even easy to ascertain whether it will be positive or negative. The mean field theory of differential rotation also suffers from similar uncertainties.

Because of these limitations of the kinematic approach with mean field equations, one may try to attack the problem by numerically solving the full MHD equations. In such an approach, the turbulence is produced by the computer simulation and the velocity field is calculated self-consistently with the effect of the Lorentz force incorporated. In the mid-1970s, Gilman started developing an ambitious computer code to model the physics of the solar convection zone by incorporating the convective turbulence, the differential rotation and the dynamo action together in a self-consistent scheme. This monumental code has been giving results since the early 1980s. Many predictions of this code, however, seem to contradict observational data flatly (Gilman

1983). It is still not very clear why Gilman's code has failed to match observations.

To sum up, the dynamo theory has given us the clue to understand why magnetic fields are ubiquitous in the astrophysical Universe. We are, however, still unable to make a very detailed quantitative model of the dynamo process in any astrophysical system due to our lack of understanding of turbulence.

Exercises

16.1   Show that λ_T for isotropic turbulence as defined in (16.17) is given by (16.19).

16.2   Let us consider the dynamo problem in a situation where all the mean quantities are axisymmetric. Using spherical coordinates, we write the magnetic field as

$$\mathbf{B} = B(r, \theta)\hat{\mathbf{e}}_\phi + \mathbf{B}_\mathrm{p},$$

where $B(r, \theta)$ is the toroidal component and $\mathbf{B}_\mathrm{p}$ is the poloidal component, which we further write as

$$\mathbf{B}_\mathrm{p} = \nabla \times [A(r, \theta)\hat{\mathbf{e}}_\phi].$$

The velocity field due to differential rotation can be written as

$$\mathbf{v} = \omega(r, \theta)r \sin \theta \hat{\mathbf{e}}_\phi,$$

where $\omega(r, \theta)$ is the angular velocity. Show that the dynamo equation (16.22) leads to two scalar equations

$$\frac{\partial B}{\partial t} = r \sin \theta (\mathbf{B}_\mathrm{p} \cdot \nabla)\omega + \hat{\mathbf{e}}_\phi \cdot \left[\nabla \times (\alpha \mathbf{B}_\mathrm{p})\right] + \lambda \left(\nabla^2 - \frac{1}{r^2 \sin^2 \theta}\right) B,$$

$$\frac{\partial A}{\partial t} = \alpha B + \lambda \left(\nabla^2 - \frac{1}{r^2 \sin^2 \theta}\right) A.$$

16.3   Consider an α^2 dynamo in Cartesian geometry so that the term GB_x in (16.25) can be neglected. Find the critical value of α for the dynamo generation and show that the generated magnetic field is non-oscillatory.

16.4   The solar dynamo has a period of 22 years and its half wavelength corresponds to about 40° in latitude. Assuming the solar dynamo to be marginally sustained, make a rough estimate of the quantity αG and the turbulent diffusion coefficient λ_T.

Epilogue

Epilogue

17

We have come to the end of a long journey. Before saying a final goodbye to the reader, we wish to present an assortment of mixed fares in this final chapter. The main purpose of the book has been to develop the fundamentals fully. We now give a glimpse of what lies beyond the horizon.

Often it happens that one does not know the details of the physical conditions inside an astrophysical system, but can make rough estimates of different kinds of energies contained in the system (kinetic, potential, magnetic, etc.). To handle such situations, one can suitably integrate the basic equations to obtain an equation connecting different types of total energies of the system. This equation is known as the *virial equation*. Since this approach is very general, we present a discussion of it in this final chapter, when the reader should be in a position to possess a broad overview of the whole field. While applying hydrodynamics and magnetohydrodynamics to astrophysical systems, often it becomes necessary to incorporate relativistic corrections or to include the effects of radiation pressure. The subjects of relativistic hydrodynamics and radiation hydrodynamics respectively deal with these problems. These are vast fields of study, and we cannot provide proper treatments of these two fields in this elementary book. Just to give an idea of how one proceeds, we discuss the basic equations of these two fields in §17.2 and §17.3. Compared to the usual style of presentation in this book, these two sections would appear like skeletons without flesh and blood. A mere knowledge of basic principles does not usually give one a deep insight into a subject. A person knowing only Newton's laws of motion cannot claim much competence in mechanics. However, we do not want to leave the reader totally unacquainted with the important subjects of relativistic hydrodynamics and radiation hydrodynamics, although we are not in a position to develop these subjects fully. Finally, we make some general comments on different astrophysical fluid and plasma systems in §17.4, pointing out that one can acquire elementary knowledge about these systems by looking up selected sections of this book.

17.1 Virial theorem

Partial differential equations have played key roles in most of the discussions in this book. If one can solve the appropriate partial differential equation in a particular situation, then one may be able to get an elaborate solution giving the details of the system. Often, however, the partial differential equations are very difficult to solve, and we may have such incomplete knowledge of our system that it may not be worthwhile to take the pains of working out an elaborate solution. In such situations, it is often possible to make important conclusions in a relatively painless manner if we know some global relationships amongst the different forms of energy in the system. Clausius (1870) noted that the total kinetic energy of a gas is related to some integral of force on the bounding surface, which he termed the *virial* of the system. See Goldstein (1980, §3-4) for a modern treatment of this problem. The approach pioneered by Clausius (1870) can be generalized to a form that includes virtually all types of fluid and plasma systems.

Let us consider an ideal magnetofluid distributed within a limited region of space. The gravitational potential at a point $\mathbf{x}$ is

$$\phi(\mathbf{x}) = -G \int \frac{\rho(\mathbf{x}') \, dV'}{|\mathbf{x} - \mathbf{x}'|}, \tag{17.1}$$

where the prime denotes source points over which the integration has to be done. The equation of motion for the system is the Euler equation with the magnetic force as given by (14.7), which we write again

$$\rho \frac{dv_i}{dt} = -\frac{\partial p}{\partial x_i} - \frac{\partial M_{ij}}{\partial x_j} - \rho \frac{\partial \phi}{\partial x_i}, \tag{17.2}$$

where we have taken P_{ij} to be diagonal because of the neglect of viscosity, and the body force is assumed to be gravitational so that $F_i = -\partial \phi / \partial x_i$ with $\phi(\mathbf{x})$ given by (17.1).

We now multiply (17.2) by x_k and integrate over all volume. This gives

$$\int \rho \frac{dv_i}{dt} x_k \, dV = -\int x_k \frac{\partial p}{\partial x_i} \, dV - \int x_k \frac{\partial M_{ij}}{\partial x_j} \, dV - \int \rho \frac{\partial \phi}{\partial x_i} x_k \, dV. \tag{17.3}$$

Let us manipulate the various terms of this equation suitably. Since

$$\rho \frac{dv_i}{dt} = \frac{\partial}{\partial t}(\rho v_i) + \frac{\partial}{\partial x_j}(\rho v_i v_j)$$

on using the equation of continuity, the L.H.S. of (17.3) is equal to

$$\int \rho \frac{dv_i}{dt} x_k \, dV = \frac{d}{dt} \int \rho v_i x_k \, dV + \int x_k \frac{\partial}{\partial x_j}(\rho v_i v_j) \, dV. \qquad (17.4)$$

Noting that

$$\int x_k \frac{\partial}{\partial x_j}(\rho v_i v_j) \, dV = \int \frac{\partial}{\partial x_j}(\rho v_i v_j x_k) \, dV - \int \rho v_i v_k \, dV,$$

and using (A.18), we obtain from (17.4)

$$\int \rho \frac{dv_i}{dt} x_k \, dV = \frac{d}{dt} \int \rho v_i x_k \, dV - 2T_{ik} + \oint \rho v_i v_j x_k \, dS_j, \qquad (17.5)$$

where

$$T_{ik} = \frac{1}{2} \int \rho v_i v_k \, dV \qquad (17.6)$$

is the kinetic energy tensor and the last term in (17.5) is an integral over the surface bounding the system. The first two terms on the R.H.S. of (17.3) can now be integrated by parts using the results (A.14) and (A.18) to transform the volume integrals into surface integrals. This gives

$$-\int x_k \frac{\partial p}{\partial x_i} \, dV = \delta_{ik} \int p \, dV - \oint x_k p \, dS, \qquad (17.7)$$

$$\int x_k \frac{\partial M_{ij}}{\partial x_j} \, dV = \int M_{ik} \, dV - \oint x_k M_{ij} \, dS_j, \qquad (17.8)$$

For astrophysical systems of finite size, the surface integrals in (17.5), (17.7) and (17.8) would vanish if the bounding surface is chosen far away where all the physical quantities are zero. We now turn to the last term in (17.3), which is called the potential energy tensor and denoted by W_{ik}, i.e.

$$W_{ik} = -\int \rho \frac{\partial \phi}{\partial x_i} x_k \, dV. \qquad (17.9)$$

Substituting (17.5), (17.7), (17.8) and (17.9) in (17.3), and setting the surface integrals in (17.5), (17.7) and (17.8) to zero, we get

$$\frac{d}{dt} \int \rho v_i x_k \, dV = 2T_{ik} + \delta_{ik} \Pi + \int M_{ik} \, dV + W_{ik}, \qquad (17.10)$$

where

$$\Pi = \int p \, dV. \qquad (17.11)$$

We now derive an explicit expression for the potential energy tensor W_{ik} by substituting for ϕ from (17.1) into (17.9):

$$W_{ik} = \int dV \, x_k \rho(\mathbf{x}) \frac{\partial}{\partial x_i} \int \frac{G\rho(\mathbf{x}') \, dV'}{|\mathbf{x} - \mathbf{x}'|}$$

$$= -G \iint dV \, dV' \, \rho(\mathbf{x}) \rho(\mathbf{x}') \frac{x_k(x_i - x_i')}{|\mathbf{x} - \mathbf{x}'|^3}.$$

It is straightforward to see that the value of this quantity does not change if we interchange the primed and the unprimed variables, since both the variables have been integrated over. Hence the expression for W_{ik} can be replaced by half the original expression plus half the expression with the primed and unprimed variables interchanged. This gives

$$W_{ik} = -\frac{G}{2} \iint dV \, dV' \, \rho(\mathbf{x}) \rho(\mathbf{x}') \frac{(x_k - x_k')(x_i - x_i')}{|\mathbf{x} - \mathbf{x}'|^3}. \tag{17.12}$$

It should be clear from this expression that W_{ik} is symmetric in i and k. We now note that all terms on the R.H.S. of (17.10) are symmetric in the subscripts i and k (since M_{ik} as defined in (14.8) is symmetric). We interchange i and k in (17.10) and add that equation to the original equation (17.10):

$$\frac{d}{dt} \int \rho(v_i x_k + v_k x_i) dV = 4T_{ik} + 2\delta_{ik}\Pi + 2 \int M_{ik} \, dV + 2W_{ik}. \tag{17.13}$$

Let us now define the moment of inertia tensor

$$I_{ik} = \int \rho x_i x_k \, dV. \tag{17.14}$$

The time derivative of this quantity is

$$\frac{dI_{ik}}{dt} = \int x_i x_k \frac{\partial \rho}{\partial t} \, dV$$

$$= -\int x_i x_k \frac{\partial}{\partial x_j}(\rho v_j) \, dV$$

$$= -\int \frac{\partial}{\partial x_j}(x_i x_k \rho v_j) \, dV + \int \rho(v_i x_k + x_i v_k) \, dV.$$

Since the first term above can be transformed into a surface integral and set to zero, we have

$$\frac{dI_{ik}}{dt} = \int \rho(v_i x_k + x_i v_k) \, dV. \tag{17.15}$$

From (17.13) and (17.15), we finally obtain

$$\frac{1}{2} \frac{d^2 I_{ik}}{dt^2} = 2T_{ik} + \delta_{ik}\Pi + \int M_{ik} \, dV + W_{ik}. \tag{17.16}$$

This is the celebrated tensor virial equation.

The physical significance of (17.16) becomes clearer if we take the trace of the equation. Using the expression (14.8) for M_{ik}, we have

$$\frac{1}{2}\frac{d^2 I}{dt^2} = 2T + 3\Pi + \int \frac{B^2}{8\pi}\, dV + W, \tag{17.17}$$

where

$$I = I_{ii} = \int \rho r^2 \, dV \tag{17.18}$$

is the moment of inertia;

$$T = T_{ii} = \frac{1}{2}\int \rho v^2 \, dV \tag{17.19}$$

is the kinetic energy; and

$$W = W_{ii} = -\frac{G}{2}\iint dV \, dV' \frac{\rho(\mathbf{x})\rho(\mathbf{x}')}{|\mathbf{x}-\mathbf{x}'|} \tag{17.20}$$

is the total potential energy of the system. Making use of (17.1), we can write (17.20) as

$$W = \frac{1}{2}\int \rho(\mathbf{x})\phi(\mathbf{x})\, dV. \tag{17.21}$$

When a system expands, its moment of inertia I increases in general, as should be obvious from the definition (17.18). It is not difficult to see that a positive term on the R.H.S. of (17.17) leads to an increase in I. Such a positive term, therefore, corresponds to an expansive tendency. On the other hand, a negative term on the R.H.S. of (17.17) tries to make the system more compact. We now note that W, as given by (17.20), is the only negative term on the R.H.S. of (17.17), all the other terms being positive. It is no wonder that the kinetic energy or the term Π, which corresponds to the thermal energy, tends to expand the system. The effect of magnetic field is more subtle. We have seen in §14.1 that the magnetic field has tension along field lines and pressure in the two perpendicular directions, which is obvious in the expression (14.9). Since all the diagonal terms in (14.9) have the same magnitude with two positive terms, we expect the overall average effect to be that of a pressure and hence expansive. As one might have anticipated, gravity is the only force which introduces a confining tendency through the negative term W.

17.1.1 Application to gravitationally bound systems

Very often we have astrophysical systems in which gravity balances the expansive forces so that the system is in a steady state. Let us consider such a system without magnetic forces. Then the only terms surviving in (17.17) are

$$2T + 3\Pi + W = 0. \tag{17.22}$$

From (3.10), (3.20) and (3.22), we have

$$p = \tfrac{1}{3}\rho\langle|\mathbf{u} - \mathbf{v}|^2\rangle. \tag{17.23}$$

From the definition (17.11) of Π, we then have

$$3\Pi = 2T_t,$$

where

$$T_t = \tfrac{1}{2} \int \rho\langle|\mathbf{u} - \mathbf{v}|^2\rangle \, dV \tag{17.24}$$

is the total thermal energy. We can thus write (17.22) as

$$2(T + T_t) + W = 0. \tag{17.25}$$

We note that the bulk kinetic energy T and the thermal energy T_t, which is essentially the kinetic energy at the microscopic level, enter (17.25) in exactly the same fashion.

For a star in static equilibrium without any bulk motion, the virial theorem takes the very simple form

$$2T_t + W = 0. \tag{17.26}$$

This is an important global relation for a star and can be derived from the condition of hydrostatic equilibrium (see, for example, Schwarzschild 1958, §5).

Let us now consider a galaxy or a cluster made up of N stars having average mass m. The mass of the whole system is then $Nm = M$. If R is a measure of the size of the system, it follows from (17.20) that the approximate value of W is given by

$$W \approx -\frac{G}{2}\frac{(Nm)^2}{R}. \tag{17.27}$$

The gravitational force can be balanced either by regular circular motions of the stars or their random motions. The virial equation (17.25) makes no distinction between the regular and the random motions. If u is the typical velocity of a star, then $\tfrac{1}{2}Nmu^2$ is the total kinetic energy, and it follows from (17.25) and (17.27) that

$$Nmu^2 - \frac{G}{2}\frac{(Nm)^2}{R} = 0,$$

from which

$$u^2 \approx \frac{GM}{2R}.$$ (17.28)

This very important equation is used routinely to estimate the mass of a stellar system from the measurements of size and stellar velocities.

We now make an estimate of the relaxation time in a stellar system. Since gravity obeys an inverse-square law just like the electrostatic force, the collision frequency of a system of gravitationally attracting particles can be estimated in exactly the same way we estimated the collision frequency in a plasma in §13.3. Replacing m_e by the mass m of a star and replacing e^2 by Gm^2, it follows from (13.18) that the collision frequency in a stellar system is of the order

$$\nu_c \approx \frac{(Gm)^2 n}{u^3},$$

where $n \approx N/R^3$ is the number density per unit volume of the stars. The inverse of ν_c gives the relaxation time

$$\tau_{rel} \approx \frac{u^3 R^3}{G^2 m^2 N} \approx \frac{u^3 R^3 N}{G^2 M^2}.$$ (17.29)

Combining (17.28) and (17.29), we get

$$\tau_{rel} \approx \frac{R}{u} N.$$

Now R/u is the typical time τ_{cr} taken by a star to cross the stellar system. Hence

$$\frac{\tau_{rel}}{\tau_{cr}} \approx N.$$ (17.30)

In other words, if there are N stars in a gravitationally bound stellar system, then the collisional relaxation time of the system is N times the system crossing time τ_{cr}. It may be noted that a fuller analysis shows τ_{rel}/τ_{cr} to be smaller than N by one or two orders of magnitude (see, for example, Binney and Tremaine 1987, §4.1).

Our Galaxy has about 10^{11} stars, whereas the crossing time for a star is about 10^8 years. It is easy to see that the collisional relaxation time is of the order of 10^{18} years, which is many orders larger than the age of the Universe. Hence collisions can be neglected while considering the stellar dynamics of our Galaxy. On the other hand, a globular cluster with typically 10^5 stars has a crossing time of 10^5 years, leading to a collisional relaxation time of about 10^9 years. This is somewhat less than the estimated age of a globular cluster, leading to the conclusion that the globular clusters, in contrast to galaxies, are collisionally relaxed.

17.1.2 Some comments on global methods

Our discussion merely gives an indication of the power of the virial approach. One should look at the classic volume by Chandrasekhar (1969) to appreciate the full potential of this approach. We have multiplied (17.2) by x_k to obtain the usual virial equation. Chandrasekhar (1969) shows that higher-order equations can be obtained by multiplying (17.2) by higher powers such as $x_k x_l$, $x_k x_l x_m$, and then integrating. These higher-order equations often turn out to be useful. In §9.3 we briefly discussed the difficult subject of self-gravitating rotating fluid masses. Chandrasekhar (1969) uses the virial theorem to analyze this subject elegantly by cutting through the maze of mathematical equations and derives many important results *exactly*.

Global methods often prove very powerful in stability analysis as well. We saw in Chapter 7 that the linear perturbation technique can be applied to study the stability of a system. This approach, however, becomes very complicated for realistic systems, especially if magnetic fields are present. One property of a stable equilibrium is that it is at the minimum of the potential energy. Any perturbations around the equilibrium ought to increase the total potential energy. Hence, to determine if an equilibrium is stable, one finds out if all types of perturbations increase the potential energy of the system. Such an approach for MHD problems was systematically developed by Bernstein *et al.* (1958).

17.2 Relativistic hydrodynamics

Relativistic effects can enter hydrodynamics either through microscopic considerations or through macroscopic considerations. Special relativity becomes important at the microscopic level if a gas is so hot that the gas particles move around at relativistic speeds. On the other hand, relativistic considerations enter at the macroscopic level (i) if the bulk velocities of the fluids are relativistic (special relativity), or (ii) if fluids are in a very strong gravitational field such that the gravitational potential energy $m\phi$ of a mass is comparable to its rest mass energy mc^2 (general relativity).

Let us first consider an ultra-hot relativistic gas such that all the gas particles have random velocities close to c. The pressure of a classical gas is given by (17.23). If all the particles are moving randomly with speed c, then the relativistic expression for pressure must be

$$p = \tfrac{1}{3}\rho c^2. \tag{17.31}$$

On applying (6.12) to this equation of state, it readily follows that the sound speed in a relativistic gas is given by

$$c_s = \frac{1}{\sqrt{3}} c,$$ (17.32)

one of the famous results for a relativistic gas.

We now mainly focus our attention to situations in which relativity enters at the macroscopic level, either because of the large bulk velocities or because of the strong gravitational fields. The general relativistic framework provides a completely general treatment of hydrodynamics at the macroscopic level. If we simplify the equations by assuming weak gravity, then we would get the appropriate special relativistic equations. We now discuss the full general relativistic equations. Since we do not expect all readers of this book to be familiar with general relativity, our discussion will be at a very elementary level. We shall try to proceed in a way such that even readers without any knowledge of relativity should have an idea how things go, although they may not be able understand everything fully.

Since many popular books and articles have been written on general relativity, even readers without any technical knowledge of the subject would know that general relativity is a new theory of gravity. In classical physics, gravity is regarded as a force. On the other hand, gravity in general relativity is no longer a force, but a curvature of space-time caused by the presence of matter-energy. The basic equation describing how the curvature of space-time is produced by matter-energy is Einstein's equation

$$G^{\alpha\beta} = \frac{8\pi G}{c^4} \mathscr{T}^{\alpha\beta}$$ (17.33)

(see, for example, Landau and Lifshitz 1975, §95). Here $G^{\alpha\beta}$, which is a complicated tensor known as the Einstein tensor, is a measure of the curvature of space-time. The tensor $\mathscr{T}^{\alpha\beta}$ on the R.H.S. is known as the energy-momentum tensor and is essentially a measure of matter-energy density. We shall see below how one obtains the expression of $\mathscr{T}^{\alpha\beta}$. So far in the book, we have not considered the distinction between contravariant and covariant tensors, because this distinction disappears in flat Cartesian space. Now, one has to keep that distinction in mind. One standard result of tensor analysis is that the divergence of the Einstein tensor $G^{\alpha\beta}$ is zero, i.e.

$$\frac{DG^{\alpha\beta}}{Dx^\beta} = 0.$$ (17.34)

where D/Dx^β indicates a special kind of derivative known as the covariant derivative, which is a generalization of the ordinary derivative appropriate for curved space (see, for example, Landau and Lifshitz 1975, §85). If we take a divergence of (17.33), then the R.H.S. should give zero as well for the sake of consistency. Hence

$$\frac{D\mathscr{T}^{\alpha\beta}}{Dx^\beta} = 0. \tag{17.35}$$

This is the compact-looking equation within which the hydrodynamic equations are hidden.

For the time being, let us forget about relativity and show that the classical hydrodynamic equations can be put in the form that the divergence of a second-rank tensor is zero. Then we shall consider how to generalize it to general relativity. Let us write x^0 for ct and x^1, x^2, x^3 for the three spatial coordinates. The Greek indices α, β, ... will run over the values 0, 1, 2, 3, whereas the Roman indices i, j, ... will run over only 1, 2, 3. It is easy to see that the continuity equation can be written in the form

$$\frac{\partial S^\alpha}{\partial x^\alpha} = 0, \tag{17.36}$$

where S^α is a four-vector with components $(\rho c, \rho v^1, \rho v^2, \rho v^3)$ and the index α repeated twice implies that we are summing over 0, 1, 2, 3. We also consider the equation of motion in the form (4.23) with the indices written above, i.e.

$$\frac{\partial}{\partial t}(\rho v^i) + \frac{\partial T^{ij}}{\partial x^j} = 0, \tag{17.37}$$

where

$$T^{ij} = p\,\delta^{ij} + \rho v^i v^j. \tag{17.38}$$

It is not difficult to note that (17.36) and (17.37) can be combined in the compact form

$$\frac{\partial(\mathscr{T}_{NR})^{\alpha\beta}}{\partial x^\beta} = 0, \tag{17.39}$$

where $(\mathscr{T}_{NR})^{\alpha\beta}$ is the non-relativistic four-dimensional energy-momentum tensor of which the various components are given by

$$(\mathscr{T}_{NR})^{00} = \rho c^2, \quad (\mathscr{T}_{NR})^{0i} = (\mathscr{T}_{NR})^{i0} = \rho c v^i, \quad (\mathscr{T}_{NR})^{ij} = T^{ij}. \tag{17.40}$$

It should be noted that we have not invoked relativity in obtaining (17.39). Writing x^0 for ct has been merely a matter of notation. The equation (17.39) combines the equations of continuity and motion of classical hydrodynamics. We now have to generalize it to obtain the basic equation of relativistic hydrodynamics.

Let us first consider how we generalize the concept of velocity in general relativity. Suppose a particle has positions x^α and $x^\alpha + dx^\alpha$ before and after an infinitesimal interval. The difference dx^α is a four-vector and the quotient obtained by dividing it by a scalar will be a four-vector as well. According to general relativity, the quantity

$$ds^2 = g_{\alpha\beta}\, dx^\alpha dx^\beta \qquad (17.41)$$

is a scalar if $g_{\alpha\beta}$ is the metric tensor. We use the convention that the metric tensor $g_{\alpha\beta}$ reduces to the following special relativistic metric in the limit of weak gravity

$$g_{\alpha\beta} \to \eta_{\alpha\beta} = \begin{pmatrix} -1 & 0 & 0 & 0 \\ 0 & 1 & 0 & 0 \\ 0 & 0 & 1 & 0 \\ 0 & 0 & 0 & 1 \end{pmatrix}. \qquad (17.42)$$

We now write

$$ds^2 = -c^2 d\tau^2. \qquad (17.43)$$

It should be clear from (17.41–17.43) that τ has the dimension of time and $d\tau \to dt$ in the non-relativistic limit. Since $d\tau$ as introduced in (17.43) must be a scalar, dx^α divided by $d\tau$ should give us a four-vector. We now define the velocity four-vector as

$$u^\alpha = \frac{1}{c}\frac{dx^\alpha}{d\tau}. \qquad (17.44)$$

In the non-relativistic limit, this clearly reduces to

$$u^\alpha \to \left(1, \frac{v^1}{c}, \frac{v^2}{c}, \frac{v^3}{c}\right).$$

We now define the energy-momentum tensor

$$\mathscr{T}^{\alpha\beta} = \rho c^2 u^\alpha u^\beta + p(g^{\alpha\beta} + u^\alpha u^\beta). \qquad (17.45)$$

We leave it as an exercise for the reader to verify that this reduces in the non-relativistic limit to the non-relativistic tensor $(\mathscr{T}_{NR})^{\alpha\beta}$ as given by (17.40). The quantities like ρ and p are defined with respect to the rest frame of the fluid.

Since $\mathscr{T}^{\alpha\beta}$ is a proper relativistic second-rank tensor and reduces to the non-relativistic expression in the appropriate limit, we take it to be relativistic generalization of the energy-momentum tensor. Just as we saw in (17.39) that the divergence of $(\mathscr{T}_{NR})^{\alpha\beta}$ is zero, similarly the divergence of $\mathscr{T}^{\alpha\beta}$ must be zero, provided we replace the ordinary derivatives by covariant derivatives. In other words, equation (17.35) must hold with $\mathscr{T}^{\alpha\beta}$ given by (17.45). Just as (17.39) contains the equations of classical hydrodynamics, we expect (17.35) with $\mathcal{T}^{\alpha\beta}$ defined by (17.45) to contain the equations of relativistic hydrodynamics. In classical hydrodynamics, the equation of energy (see §4.1.2) appears as one of the fundamental equations in addition to the equations of continuity and motion. One wonders what happened to that equation in the relativistic treatment. We refer the reader to Landau and Lifshitz (1987, Chapter XV) or Mihalas and Mihalas (1984, §4.2) for a discussion of this point. Basically, the component $\mathcal{T}^{00}$ contains all forms of energy and the equation of energy is hidden within (17.35).

If we are considering a system in which velocities of the order of c occur without the gravitational field being very strong, then we have to incorporate only special relativity and not general relativity. In a general relativistic treatment, gravity appears in the curvature of space-time and one does not have to incorporate the gravitational force separately. In a special relativistic treatment, however, it is necessary to write any extra force such as gravity in the form of a four-force f^{α}. See Goldstein (1980, §7-6) for a discussion of the concept of four-force. Then, instead of (17.35), we write

$$\frac{\partial \mathcal{T}^{\alpha\beta}}{\partial x^{\beta}} = f^{\alpha} \qquad (17.46)$$

for special relativistic situations. Note that the derivative in (17.46) is an ordinary derivative, unlike the covariant derivative in (17.35). At first sight, it may seem surprising that the general relativistic equation (17.35) appears simpler than the special relativistic equation (17.46). But this is misleading. Calculating a covariant derivative in curved space-time is usually an immensely complicated problem, and (17.46) is actually a simpler equation than (17.35). In fact, (17.46) follows from (17.35) in the limit of weak gravitational field.

The above discussion should give the reader some idea how hydrodynamics is adapted to relativistic situations. Needless to say, just an acquaintance with the basic equation usually does not make one capable of solving practical problems. Only when one works out some important consequences of the basic equation, does one start to have a feeling for the subject. But such a development of relativistic hydrodynamics is beyond the scope of this elementary book. We also do not discuss how MHD equations are generalized to relativistic situations. The reader may look up Misner, Thorne and Wheeler (1973, §22.3-4) for a discussion of hydrodynamics and electrodynamics in curved space-time.

17.3 Radiation hydrodynamics

We now briefly consider the situation in which the dynamics of a fluid is influenced by the presence of a strong radiation field. In this book, we have neglected the interaction between fluids and radiation. One straightforward interaction between fluids and radiation is that a fluid can gain thermal energy by absorbing radiation and can lose thermal energy by emitting radiation. Such gains and losses can be incorporated in the energy equation (4.13), as pointed out in §4.1.2. A much more non-trivial interaction between fluids and radiation is when the radiation pressure causes fluid motions. The radiation pressure in the atmospheres of very massive stars is so strong that it can drive an outward flow of gas in the form of a radiation-driven wind. The study of such situations is the subject of *radiation hydrodynamics*.

We shall see below that one can introduce an energy-momentum tensor $R^{\alpha\beta}$ for the radiation field just like the energy-momentum tensor $\mathcal{T}^{\alpha\beta}$ of the fluid introduced by us in the previous section. In the presence of a radiation field with the energy-momentum tensor $R^{\alpha\beta}$, the equation (17.46) is then modified to

$$\frac{\partial}{\partial x^{\beta}}(\mathcal{T}^{\alpha\beta} + R^{\alpha\beta}) = f^{\alpha}. \tag{17.47}$$

This can also be written in the form

$$\frac{\partial \mathcal{T}^{\alpha\beta}}{\partial x^{\beta}} = f^{\alpha} + G^{\alpha}, \tag{17.48}$$

where

$$G^{\alpha} = -\frac{\partial R^{\alpha\beta}}{\partial x^{\beta}} \tag{17.49}$$

is the four-force due to the radiation field on the fluid. Our job now is to express this force G^{α} in terms of various physical variables.

Just as we introduced a distribution function for a gas of particles in §1.5, we can introduce a similar distribution function for a gas of photons. In the case of particles, the distribution function is a function of $\mathbf{x}$, $\mathbf{u}$ and t. In the case of photons, the distribution function certainly should depend on $\mathbf{x}$ and t. However, instead of $\mathbf{u}$ as in the case of material particles, a photon is characterized by its frequency v and direction of motion $\mathbf{n}$. Hence we write the photon distribution function as $f_R(\mathbf{x}, \mathbf{n}, v, t)$. Instead of this photon distribution function, it is often customary to use the specific intensity $I(\mathbf{x}, \mathbf{n}, v, t)$ which is related to the distribution function as follows

$$I(\mathbf{x}, \mathbf{n}, v, t) = (chv)f_R(\mathbf{x}, \mathbf{n}, v, t). \tag{17.50}$$

It may be noted that the specific intensity $I(\mathbf{x}, \mathbf{n}, v, t)$ can be introduced from purely macroscopic considerations without introducing the concept of a photon gas. We refer the reader to Mihalas (1978, §1-1) for the macroscopic definition of specific intensity and its relation to the photon distribution function.

We now consider the quantity

$$R^{\alpha\beta} = \frac{1}{c} \int dv \int d\Omega \, I \, n^{\alpha} n^{\beta}, \qquad (17.51)$$

where n^{α} is a four-dimensional vector with components ($n^0 = 1$, n^1, n^2, n^3) and $d\Omega$ is the element of solid angle. So the expression (17.51) involves integration over all solid angles and all frequencies of radiation. Substituting (17.50) into (17.51), it should be obvious that R^{00} is the energy density of the radiation field and R^{0i} is c times its momentum density (keeping in mind that the momentum of a photon is hv/c). Now note from (17.40) that $\mathcal{T}^{00}$ gives the energy density of the fluid and $\mathcal{T}^{0i}$ gives c times its momentum density. Based on such comparison, one concludes that $R^{\alpha\beta}$ introduced through (17.51) is the energy-momentum tensor of the radiation field, which should be used in the basic equation (17.47) of radiation hydrodynamics. To use $R^{\alpha\beta}$ in a relativistic formulation, one, however, has to justify that it transforms as a second-rank tensor under Lorentz transformation. We again refer the reader to Mihalas (1978, §14-3) for a discussion of this point.

Finally we discuss how to find an expression for the four-force G^{α} given by (17.49). For this purpose, we have to consider the Boltzmann equation for the photon distribution function $f_R(\mathbf{x}, \mathbf{n}, v, t)$. Let us first consider the situation where there is no matter present, so that the photon gas can be considered collisionless. Keeping in mind that photons move with velocity $\mathbf{u} = c\mathbf{n}$, we substitute $\dot{\mathbf{x}} = \mathbf{u} = c\mathbf{n}$ and $\dot{\mathbf{u}} = 0$ in the collisionless Boltzmann equation (1.19), which gives us

$$\frac{1}{c} \frac{\partial f_R}{\partial t} + \mathbf{n} \cdot \nabla f_R = 0.$$

Because of the relation (17.50) between the specific intensity $I(\mathbf{x}, \mathbf{n}, v, t)$ and the photon distribution function $f_R(\mathbf{x}, \mathbf{n}, v, t)$, we may as well write the collisionless Boltzmann equation in terms of the specific intensity:

$$\frac{1}{c} \frac{\partial I}{\partial t} + \mathbf{n} \cdot \nabla I = 0. \qquad (17.52)$$

This is the equation satisfied by the specific intensity in free space. When matter is present, we have to add appropriate terms to this equation representing the interaction between radiation and matter. We saw in §2.1 that collisions between material particles are taken care of by adding two terms in the Boltzmann equation corresponding to particles entering and leaving a volume element of phase space. In the same fashion, we have to add two terms on the R.H.S. of (17.52) corresponding to radiation being added or removed by matter. The effect of emission of radiation by matter is incorporated by adding a term $\eta(\mathbf{x}, \mathbf{n}, v, t)$ on the R.H.S. of (17.52). This term $\eta(\mathbf{x}, \mathbf{n}, v, t)$ is referred to as the emission coefficient of matter. The absorption of radiation by matter is proportional to the amount of radiation present. Hence the effect of absorption is included though a term $-\chi(\mathbf{x}, \mathbf{n}, v, t)I(\mathbf{x}, \mathbf{n}, v, t)$, where $\chi(\mathbf{x}, \mathbf{n}, v, t)$ is known as the absorption coefficient. On including these terms, (17.52) becomes

$$\frac{1}{c}\frac{\partial I}{\partial t} + \mathbf{n} \cdot \nabla I = \eta - \chi I. \tag{17.53}$$

In many astrophysical problems, one has to study a time-independent radiation field in the presence of matter. For such situations, one has to use the standard *radiative transfer equation*, which follows on setting $\partial I / \partial t = 0$ in (17.53) (see, for example, Mihalas 1978, Chapter 2; Rybicki and Lightman 1979, Chapter 1).

It may be noted that equation (17.53) is usually written in this form only when we are in the rest frame of the fluid. If the fluid is moving with respect to our frame of reference, then we have to carry out appropriate Lorentz transformations (see, for example, Mihalas 1978, §15-3; Mihalas and Mihalas 1984, Chapter 7). In this elementary treatment, however, we neglect those extra complications and work with (17.53). On integrating (17.53) over all solid angles and all frequencies, we get

$$\frac{\partial R^{00}}{\partial x^0} + \frac{\partial R^{0i}}{\partial x^i} = \frac{1}{c}\int dv \int d\Omega \left[\eta(\mathbf{x}, \mathbf{n}, v, t) - \chi(\mathbf{x}, \mathbf{n}, v, t)I(\mathbf{x}, \mathbf{n}, v, t)\right]. \tag{17.54}$$

If we first multiply (17.53) by n^i and then carry on the integration, we arrive at

$$\frac{\partial R^{i0}}{\partial x^0} + \frac{\partial R^{ij}}{\partial x^j} = \frac{1}{c}\int dv \int d\Omega \left[\eta(\mathbf{x}, \mathbf{n}, v, t) - \chi(\mathbf{x}, \mathbf{n}, v, t)I(\mathbf{x}, \mathbf{n}, v, t)\right]n^i. \tag{17.55}$$

We now combine (17.54) and (17.55) in the compact form

$$\frac{\partial R^{\alpha\beta}}{\partial x^\beta} = \frac{1}{c}\int dv \int d\Omega \left[\eta(\mathbf{x}, \mathbf{n}, v, t) - \chi(\mathbf{x}, \mathbf{n}, v, t)I(\mathbf{x}, \mathbf{n}, v, t)\right]n^\alpha. \tag{17.56}$$

From (17.49) and (17.56), we conclude that the four-force due to the radiation on the fluid is

$$G^\alpha = -\frac{1}{c}\int dv \int d\Omega \left[\eta(\mathbf{x}, \mathbf{n}, v, t) - \chi(\mathbf{x}, \mathbf{n}, v, t)I(\mathbf{x}, \mathbf{n}, v, t)\right]n^\alpha. \tag{17.57}$$

Thus the dynamics of a fluid in the presence of a radiation field is governed by the basic equation (17.48) with G^α connected to the emission and absorption properties of the fluid through (17.57).

As in the previous section, we stop here after arriving at the basic equation. Readers wishing to acquire a deeper insight into the subject are urged to look at Mihalas and Mihalas (1984, Chapter 8) to learn how the basic equations are applied to practical problems of astrophysical interest.

17.4 A guided tour through the world of astrophysical fluids and plasmas

Our aim in this book has been to develop the basic physics of fluids and plasmas with applications to astrophysical systems as illustrations of our theory. For students wishing to specialize in this subject, this book should ideally be followed by a second book which discusses one by one those astrophysical systems in which fluid and plasma processes play important roles. In lieu of such a second volume, we offer only a small section of general comments. It will be found that the discussions of any particular astrophysical system are scattered in different portions of this book, wherever that particular system seemed convenient to illustrate some point made in that portion of the book. After learning the material of this book, a reader can form an over-all idea of the different fluid and plasma processes in a particular astrophysical system by looking up those sections of the book in which that system is discussed. Along with a few general comments, we give below a list of various sections which one should look up together to form an idea about a given astrophysical system. To learn more about a system, the reader should consult the references given at the end of the *Suggestions for further reading*.

17.4.1 Stellar convection, rotation and oscillations

In elementary textbooks on stellar astronomy, a star is usually mod-elled as a non-rotating, non-magnetic object of spherical symmetry within which static equilibrium prevails, unless some regions are un-stable to convection. To study such an idealized model of a star, one does not require much sophisticated hydrodynamics. For the regions in static equilibrium, the hydrostatic equilibrium equation is the only hydrodynamic input required. Read §4.4 and work out Exercise 4.3 to learn about hydrostatic equilibrium in a spherically symmetric system. The condition for stability against convection is discussed in §7.2. If a region of the star is convectively unstable, then one has to use some model of convective turbulence as pointed out in §8.6. The layers just underneath the solar surface are unstable to convection, as confirmed by the observations of granulation discussed in §7.3.

If a star is rotating, then it becomes an immensely complicated hydrodynamic system. The theory of rotating stars is still one of the ill-understood areas of astrophysics. The theory of incompressible rotating fluids, of which some elementary results are presented in §9.3, is complicated enough. If the fluid is compressible and inhomogeneous, then the analysis becomes formidable. Some of the difficulties are pointed out in §9.4. Some comments are made on the subject of stellar oscillations in §7.6. With the study of solar oscillations with increasingly sophisticated instruments, this subject has suddenly blossomed into one of the very active areas in modern astrophysics. Solar oscillations have provided a wealth of information about the interior of the Sun. Perhaps the most dramatic result in this area is the production of a map of angular velocity distribution presented in Figure 7.8. So there is at least one star within which the angular velocity distribution has been mapped observationally. But our theoretical understanding of rotating stars is still not good enough to provide a proper explanation of this map.

17.4.2 Magnetohydrodynamics of the Sun and the solar system

Some of the most intriguing theoretical predictions of MHD are for systems with large magnetic Reynolds numbers. Since such large magnetic Reynolds numbers cannot be reached in laboratory experiments, the Sun is the best laboratory for studying some of these magnetohydrodynamic phenomena. There is growing evidence that similar phenomena take place in the distant stars as well. In fact, some stars seem to be much more strongly magnetic than the Sun. But it is not possible to study these phenomena in distant stars in detail, as we can in the case of the Sun.

The origin of the solar magnetic fields by the dynamo process is discussed in Chapter 16. The observational data on the statistics of sunspots are summarized in §16.1, whereas §16.6 shows how a simple dynamo solution can explain much of these data. After the question of origin, the second question is to understand why the solar magnetic fields appear concentrated in sunspots. The theory of magnetoconvection discussed in §14.6 throws light on this question. After being produced in the interior of the Sun, the solar magnetic fields rise to the surface due to magnetic buoyancy, as pointed out in

§14.7. The solar corona is full of magnetic fields. The solar flares taking place above the surface are presumably caused by the conversion of magnetic energy to other forms due to fast reconnection processes outlined in §15.2. In §15.5 we discuss how the magnetic fields in the corona may be responsible for the high temperature of the corona. It is pointed out in §4.4 that this high temperature makes it impossible for the corona to be in static equilibrium and drives the solar wind. An idealized model of the solar wind is presented in §6.8. We point out in §14.9.2 that the presence of magnetic fields makes the theoretical investigation of the wind more complicated.

Most of the planets have dipole-type magnetic fields associated with them. We point out in §10.4 that charged particles may be trapped by a planetary magnetic field, giving rise to a magnetosphere around the planet. The solar wind flowing through the solar system impinges on the planetary magnetospheres. Due to the presence of the strong magnetic field, the solar wind cannot easily penetrate into the magnetosphere, but flows around it after producing a shock in front of the magnetosphere. The interaction between the magnetosphere and the solar wind makes the magnetosphere highly asymmetric in the forward and backward directions, as sketched in Figure 17.1.

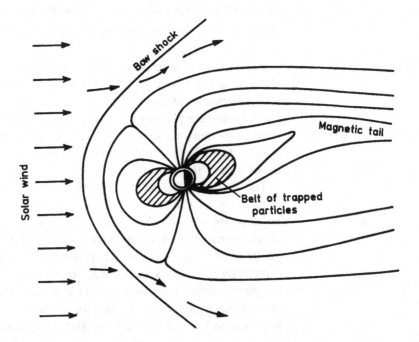

Figure 17.1 A sketch of a planetary magnetosphere.

17.4.3 Neutron stars and pulsars

For a long time, astrophysicists regarded neutron stars as hypothetical objects. After the discovery of pulsars, however, it was realized very quickly that pulsars have to be rotating neutron stars. The main arguments for identifying pulsars as rotating neutron stars are summarized in §9.4. Pulsars are also known to have strong magnetic fields for reasons pointed out in §14.2. We discuss in §13.7 that the strong magnetic field of the pulsar should create a magnetosphere of plasma around the pulsar. Presumably, complicated plasma processes in this magnetosphere are responsible for the emission of radiation from a pulsar. Since the low-frequency radio emission from a pulsar is affected by the interstellar medium during its passage, one can obtain information about the interstellar medium by analyzing pulsar signals, as discussed in §12.6.

Within a decade of the discovery of pulsars, a second evidence came for the existence of neutron stars. With the advent of X-ray astronomy, it was found that some close binary stars are strong emitters of X-radiation. The most plausible explanation for this phenomenon is that one member of the binary is a neutron star accreting matter from the other giant star. The theory of accretion disks is developed in §5.7, whereas we present a discussion of spherical accretion in §6.8.

17.4.4 Interstellar medium

The space between stars is filled with the interstellar medium threaded by magnetic fields. We point out in §12.6 that pulsar signals can be analyzed to provide valuable information about the interstellar medium including the interstellar magnetic field. This magnetic field is presumably produced by the dynamo process discussed in Chapter 16. Some observations of galactic magnetic fields are described in §16.1. The magnetic field causes the interstellar medium to break into clumps due to the Parker instability, as pointed out in §14.8.

New stars are born out of the interstellar medium and the massive old stars give back some of their material to the interstellar medium during supernova explosions. The star formation process is triggered by the Jeans instability discussed in §7.5. In order for the gravitational collapse to proceed, considerable amounts of angular momentum and magnetic flux have to be removed from the collapsing gas mass. The removal of angular momentum and the removal of magnetic flux are respectively discussed in §14.9.1 and §14.10. A supernova explosion sends out a blast wave in the interstellar medium. The dynamics of such a blast wave is studied in §6.6. We point out in §7.4.2 how Rayleigh–Taylor instabilities may develop in supernova remnants.

The interstellar medium is also full of cosmic ray particles, which are probably produced in the shock fronts of supernova explosions. The important subject of particle acceleration in shock fronts is discussed in §10.5.

17.4.5 Active galactic nuclei

There are many galaxies which are not merely simple collections of stars, but seem to possess central powerhouses emitting huge amounts of energy. Quasars are the most extreme examples of such active galaxies. Often the energy is emitted from a very compact nucleus in the central region of the galaxy. The most obvious explanation for such galactic activity is that an active galaxy has a massive black hole in its nucleus and the power is released by the accretion of matter onto this black hole. We present the theory of accretion disks in §5.7 and the theory of spherical accretion in §6.8.

One intriguing aspect of galactic activity is that often these active nuclei produce plasma jets of gigantic size going out of the parent galaxies. A hydrodynamic model of jet production is discussed in §6.7. It appears that magnetic fields probably have an important role in the production of jets. We discuss a hydromagnetic model in §14.9.3, pointing out how the magnetic field may be responsible for driving the jet from the accretion disk around the galactic nucleus.

An active galactic nucleus is a system which usually necessitates the application of relativistic corrections to the usual hydrodynamic and hydromagnetic equations. We present the rudiments of relativistic hydrodynamics in §17.2. Many extragalactic jets seem to be moving with speeds close to c. One has to apply special relativity to study the dynamics of such jets. On the other hand, the gravitational field is very strong in the inner regions of the accretion disk around the black hole in the galactic nucleus. Hence general relativity becomes essential in the study of such galactic accretion disks.

17.4.6 Stellar dynamics

The stars in a galaxy or a cluster can be thought to make up a stellar fluid just like the particles in an ordinary gas or a plasma. We present a comparison of these systems in §3.7 and point out that a fluid of stars is, in some ways, profoundly different from an ordinary gas or a plasma. The stars interact through long-range gravitational forces which cannot be screened (unlike electromagnetic forces). This long-range nature of interactions makes a thermodynamic equilibrium impossible for the stellar system. Hence many concepts developed in this book which tacitly assume thermodynamic equilibrium cannot be applied to stellar systems. Although it is *not* possible to get a systematic introduction to stellar dynamics from this book, a few sections here and there would give the reader some elementary idea of the subject. Read §3.2 and work out Exercise 3.3 to learn how the moment equations can be useful in stellar dynamics, even though they cannot be closed due to the lack of thermodynamic equilibrium.

Depending on whether collisions have played an important role or not, stellar dynamics is broadly divided into two parts: *collisionless* and *collisional* stellar dynamics. The virial theorem discussed in §17.1.1 is applicable to both types of stellar dynamical systems. The collisional relaxation time for a stellar system is also estimated in §17.1.1 by using some results obtained in §13.3 for collisions between particles having an inverse-square law of force. Some techniques for treating collisions in plasmas can be applied to collisional stellar dynamics as well. Read discussions of the BBGKY hierarchy in §11.4 and the Fokker–Planck equation in §11.6 to learn about some ideas which are fruitful in stellar dynamics. The very important concept of dynamical friction is discussed in §11.6.

17.5 Final goodbye to the reader

The previous section should make it clear that fluid mechanics and plasma physics are used in studying diverse types of astrophysical systems. The extensive use of fluid mechanics and plasma physics in astrophysical problems, however, began only after World War II. Until the time of World War II, nearly all of our knowledge about the astronomical Universe was gathered through observations at optical (i.e. visible) wavelengths. It may be noted that black bodies with temperatures in the range 4000 K to 10 000 K have the peaks of their spectra within the optical wavelengths, and this temperature range roughly happens to coincide with the range of surface temperatures of the main-sequence stars. Hence astronomers mainly saw ordinary stars and collections of such stars (such as normal galaxies) through their telescopes as long as all astronomical observations were limited to the optical window of the electromagnetic spectrum. Although the evidence for the existence of interstellar matter was mounting, optical observations proved inadequate to reveal its detailed nature. As we have pointed out in §17.4.1, the study of a spherically symmetric main-sequence star does not involve any sophisticated hydrodynamics. One needs to use only the equation of hydrostatic equilibrium and some phenomenological model of convection. Hence an astronomer in the first half of the twentieth century did not require much knowledge of fluid mechanics and plasma physics to make sense of the astronomical Universe known at that time.

The situation changed drastically after World War II. New windows of electromagnetic spectrum started being opened for astronomical observation. The radar technology developed during World War II gave a tremendous boost to the nascent subject of radio astronomy. The detailed structure and distribution of the interstellar medium could be mapped with radio telescopes. Radio astronomy also unravelled mysterious objects nobody suspected before. The study of active galactic nuclei became a major area of research with the discovery of radio galaxies, quasars and extragalactic radio jets. The dramatic discovery of pulsars confirmed the existence of neutron stars. Optical astronomy also got revolutionized by the invention of many instruments which could be used in conjunction with the telescope. Although the Sun has been studied with optical telescopes ever since the time of Galileo, many of the intriguing fluid and plasma processes on and around the Sun could be studied in detail only within the last few decades. The evidence of the galactic magnetic field also first came through optical observations shortly after World War II and established the fact that magnetic fields are ubiquitous in the Universe. Another boon to astronomy was the heralding of the Space Age in the 1950s. It became possible to send instruments outside the Earth's atmosphere. *In-situ* measurements from spacecrafts led to the discovery of the Van Allen belt and the solar wind predicted theoretically a few years earlier. X-ray astronomy started taking its first halting steps with rocket-borne instruments shortly after World War II. But X-ray astronomy really came of age when sophisticated X-ray telescopes were carried in spacecrafts. The discovery of X-ray emitting binary stars rapidly led to the development of the theory of accretion disks, which is also applied to the study of active galactic nuclei. One of the last windows of astronomical observation—the infrared window—is providing us unusual glimpses of fluid flow patterns in the star-forming regions.

Even around 1970, one could think of becoming a professional astronomer without much knowledge of fluid mechanics and plasma physics. But those days are gone forever. The new astronomies have opened up a radically new vision of the Universe in which fluid and plasma processes reign supreme. One has to carry on detailed hydrodynamic and hydromagnetic modelling to understand the behaviours of many astronomical systems unravelled by the new astronomies. The previously undreamt-of computational powers made available by successive generations of computers have led to extremely rapid developments in the field of astrophysical hydrodynamics and magnetohydrodynamics. It may be noted that the theoretical foundations of plasma physics were being firmly established around the same time when the importance of astrophysical plasma processes was realized. The various fusion research programmes which started around 1950 have not yet achieved their final goal, but have given a tremendous boost to plasma physics and have provided a much deeper insight into the complex behaviours of plasmas.

When we look up at the dark, clear night sky at a place far away from the city lights, it gives us a feeling of peace, tranquility and timelessness. People of ancient civilizations used to regard the celestial sphere as an abode of calm and tranquility. We see changes around us on the surface of the Earth. We see things grow and decay. But the heavens were thought to be eternal and changeless. That is why the sudden appearance of a comet used to cause such consternation. This view of a tranquil and eternal stellar Universe survived the scientific revolution of the Renaissance, down to the early decades of the twentieth century. Only in the early decades of the twentieth century did it become clear that stars are born and die just like biological organisms. The Universe itself was born in a fiery explosion and will eventually die either through unlimited expansion or through another crunch. But the time scales of these evolutions seemed too vast by human standards. The Universe still appeared to be an ordered and peaceful place, where changes take place gracefully and imperceptibly in majestically long time scales. Only with the development of the new astronomies of the late twentieth century, the ancient image of a tranquil and ordered Universe is finally shattered. We have learnt that the Universe is quite 'messy' and full of many violent processes not suspected before. Incredible amounts of energy can be released within very compact and small regions. Rapid variabilities are often not uncommon. Fluid mechanics and plasma physics are providing the keys in our understanding of this previously unknown messy and violent Universe.

This book is *not* meant to bring you to such a level that you start understanding any odd research paper on astrophysical fluids and plasmas. Rather, the book presents the minimum which, in my opinion, every student of modern astrophysics should know. For some of you embarking on a professional career of astrophysics, this book may provide all you need to know about fluid mechanics and plasma physics. But I fondly hope that there will be some readers for whom this book will be only a starting point and who may wish to devote their lives studying astrophysical fluids and plasmas. With my best wishes to such readers as well as the others, let me say my final goodbye!

APPENDIX A

Useful vector relations

We assume that the readers of this book are familiar with the dot and cross products of vectors as well as the gradient, divergence and curl operators. Here we collect together some important relations involving the products and the operators, which are used throughout the book.

A.1 General identities

For products of three vectors $\mathbf{A}$, $\mathbf{B}$ and $\mathbf{C}$, we have the following identities:

$$\mathbf{A} \cdot (\mathbf{B} \times \mathbf{C}) = \mathbf{B} \cdot (\mathbf{C} \times \mathbf{A}) = \mathbf{C} \cdot (\mathbf{A} \times \mathbf{B}), \tag{A.1}$$

$$\mathbf{A} \times (\mathbf{B} \times \mathbf{C}) = (\mathbf{A} \cdot \mathbf{C})\mathbf{B} - (\mathbf{A} \cdot \mathbf{B})\mathbf{C}. \tag{A.2}$$

Let ϕ, ψ be two scalar fields and $\mathbf{A}$, $\mathbf{B}$ two vector fields. The different operators acting on products satisfy the following relations:

$$\nabla(\phi\psi) = \phi\,\nabla\psi + \psi\,\nabla\phi, \tag{A.3}$$

$$\nabla \cdot (\psi\mathbf{A}) = \mathbf{A} \cdot \nabla\psi + \psi\,\nabla \cdot \mathbf{A}, \tag{A.4}$$

$$\nabla \times (\psi\mathbf{A}) = \psi\nabla \times \mathbf{A} - \mathbf{A} \times \nabla\psi, \tag{A.5}$$

$$\nabla(\mathbf{A} \cdot \mathbf{B}) = \mathbf{A} \times (\nabla \times \mathbf{B}) + \mathbf{B} \times (\nabla \times \mathbf{A}) + (\mathbf{A} \cdot \nabla)\mathbf{B} + (\mathbf{B} \cdot \nabla)\mathbf{A}, \tag{A.6}$$

$$\nabla \cdot (\mathbf{A} \times \mathbf{B}) = \mathbf{B} \cdot (\nabla \times \mathbf{A}) - \mathbf{A} \cdot (\nabla \times \mathbf{B}), \tag{A.7}$$

$$\nabla \times (\mathbf{A} \times \mathbf{B}) = \mathbf{A}(\nabla \cdot \mathbf{B}) - \mathbf{B}(\nabla \cdot \mathbf{A}) + (\mathbf{B} \cdot \nabla)\mathbf{A} - (\mathbf{A} \cdot \nabla)\mathbf{B}. \tag{A.8}$$

Two operators operating in succession satisfy the relations:

$$\nabla \times \nabla\psi = 0, \tag{A.9}$$

$$\nabla \cdot (\nabla \times \mathbf{A}) = 0, \tag{A.10}$$

$$\nabla \cdot \nabla \psi = \nabla^2 \psi, \tag{A.11}$$

$$\nabla \times (\nabla \times \mathbf{A}) = \nabla(\nabla \cdot \mathbf{A}) - \nabla^2 \mathbf{A}. \tag{A.12}$$

A.2 Integral relations

If V be a volume bounded by a closed surface S, with $d\mathbf{S}$ taken positive in the outward direction,

$$\oint_S \mathbf{A} \cdot d\mathbf{S} = \int_V (\nabla \cdot \mathbf{A}) \, dV, \tag{A.13}$$

$$\oint_S \psi \, d\mathbf{S} = \int_V (\nabla \psi) \, dV, \tag{A.14}$$

where (A.13) is Gauss's theorem.

If an open surface S is bounded by a contour C, of which $d\mathbf{l}$ is the line element, then Stokes's theorem states that

$$\oint_C \mathbf{A} \cdot d\mathbf{l} = \int_S (\nabla \times \mathbf{A}) \cdot d\mathbf{S}. \tag{A.15}$$

A.3 Tensorial notation

We sometimes use the tensorial notation in our analysis. When tensors are considered in Cartesian geometry, it is not necessary to distinguish between contravariant and covariant tensors. We, therefore, always indicate the components of tensors by subscripts. The components of a vector $\mathbf{A}$ are indicated by $A_i = (A_x, A_y, A_z)$. The only exception is the position vector $\mathbf{x}$ of which the components are indicated by $x_i = (x, y, z)$, i.e. not by (x_x, x_y, x_z).

The letters $i, j, k, l, \ldots$ are used as subscripts to denote the components of vectors and tensors. We use the standard summation convention: an index repeated twice in a term implies summation over x, y and z, unless otherwise specified. To give some examples,

$$A_i B_i = A_x B_x + A_y B_y + A_z B_z = \mathbf{A} \cdot \mathbf{B}, \tag{A.16}$$

$$\frac{\partial A_i}{\partial x_i} = \frac{\partial A_x}{\partial x} + \frac{\partial A_y}{\partial y} + \frac{\partial A_z}{\partial z} = \nabla \cdot \mathbf{A}. \tag{A.17}$$

We have used the summation convention only for summing over x, y and z. Any other kind of summation is indicated explicitly.

The components (T_{ix}, T_{iy}, T_{iz}) of a tensor T_{ij} can be thought to be a vector. The divergence of this vector is given by $\partial T_{ij}/\partial x_j$. We can apply Gauss's theorem (A.13) to transform a volume integral of this divergence to a surface integral, i.e.

$$\int_V \frac{\partial T_{ij}}{\partial x_j} \, dV = \oint_S T_{ij} \, dS_j, \tag{A.18}$$

since $T_{ij}dS_j$ is the dot product between the vector (T_{ix}, T_{iy}, T_{iz}) and the surface element $d\mathbf{S}$.

We often use the Kronecker delta δ_{ij} defined as

$$\delta_{xx} = \delta_{yy} = \delta_{zz} = 1,$$

$$\delta_{xy} = \delta_{yx} = \delta_{yz} = \delta_{zy} = \delta_{zx} = \delta_{xz} = 0. \tag{A.19}$$

It is obvious that

$$\delta_{ij}A_j = A_i, \tag{A.20}$$

$$\delta_{ii} = 3. \tag{A.21}$$

We also assume the reader to be familiar with the Levi-Civita symbol ϵ_{ijk} defined as

$$\epsilon_{xyz} = \epsilon_{yzx} = \epsilon_{zxy} = 1,$$

$$\epsilon_{xzy} = \epsilon_{yxz} = \epsilon_{zyx} = -1, \tag{A.22}$$

$$\text{all other } \epsilon_{ijk} = 0.$$

The cross product $\mathbf{C} = \mathbf{A} \times \mathbf{B}$ of a vector can be written as

$$C_i = (\mathbf{A} \times \mathbf{B})_i = \epsilon_{ijk}A_jB_k, \tag{A.23}$$

whereas the curl of a vector can be written as

$$(\nabla \times \mathbf{A})_i = \epsilon_{ijk}\frac{\partial A_k}{\partial x_j}. \tag{A.24}$$

One important relation to note is

$$\epsilon_{ijk}\epsilon_{pqk} = \delta_{ip}\delta_{jq} - \delta_{iq}\delta_{jp}. \tag{A.25}$$

APPENDIX B

Integrals in kinetic theory

In kinetic theory, one often encounters integrals of the type

$$I_n = \int_0^\infty e^{-\alpha x^2} x^n dx, \tag{B.1}$$

where n is a non-negative integer. We note that

$$I_{n+2} = -\frac{\partial I_n}{\partial \alpha}. \tag{B.2}$$

Using this relation, all integrals of the type (B.1) can be evaluated from the first two integrals

$$I_0 = \frac{1}{2}\sqrt{\frac{\pi}{\alpha}}, \tag{B.3}$$

$$I_1 = \frac{1}{2\alpha}. \tag{B.4}$$

Evaluating I_1 is straightforward. Methods of evaluating I_0 are discussed in many standard textbooks (see, for example, Reif 1965, pp. 606–9). We list below some more of the integrals obtained by applying (B.2) on (B.3) and (B.4):

$$I_2 = \frac{1}{4}\sqrt{\frac{\pi}{\alpha^3}}, \tag{B.5}$$

$$I_3 = \frac{1}{2\alpha^2}. \tag{B.6}$$

APPENDIX C

Formulae and equations in cylindrical and spherical coordinates

C.1 Vector formulae in cylindrical coordinates

If ψ is a scalar field and $\mathbf{A} = A_r\hat{\mathbf{e}}_r + A_\theta\hat{\mathbf{e}}_\theta + A_z\hat{\mathbf{e}}_z$ is a vector field, then

$$\nabla\psi = \frac{\partial\psi}{\partial r}\hat{\mathbf{e}}_r + \frac{1}{r}\frac{\partial\psi}{\partial\theta}\hat{\mathbf{e}}_\theta + \frac{\partial\psi}{\partial z}\hat{\mathbf{e}}_z, \tag{C.1}$$

$$\nabla\cdot\mathbf{A} = \frac{1}{r}\frac{\partial}{\partial r}(rA_r) + \frac{1}{r}\frac{\partial A_\theta}{\partial\theta} + \frac{\partial A_z}{\partial z}, \tag{C.2}$$

$$\nabla\times\mathbf{A} = \left(\frac{1}{r}\frac{\partial A_z}{\partial\theta} - \frac{\partial A_\theta}{\partial z}\right)\hat{\mathbf{e}}_r + \left(\frac{\partial A_r}{\partial z} - \frac{\partial A_z}{\partial r}\right)\hat{\mathbf{e}}_\theta + \frac{1}{r}\left[\frac{\partial}{\partial r}(rA_\theta) - \frac{\partial A_r}{\partial\theta}\right]\hat{\mathbf{e}}_z, \tag{C.3}$$

$$\nabla^2\psi = \frac{1}{r}\frac{\partial}{\partial r}\left(r\frac{\partial\psi}{\partial r}\right) + \frac{1}{r^2}\frac{\partial^2\psi}{\partial\theta^2} + \frac{\partial^2\psi}{\partial z^2}. \tag{C.4}$$

C.2 The Navier–Stokes equation in cylindrical coordinates

The components of the Navier–Stokes equation (5.10) in cylindrical coordinates are

$$\frac{\partial v_r}{\partial t} + v_r\frac{\partial v_r}{\partial r} + \frac{v_\theta}{r}\frac{\partial v_r}{\partial\theta} + v_z\frac{\partial v_r}{\partial z} - \frac{v_\theta^2}{r} = -\frac{1}{\rho}\frac{\partial p}{\partial r}$$
$$+ \nu\left(\frac{\partial^2 v_r}{\partial r^2} + \frac{1}{r^2}\frac{\partial^2 v_r}{\partial\theta^2} + \frac{\partial^2 v_r}{\partial z^2} + \frac{1}{r}\frac{\partial v_r}{\partial r} - \frac{2}{r^2}\frac{\partial v_\theta}{\partial\theta} - \frac{v_r}{r^2}\right) + F_r, \tag{C.5}$$

$$\frac{\partial v_\theta}{\partial t} + v_r\frac{\partial v_\theta}{\partial r} + \frac{v_\theta}{r}\frac{\partial v_\theta}{\partial\theta} + v_z\frac{\partial v_\theta}{\partial z} + \frac{v_r v_\theta}{r} = -\frac{1}{\rho r}\frac{\partial p}{\partial\theta}$$
$$+ \nu\left(\frac{\partial^2 v_\theta}{\partial r^2} + \frac{1}{r^2}\frac{\partial^2 v_\theta}{\partial\theta^2} + \frac{\partial^2 v_\theta}{\partial z^2} + \frac{1}{r}\frac{\partial v_\theta}{\partial r} + \frac{2}{r^2}\frac{\partial v_r}{\partial\theta} - \frac{v_\theta}{r^2}\right) + F_\theta, \tag{C.6}$$

$$\frac{\partial v_z}{\partial t} + v_r \frac{\partial v_z}{\partial r} + \frac{v_\theta}{r} \frac{\partial v_z}{\partial \theta} + v_z \frac{\partial v_z}{\partial z}$$

$$= -\frac{1}{\rho} \frac{\partial p}{\partial z} + v \left(\frac{\partial^2 v_z}{\partial r^2} + \frac{1}{r^2} \frac{\partial^2 v_z}{\partial \theta^2} + \frac{\partial^2 v_z}{\partial z^2} + \frac{1}{r} \frac{\partial v_z}{\partial r} \right) + F_z. \quad \text{(C.7)}$$

C.3 Vector formulae in spherical coordinates

If ψ is a scalar field and $\mathbf{A} = A_r \hat{\mathbf{e}}_r + A_\theta \hat{\mathbf{e}}_\theta + A_\phi \hat{\mathbf{e}}_\phi$ is a vector field, then

$$\nabla \psi = \frac{\partial \psi}{\partial r} \hat{\mathbf{e}}_r + \frac{1}{r} \frac{\partial \psi}{\partial \theta} \hat{\mathbf{e}}_\theta + \frac{1}{r \sin \theta} \frac{\partial \psi}{\partial \phi} \hat{\mathbf{e}}_\phi, \quad \text{(C.8)}$$

$$\nabla \cdot \mathbf{A} = \frac{1}{r^2} \frac{\partial}{\partial r} (r^2 A_r) + \frac{1}{r \sin \theta} \frac{\partial}{\partial \theta} (\sin \theta A_\theta) + \frac{1}{r \sin \theta} \frac{\partial A_\phi}{\partial \phi} \quad \text{(C.9)}$$

$$\nabla \times \mathbf{A} = \frac{1}{r \sin \theta} \left[\frac{\partial}{\partial \theta} (\sin \theta A_\phi) - \frac{\partial A_\theta}{\partial \phi} \right] \hat{\mathbf{e}}_r + \left[\frac{1}{r \sin \theta} \frac{\partial A_r}{\partial \phi} \right.$$

$$\left. - \frac{1}{r} \frac{\partial}{\partial r} (r A_\phi) \right] \hat{\mathbf{e}}_\theta + \frac{1}{r} \left[\frac{\partial}{\partial r} (r A_\theta) - \frac{\partial A_r}{\partial \theta} \right] \hat{\mathbf{e}}_\phi \quad \text{(C.10)}$$

$$\nabla^2 \psi = \frac{1}{r^2} \frac{\partial}{\partial r} \left(r^2 \frac{\partial \psi}{\partial r} \right) + \frac{1}{r^2 \sin \theta} \frac{\partial}{\partial \theta} \left(\sin \theta \frac{\partial \psi}{\partial \theta} \right) + \frac{1}{r^2 \sin^2 \theta} \frac{\partial^2 \psi}{\partial \phi^2}. \quad \text{(C.11)}$$

C.4 The Navier–Stokes equation in spherical coordinates

The components of the Navier–Stokes equation (5.10) in spherical coordinates are

$$\frac{\partial v_r}{\partial t} + v_r \frac{\partial v_r}{\partial r} + \frac{v_\theta}{r} \frac{\partial v_r}{\partial \theta} + \frac{v_\phi}{r \sin \theta} \frac{\partial v_\phi}{\partial \phi} - \frac{v_\theta^2 + v_\phi^2}{r}$$

$$= -\frac{1}{\rho} \frac{\partial p}{\partial r} + v \left[\frac{1}{r} \frac{\partial^2}{\partial r^2} (r v_r) + \frac{1}{r^2} \frac{\partial^2 v_r}{\partial \theta^2} + \frac{1}{r^2 \sin^2 \theta} \frac{\partial^2 v_r}{\partial \phi^2} + \frac{\cot \theta}{r^2} \frac{\partial v_r}{\partial \theta} \right.$$

$$\left. - \frac{2}{r^2} \frac{\partial v_\theta}{\partial \theta} - \frac{2}{r^2 \sin \theta} \frac{\partial v_\phi}{\partial \phi} - \frac{2 v_r}{r^2} - \frac{2 \cot \theta}{r^2} v_\theta \right] + F_r, \quad \text{(C.12)}$$

$$\frac{\partial v_\theta}{\partial t} + v_r \frac{\partial v_\theta}{\partial r} + \frac{v_\theta}{r} \frac{\partial v_\theta}{\partial \theta} + \frac{v_\phi}{r \sin \theta} \frac{\partial v_\theta}{\partial \phi} + \frac{v_r v_\theta}{r} - \frac{v_\phi^2 \cot \theta}{r}$$

$$= -\frac{1}{\rho r} \frac{\partial p}{\partial \theta} + v \left[\frac{1}{r} \frac{\partial^2}{\partial r^2} (r v_\theta) + \frac{1}{r^2} \frac{\partial^2 v_\theta}{\partial \theta^2} + \frac{1}{r^2 \sin^2 \theta} \frac{\partial^2 v_\theta}{\partial \phi^2} + \frac{\cot \theta}{r^2} \frac{\partial v_\theta}{\partial \theta} \right.$$

$$\left. - \frac{2 \cos \theta}{r^2 \sin^2 \theta} \frac{\partial v_\phi}{\partial \phi} + \frac{2}{r^2} \frac{\partial v_r}{\partial \theta} - \frac{v_\theta}{r^2 \sin^2 \theta} \right] + F_\theta, \quad \text{(C.13)}$$

$$\frac{\partial v_\phi}{\partial t} + v_r \frac{\partial v_\phi}{\partial r} + \frac{v_\theta}{r} \frac{\partial v_\phi}{\partial \theta} + \frac{v_\phi}{r \sin \theta} \frac{\partial v_\phi}{\partial \phi} + \frac{v_r v_\phi}{r} + \frac{v_\theta v_\phi \cot \theta}{r}$$

$$= -\frac{1}{\rho r \sin \theta} \frac{\partial p}{\partial \phi} + v \left[\frac{1}{r} \frac{\partial^2}{\partial r^2}(rv_\phi) + \frac{1}{r^2} \frac{\partial^2 v_\phi}{\partial \theta^2} + \frac{1}{r^2 \sin^2 \theta} \frac{\partial^2 v_\phi}{\partial \phi^2} \right.$$

$$\left. + \frac{\cot \theta}{r^2} \frac{\partial v_\phi}{\partial \theta} + \frac{2}{r^2 \sin \theta} \frac{\partial v_r}{\partial \phi} + \frac{2 \cos \theta}{r^2 \sin^2 \theta} \frac{\partial v_\theta}{\partial \phi} - \frac{v_\phi}{r^2 \sin^2 \theta} \right] + F_\phi.$$

$$(C.14)$$

Values of various quantities

D.1 Physical constants

Speed of light	$c = 3.00 \times 10^{10}$ cm s^{-1}
Gravitation constant	$G = 6.67 \times 10^{-8}$ dyne cm^2 g^{-2}
Planck constant	$h = 6.63 \times 10^{-27}$ erg s
Boltzmann constant	$\kappa_B = 1.38 \times 10^{-16}$ erg deg^{-1}
Charge of electron	$e = 4.80 \times 10^{-10}$ e.s.u.
Mass of electron	$m_e = 9.11 \times 10^{-28}$ g
Mass of hydrogen atom	$m_H = 1.67 \times 10^{-24}$ g
Stefan–Boltzmann constant	$\sigma = 5.67 \times 10^{-5}$ erg cm^{-2} s^{-1} deg^{-4}
Constant in Wien's law	$\lambda_m T = 0.290$ cm deg^{-1}
Standard atmospheric pressure	$= 1.01 \times 10^6$ dyn cm^{-2}
1 electron volt	eV $= 1.60 \times 10^{-12}$ erg
1 angstrom	Å $= 10^{-8}$ cm
1 calorie	$= 4.19 \times 10^7$ erg

D.2 Astronomical constants

1 astronomical unit	AU $= 1.50 \times 10^{13}$ cm
1 parsec	pc $= 3.09 \times 10^{18}$ cm
Mass of Sun	$M_\odot = 1.99 \times 10^{33}$ g
Radius of Sun	$R_\odot = 6.96 \times 10^{10}$ cm
Luminosity of Sun	$L_\odot = 3.90 \times 10^{33}$ erg s^{-1}
Mass of Earth	$M_\oplus = 5.98 \times 10^{27}$ g
Radius of Earth	$R_\oplus = 6.37 \times 10^8$ cm

D.3 Characteristics of some common fluids at 15 °C and 1 atm

ρ = density in g cm^{-3}, c_p = specific heat in joule g^{-1} deg^{-1}, μ = viscosity in g cm^{-1} s^{-1}, v = kinematic viscosity in cm^2 s^{-1}, K = thermal conductivity in joule cm^{-1} s^{-1} deg^{-1}, κ = thermometric conductivity in cm^2 s^{-1}.

	Air	Water	Ethyl alcohol	Glycerine	Mercury
ρ	0.00123	0.999	0.79	1.26	13.6
c_p	1.01	4.19	2.34	2.34	0.140
μ	0.000178	0.0114	0.0134	23.3	0.0158
v	0.145	0.0114	0.0170	18.5	0.00116
K	0.000253	0.0059	0.00183	0.0029	0.080
κ	0.202	0.00140	0.00099	0.00098	0.042

Electrical conductivity of mercury at 15 °C, $\sigma = 9.4 \times 10^{15}$ e.s.u.
Magnetic diffusivity of mercury at 15 °C, $\lambda = 7.6 \times 10^3$ cm^2 s^{-1}

Basic parameters pertaining to plasmas

B = magnetic field in G, n = number density in cm^{-3}, T = temperature in K.

Electron gyrofrequency	$\omega_{\mathrm{c}} =$	$Be/m_{\mathrm{e}}c$	$= 1.76 \times 10^7 B \ \mathrm{s}^{-1}$
Electron plasma frequency	$\omega_{\mathrm{p}} =$	$(4\pi n e^2/m_{\mathrm{e}})^{1/2}$	$= 5.64 \times 10^4 n^{1/2} \ \mathrm{s}^{-1}$
Debye length	$\lambda_{\mathrm{D}} =$	$(\kappa_{\mathrm{B}} T/8\pi n e^2)^{1/2}$	$= 4.9(T/n)^{1/2} \ \mathrm{cm}$
Plasma parameter	$g =$	$1/n\lambda_{\mathrm{D}}^3$	$= 8.6 \times 10^{-3} n^{1/2} T^{-3/2}$

Approximate values of various transport coefficients for a fully ionized, unmagnetized hydrogen plasma (source for thermal conductivity and viscosity: Chapman 1954):

Electrical conductivity	$\sigma \approx 10^7 \ T^{3/2}$	e.s.u.
Thermal conductivity	$K \approx 10^{-5} \ T^{5/2}$	erg cm^{-1} s^{-1} deg^{-1}
Viscosity	$\mu \approx 10^{-17} \ T^{5/2}$	g cm^{-1} s^{-1}

Suggestions for further reading

No attempt at completeness is made here. It is not possible for an individual to be acquainted with everything written on all the topics covered in this book. I have mainly included those references which I myself have found useful and which, I believe, will be useful for those readers of the book who want to go beyond what is covered here. I have purposely refrained from citing literature of a very technical nature, which may not be understandable to a person whose level of knowledge is at the level of the present book. Some suitable references may not have been included merely due to the accident of my not knowing about them. I apologize to those authors who may feel that something written by them ought to have been referenced.

After giving some references on the background assumed for the reader and a few general references, I provide references on the materials covered in different chapters. References on various astrophysical fluid and plasma systems are put together at the end. So, even if an astrophysical system is discussed in a particular chapter, the reference for it may not appear under that chapter, but rather at the end. While referencing review articles—especially on astrophysical topics—I have tried to select articles which present the fundamentals clearly rather than articles which summarize recent research. So some of the review articles referenced here were written long ago (maybe 10 or 15 years ago). Although the more technical parts of these articles may be outdated, most of these articles have not yet been superseded in the presentation of fundamentals. On the other hand, review articles focussing on current research become outdated quickly, as research frontiers move on. Hence I have not felt compelled to provide references to such review articles, which are likely to be superseded soon by newer reviews. This also explains the almost complete lack of references to conference proceedings.

Background

In this book, we assume the reader to have a background of classical mechanics, electromagnetic theory, thermal physics and the standard mathematical tools often known collectively as the methods of mathematical physics. For the necessary background of classical mechanics, we recommend Goldstein (1980), Landau and Lifshitz (1976) or Raychaudhuri (1983). The electromagnetic theory background is covered in Panofsky and Phillips (1962), Jackson (1975) or Landau and Lifshitz (1975). We recommend Fermi (1937) or Pippard (1957) for thermodynamics, whereas Zemansky and Dittman (1981) or Saha and Srivastava (1965) are good textbooks on general thermal physics. Some of the good modern textbooks on mathematical methods are Arfken (1985), Mathews and Walker (1970) and Bender and Orszag (1978). No previous background of astrophysics is assumed, but a little acquaintance with astrophysical phenomenology may be helpful in appreciating certain discussions. Shu (1982) and Abell (1982) are excellent, though rather elementary, undergraduate textbooks meant for readers without any knowledge of calculus. Unsöld (1977) is a more advanced book providing a general introduction to astrophysics. A compact and convenient glossary of astrophysical terms has been compiled by Hopkins (1980).

General references

Before discussing the specific references for different parts of the book, we mention the book by Shu (1992) written with the same aim as the present book. The reader of this book may find it interesting to look at Shu's book, which covers roughly the same ground with a very different perspective and a very different style. Another book at the same level is by Shore (1993). It may not be good manners to make derogatory remarks about a predecessor, but unfortunately I do not have anything positive to say about that book. The recent book by Battaner (1996) attempts an extremely ambitious coverage in a very limited number of pages. Consequently, the basics of fluid mechanics and plasma physics are presented in a rather sketchy fashion.

Chapters 2–3

Our treatment follows that of Huang (1987). The first edition of Huang (published in 1963) presents kinetic theory of gases more fully than the second edition (Huang 1987). Chapters 3–5 of Huang

(1987) constitute a good introduction to the subject. Another good introduction to the kinetic theory of gases is Chapters 12–14 of the very clear and well-known textbook by Reif (1965). For a rigorous, but somewhat dry, treatment of the subject, see Lifshitz and Pitaevskii (1981, Chapter I). None of these books treats the very difficult subject of Chapman–Enskog theory fully. The classic reference on that subject is the monumental volume of Chapman and Cowling (1970). The central part of the book, where the authors present the Chapman–Enskog theory (Chapters 7–10), is difficult reading due to the intrinsic difficulty of the subject, although the authors present the subject as clearly as possible. For readers not brave enough to plunge into this part of the book, Chapters 12–14 provide a summary of important results along with comparisons with experimental data.

Chapters 4–5

The textbooks on hydrodynamics are too numerous to enlist fully. Some of the textbooks are written from the physics student's point of view, some from the engineering student's point of view, and some from the mathematics student's point of view. For somebody wishing to learn hydrodynamics as a branch of physics, Landau and Lifshitz (1987) is a good introduction. It is the Volume 6 in the celebrated *Course of Theoretical Physics*, in which several volumes (notably *Mechanics*, *Classical Theory of Fields*, *Statistical Mechanics*) raised textbook writing to the height of creative science. Although the volume on *Fluid Mechanics* is not quite of the same class as some of these other volumes, it is nevertheless a useful textbook. Another standard textbook is Batchelor (1967). The title of the book (*An Introduction to Fluid Dynamics*) may convey the misleading impression of the book being comprehensive; but the book actually devotes some 600 pages to just the material covered in Chapters 4–5 of our book. The presentation is masterly and encyclopaedic, although the book may not appeal to those readers who wish to have a broad overview of the whole of hydrodynamics. We recommend Tritton (1988) as another comprehensive introduction emphasizing the physics aspect, whereas Milne-Thomson (1960) is one of the standard older works. Amongst the many engineering textbooks, Vennard and Street (1976) and Duncan, Thom and Young (1970) are widely used.

Serious students of the subject should make an acquaintance with the classics on the subject. Perhaps the honour of being the greatest treatise on hydrodynamics ever written should go to Lamb (1932). The first edition of the book (published in 1879) was written well before the development of the vector notation. Those of us who are raised on the vector notation from the beginning often feel horrified by the appearance of vector equations written in the long notation. Once this initial repulsion is overcome, reading Lamb (1932) is a truly rewarding experience. Written at a time when footnotes were fashionable, the book abounds in delightful footnotes giving the historical origins of many important concepts in hydrodynamics. Amongst other classics in the field, one should mention the two small volumes by Prandtl and Tietjens (1934a, 1934b), based on the notes taken by Tietjens during the series of lectures delivered in Göttingen by the great Prandtl whose research paved the way towards bridging the gap between pure and applied hydrodynamics.

Finally, we recommend in highest terms the collection of photographs gathered by Van Dyke (1982) for the sheer visual pleasure this volume can provide. Anybody who has succumbed to the charms of this beautiful volume ought to remain a lover of hydrodynamics forever.

Chapter 6

Some of the standard hydrodynamics textbooks, such as Landau and Lifshitz (1987) mentioned already, cover much of the material presented in this chapter. Liepmann and Roshko (1956) is the standard pedagogical textbook on gas dynamics. It builds up the subject systematically from the basics to engineering applications. Different aspects of the mathematical theory of shock waves are discussed in Witham (1974). Although written from a mathematician's viewpoint, the clarity and lucidity of presentation make it a delight to read this book. The classic treatise on the application of the method of characteristics to gas dynamics problems is Courant and Friedrichs (1948). Another clear introduction on gas dynamics is Zeldovich and Raizer (1966), which covers some unusual topics not usually discussed in other books.

Chapter 7

The outstanding textbook on the subject of hydrodynamic waves is Lighthill (1978). Although only limited varieties of waves are discussed, most of the important theoretical techniques are developed in a brilliant and masterly fashion.

One of the first books on the subject of hydrodynamic instability is the small volume by Lin (1955). The book by Chandrasekhar (1961) is much bigger and comes closest to being a pedagogical textbook amongst Chandrasekhar's several books. Although Chandrasekhar's volumes on *Stellar Structure* or *Stellar Dynamics* are regarded as classics giving his personal perspectives on these subjects, no sensible instructor would use them as basic textbooks in modern courses. In contrast, Chandrasekhar (1961) presents standard material in a standard way and is still the best introduction for somebody wishing to learn the subject. Another good introduction is Drazin and Reid (1981). It covers shear flow instabilities not treated by Chandrasekhar (1961) and has a chapter introducing the nonlinear theory.

Chapter 8

One of the best pedagogical introductions to the difficult subject of turbulence is Tennekes and Lumley (1972). It teaches the reader how to make calculations in many practical situations, even though we do not have a full theory of turbulence. The brilliant small volume by Batchelor (1953), based on his Adams Prize essay, presents many of the important theoretical concepts very lucidly, although the book is limited in scope only to homogeneous turbulence. An indispensable reference for the serious student is the two massive volumes by Monin and Yaglom (1971, 1975) translated from Russian. The sheer sizes of these volumes may appear forbidding, but they provide pleasurable reading because of a relaxed style of scientific writing which is fast disappearing in today's world. Written at the end of a fruitful period of turbulence research stimulated by Kolmogorov's ideas, these volumes present an exhaustive account of the research in that period.

Chapter 9

The standard work on hydrodynamics in rotating frames is Greenspan (1968). An excellent short introduction to the subject of self-gravitating rotating fluid masses can be found in Chapter XII of Lamb (1932). Chandrasekhar (1969) presents a sophisticated treatment of the subject based on the virial principle. The standard textbook on geophysical fluid dynamics by Pedlosky (1987) discusses the large-scale oceanic and atmospheric circulations in the rotating frame of the Earth.

Chapter 10

Some of the older textbooks on plasma physics cover the plasma orbit theory more fully than the modern textbooks. There are textbooks by the two great astrophysicists Chandrasekhar (1960) and Spitzer (1962) from the early years of plasma research. Amongst modern textbooks, Chen (1984) gives a good introduction to the subject in Chapter 2. Readers desirous of having a first-hand account from the creator of plasma orbit theory should look up Chapter 2 of Alfvén (1950).

See Parks (1991) for a discussion on the role of plasma orbit theory and its applications to the study of space plasmas. The subject of particle acceleration in astrophysics is one of the central themes in the superb book by Longair (1994).

Chapter 11

The standard reference on the subject of plasma kinetic theory is Montgomery and Tidman (1964). Written towards the end of the golden period ($\approx$ 1955–1965) of plasma kinetic theory research, this book provides an insider's account of the achievements of that period. Amongst the standard textbooks on plasma physics, Nicholson (1983) gives an exceptionally clear introduction to plasma kinetic theory in §3.1–7.2. The comprehensive textbook by Clemmow and Dougherty (1969) also should be of interest to the reader.

Chapter 2 of Krall and Trivelpiece (1973) provides an introduction to the statistical mechanics of equilibrium plasmas through the Gibbs ensemble approach—a topic not covered in this book. The famous review article by Chandrasekhar (1943a) gives a thorough analysis of how long-range collisions can be treated as stochastic processes and lead to the Fokker–Planck equation.

Chapters 12–13

The materials of these chapters constitute the core of modern textbooks on plasma physics. There are too many such textbooks. We mention here only some of the most widely used ones. Krall and Trivelpiece (1973) has remained the rigorous and authoritative textbook for a long time. Although mature students benefit by consulting it, complete beginners are often lost in this big book. A more user-friendly book at a less advanced level is Chen (1984). It provides a coherent and pedagogical account of the basic topics of plasma physics in a way suitable for the beginning student. Amongst the elementary textbooks, Nicholson (1983) has the most thorough and complete coverage of the formal foundations. It is a book of matchless clarity in presentation and deserves to be much more widely used than it is today. It is a favourite book of the present author. Chapters III–IV of Lifshitz and Pitaevskii (1981) cover many topics in plasma physics.

Parks (1991) develops the basics of plasma physics very clearly for students wishing to specialize on space plasmas. The standard monograph on the subject of plasma waves is by Stix (1992). It tries to make sense of a rather messy and complex subject.

Chapters 14–15

There is a lack of a comprehensive modern textbook on magneto-hydrodynamics. The small volume by Cowling (1976) is one of the best compact introductions. The book is restricted only to elementary topics in MHD, but these are covered with extreme lucidity. The old book by Ferraro and Plumbton (1966) has a fascinating historical introduction and is still worth reading because of the very clear style. Another standard older reference is Roberts (1967). Biscamp (1993) has put together several important advanced topics, which are usually not found treated within one volume. Despite the attractive coverage of topics, the book may not be very easily understandable to a beginner. There is a superb review article on the subject of MHD waves by Roberts (1985). Chandrasekhar (1961) discusses how the presence of a magnetic field affects various hydrodynamic instabilities. A famous reference on the subject of fusion research is Rose and Clark (1961). Although many technical details given in the later chapters are hopelessly out of date, the book provides one of the clearest statements on the goals of fusion research and captures the euphoria of the early post-war years. The problem of plasma relaxation in various laboratory devices has been reviewed by Taylor (1986).

We also mention the classics in the field for serious students. The famous monograph by Alfvén (1950) played a significant role in establishing this field of research. Perhaps the first edition is more valuable as a historical document of great importance than the second revised edition (Alfvén and Fälthammar 1963). The monumental volume by Parker (1979) has some 800 pages densely packed with equations and is not exactly suited for bedtime reading. It is a book of deep scholarship and profound originality by perhaps the greatest scientist in the field of astrophysical plasmas. Parker has a unique style of combining physical insight with mathematical analysis, so that one never loses sight of the (astro)physical world in the maze of mathematical equations.

Chapter 16

The small book by Moffatt (1978) is an extremely clear introduction to dynamo theory with chapters on solar and terrestrial magnetic fields. The first chapter traces the tortuous historical path through which the modern dynamo theory developed. A more formal introduction by the creators of mean field magnetohydrodynamics is Krause and Rädler (1980). This book throws light on many subtleties in the mean field theory of turbulence. Zeldovich, Ruzmaikin and Sokoloff (1983) discuss the origin of magnetic fields in diverse astrophysical systems and survey the Russian contributions on the subject.

Astrophysical fluid and plasma systems

Throughout the book, we have illustrated various topics with applications to different astrophysical systems from the Sun to the active galactic nuclei. We now mention some basic references on these astrophysical systems covering the fluid and plasma aspects.

Stellar convection, rotation and oscillations. Amongst the many authors of stellar interiors textbooks, Kippenhahn and Weigert (1990) give very clear introductions to all these subjects. The textbook on solar physics by Stix (1989) discusses convection, rotation and oscillations in the Sun. The theory of rotating stars forms the subject of a monograph by Tassoul (1978). Rüdiger (1989) presents an analysis of the very difficult subject of differential rotation. There is an excellent review on helioseismology by Brown, Mihalas and Rhodes (1986).

Magnetohydrodynamics of the Sun and the solar system. The outstanding textbook by Priest (1982) encompasses all aspects of solar MHD. Selected parts of the book can be used as an introduction to MHD as well, since the author develops the subject without assuming any previous background of MHD. A good review on the solar dynamo is by Gilman (1986), whereas we recommend the review by Thomas and Weiss (1992) for readers wanting to learn about sunspots. For the complex subject of solar flares, not much discussed in our book, see the monograph by Tandberg-Hansen and Emslie (1988). Parks (1991) develops various plasma physics concepts necessary to study solar system plasmas and planetary magnetospheres.

Neutron stars and pulsars. The comprehensive graduate textbook by Shapiro and Teukolsky (1983) covers different aspects of neutron stars and pulsars, including the fluid and plasma topics. This book provides an introduction to both spherical accretion and disk accretion (Chapter 14). Amongst other good introductions to accretion disks, one may mention the superb review article by Pringle (1981) or the monograph by Frank, King and Raine (1984).

Interstellar medium. The standard graduate textbook on the interstellar medium is by Spitzer (1978). Chapters 10–13 of this book provide a comprehensive introduction to various hydrodynamic and MHD problems pertaining to the interstellar medium. There is a review article on astrophysical blast waves by Ostriker and McKee (1988), whereas the complex subject of star formation has been reviewed by Shu, Adams and Lizano (1987). Ruzmaikin, Shukurov and Sokoloff (1988) discuss the dynamo process in the interstellar medium giving rise to the galactic magnetic field.

Active galactic nuclei. There is a lack of a good pedagogically clear textbook on fluid and plasma problems pertaining to the active galactic nuclei. The long and comprehensive review article by Begelman, Blandford and Rees (1984) is truly a review article, in the sense that even the basics are merely summarized and not worked out. The Saas-Fee lecture course by Blandford in Blandford, Netzer and Woltjer (1990) would be more easily accessible to a beginner. The review article by Rees (1984) discusses various fluid and plasma processes around black holes.

Stellar dynamics. The standard textbook by Binney and Tremaine (1987) gives an introduction to both collisionless and collisional stellar dynamics. The phenomenological background of the subject is provided by Mihalas and Binney (1981). The old review article by Oort (1965) is a masterpiece and should be read by anybody willing to learn how theory and observations can go hand-in-hand in this field. The collisional dynamics of star clusters is treated in the monograph by Spitzer (1987).

References

Each reference is cited in the portions of the book indicated within square brackets. S. f. r. stands for *Suggestions for further reading*.

Abell, G. 1982, *The Exploration of the Universe*, 4th edn. Saunders. [S. f. r.]

Abramowitz, M. and Stegun, I. A. 1964, *Handbook of Mathematical Functions*. National Bureau of Standards. Reprinted by Dover. [§14.3]

Alfvén, H. 1940, *Ark. f. Mat. Astr. o. Fysik* **27A**, No. 22. [§10.1]

Alfvén, H. 1942a, *Ark. f. Mat. Astr. o. Fysik* **29B**, No. 2. [§14.2, §14.5]

Alfvén, H. 1942b, *Nature* **150**, 405. [§14.5]

Alfvén, H. 1950, *Cosmical Electrodynamics*. Oxford University Press. [Quot. Part II, §10.1, Fig. 10.1, S. f. r.]

Alfvén, H. and Fälthammar, C.-G. 1963, *Cosmical Electrodynamics*, 2nd edn. Oxford University Press. [S. f. r.]

Appleton, E. V. 1927, *Proc. Union Radio Scientifique Internationale*, Washington. [§12.5]

Arfken, G. 1985, *Mathematical Methods for Physicists*, 3rd edn. Academic Press. [§4.8, §5.6, §5.7, §14.3, S. f. r.]

Ashcroft, N. W. and Mermin, N. D. 1976, *Solid State Physics*. Saunders. [§11.3, §13.4]

Axford, W. I., Leer, E. and Skadron, G. 1977, *Proc. 15th International Cosmic Ray Conf.* **11**, 132. [§10.5]

Baade, W. and Zwicky, F. 1934, *Phys. Rev.* **45**, 138. [§9.4]

Backer, D. C., Kulkarni, S. R., Heiles, C., Davis, M. M. and Gross, W. M. 1982, *Nature* **300**, 615. [§9.4]

Balescu, R. 1960, *Phys. Fluids* **3**, 52. [§11.4]

Batchelor, G. K. 1953, *The Theory of Homogeneous Turbulence*. Cambridge University Press. [S. f. r.]

Batchelor, G. K. 1967, *An Introduction to Fluid Dynamics*. Cambridge University Press. [§5.2, §5.5, Fig. 5.3, S. f. r.]

Battaner, E. 1996, *Astrophysical Fluid Dynamics*. Cambridge University Press. [S. f. r.]

Begelman, M. C., Blandford, R. D. and Rees, M. J. 1984, *Rev. Mod. Phys.* **56**, 255. [S. f. r.]

Bell, A. R. 1978, *Mon. Not. Roy. Astron. Soc.* **182**, 147 and 443. [§10.5]

Bénard, H. 1900, *Revue Gén. Sci. Pur. Appl.* **11**, 1261 and 1309. [§7.3]

Bender, C. M. and Orszag, S. A. 1978, *Advanced Mathematical Methods for Scientists and Engineers*. McGraw-Hill. [S. f. r.]

Bennett, W. H. 1934, *Phys. Rev.* **45**, 890. [§14.3]

Berger, M. A. and Field, G. B. 1984, *J. Fluid Mech.* **147**, 133. [§15.3]

Bernoulli, D. 1738, *Hydrodynamica*. [§4.5]

Bernstein, I. B., Frieman, E. A., Kruskal, M. D. and Kulsrud, R. M. 1958, *Proc. Roy. Soc.* **A 244**, 17. [§17.1]

Bethe, H. and Critchfield, C. H. 1938, *Phys. Rev.* **54**, 248. [§5.7]

Bhatnagar, P. L., Gross, E. P. and Krook, M. 1954, *Phys. Rev.* **94**, 511. [§3.4]

Biermann, L. 1941, *Vierteljahrsschrift Astron. Ges.* **76**, 194. [§14.6]

Biermann, L. 1948, *Zs. f. Astrophys.* **25**, 135. [§8.6]

Binney, J. and Tremaine, S. 1987. *Galactic Dynamics*. Princeton University Press. [§3.2, §3.7, §9.3, §9.5, §11.6, §17.1, S. f. r.]

Biscamp, D. 1993, *Nonlinear Magnetohydrodynamics*. Cambridge University Press. [S. f. r.]

Bjerknes, V. 1937, *Astroph. Norvegica* **2**, 263. [§9.2]

Blackett, P. M. S. 1947, *Nature* **159**, 658. [Quot. Part II]

Blandford, R. D., Netzer, H. and Woltjer, L. 1990, *Active Galactic Nuclei*. Springer-Verlag. [S. f. r.]

Blandford, R. D. and Ostriker, J. P. 1978, *Astrophys. J.* **221**, L29. [§10.5]

Blandford, R. D. and Payne, D. G. 1982, *Mon. Not. Roy. Astron. Soc.* **199**, 883. [§14.9]

Blandford, R. D. and Rees, M. J. 1974, *Mon. Not. Roy. Astron. Soc.* **169**, 395. [§6.7, §14.9]

Bodin, H. A. B. and Newton, A. A. 1980, *Nucl. Fusion* **20**, 1255. [Fig. 15.6]

Bogoliubov, N. N. 1946, *Problems of a Dynamical Theory in Statistical Physics* (in Russian). State Technical Press, Moscow. [§11.4]

Boltzmann, L. 1872, *Sitzungsber. Kaiserl. Akad. Wiss. Wien* **66**, 275. [§2.2, §2.4]

Bondi, H. 1952, *Mon. Not. Roy. Astron. Soc.* **112**, 195. [§6.8]

Born, M. and Green, H. S. 1949, *A General Kinetic Theory of Liquids*. Cambridge University Press. [§11.4]

Born, M. and Wolf, E. 1980, *Principles of Optics*, 6th edn. Pergamon. [Ex. 12.2]

Boussinesq, J. 1903, *Théorie Analytique de la Chaleur*. Gauthier-Villars. [§7.3]

Brown, T. M., Mihalas, B. W. and Rhodes, E. J. 1986, in *Physics of the Sun, Vol. I* (ed. P. A. Sturrock), p. 177. Reidel. [S. f. r.]

Brunt, D. 1927, *Quart. J. Roy. Met. Soc.* **53**, 30. [§7.2]

Chandrasekhar, S. 1943a, *Rev. Mod. Phys.* **15**, 1. [§11.6, S. f. r.]

Chandrasekhar, S. 1943b, *Astrophys. J.* **97**, 255. [§11.6]

Chandrasekhar, S. 1952, *Phil. Mag. (7)* **43**, 501. [§14.6]

Chandrasekhar, S. 1960, *Plasma Physics*. University of Chicago Press. [S. f. r.]

Chandrasekhar, S. 1961, *Hydrodynamic and Hydromagnetic Stability*. Oxford University Press. Reprinted by Dover. [Pref., §7.3, §14.6, Fig. 14.7, S. f. r.]

Chandrasekhar, S. 1969, *Ellipsoidal Figures of Equilibrium*. Yale University Press. Reprinted by Dover. [§9.3, Fig. 9.3, Fig. 9.4, §17.1, S. f. r.]

Chandrasekhar, S. and Kendall, P. C. 1957, *Astrophys. J.* **126**, 457. [§14.3]

Chapman, S. 1916, *Phil. Trans. Roy. Soc.* **A 216**, 279. [§3.4]

Chapman, S. 1954, *Astrophys. J.* **120**, 151. [§13.5, Append. E]

Chapman, S. 1957, *Smithsonian Contrib. Astrophys.* **2**, 1. [§4.4]

Chapman, S. and Cowling, T. G. 1970, *The Mathematical Theory of Non-Uniform Gases*, 3rd edn. Cambridge University Press. [§2.4, §3.4, §3.5, S. f. r.]

Chen, F. F. 1984, *Introduction to Plasma Physics and Controlled Fusion, Vol. I*, 2nd edn. Plenum Press. [Pref., S. f. r.]

Chew, G. F., Goldberger, M. L. and Low, F. E. 1956, *Proc. Roy. Soc.* **A 236**, 112. [§11.5]

Choudhuri, A. R. 1989, *Solar Phys.* **123**, 217. [§14.7]

Choudhuri, A. R., Schüssler, M. and Dikpati, M. 1995, *Astron. Astrophys.* **303**, L29. [§16.6]

Clausius, R. 1858, *Ann. d. Phys. u. Chemie* **105**, 239. [§2.1]

Clausius, R. 1870, *Sitzungsber. Niedderrheinisch. Ges. Bonn*, 114. [§17.1]

Clemmow, P. C. and Dougherty, J. P. 1969, *Electrodynamics of Particles and Plasmas*. Addison-Wesley. [S. f. r.]

Copson, E. T. 1935, *Theory of Functions of a Complex Variable*. Oxford University Press. [§4.8, §5.6]

Courant, R. and Friedrichs, K. O. 1948, *Supersonic Flow and Shock Waves*. Interscience. [§6.4, S. f. r.]

Cowling, T. G. 1934, *Mon. Not. Roy. Astron. Soc.* **94**, 39. [§16.3]

Cowling, T. G. 1941, *Mon. Not. Roy. Astron. Soc.* **101**, 367. [§7.6]

Cowling, T. G. 1976, *Magnetohydrodynamics*, 2nd edn. Adam Hilger. [Pref., S. f. r.]

Debye, P. and Hückel, E. 1923, *Phys. Zs.* **24**, 185. [§11.2]

Deubner, F.-L. 1975, *Astron. Astrophys.* **44**, 371. [§7.6]

Drazin, P. G. and Reid, W. H. 1981, *Hydrodynamic Stability*. Cambridge University Press. [§7.7, S. f. r.]

Dreher, J. W. and Feigelson, E. D. 1984, *Nature* **308**, 43. [Fig. 6.7]

Drude, P. 1900, *Ann. d. Physik* **1**, 566. [§13.5]

Duncan, W. J., Thom, A. S. and Young, A. D. 1970, *Mechanics of Fluids*, 2nd edn. Edward Arnold. [S. f. r.]

Dungey, J. W. 1953, *Phil. Mag. (7)* **44**, 725. [§15.2]

Eddington, A. S. 1918, *Mon. Not. Roy. Astron. Soc.* **79**, 177. [§7.6]

Eddington, A. S. 1925, *Observatory* **48**, 73. [§9.4]

Eddington, A. S. 1926, *The Internal Constitution of the Stars*. Cambridge University Press. [§11.3]

Emden, R. 1907, *Gaskugeln*. Teubner, Leipzig. [Ex. 4.3]

Enskog, D. 1917, *Kinetische Theorie der Vorgänge in mässig vendünnten Gasen.* Dissertation, Upsala. [§3.4]

Euler, L. 1755, *Hist. de l'Acad. de Berlin.* [§4.1]

Euler, L. 1759, *Novi Comm. Acad. Petrop.* **14**, 1. [§4.1]

Fermi, E. 1937, *Thermodynamics.* Prentice-Hall. Reprinted by Dover. [S. f. r.]

Fermi, E. 1949, *Phys. Rev.* **75**, 1169. [§10.5]

Ferraro, V. C. A. 1937, *Mon. Not. Roy. Astron. Soc.* **97**, 458. [§14.9]

Ferraro, V. C. A. and Plumbton, C. 1966, *An Introduction to Magnetofluid Mechanics*, 2nd edn. Oxford University Press. [S. f. r.]

Feynman, R. P., Leighton, R. B. and Sands, M. 1964, *Feynman Lectures on Physics, Vol. II.* Addison-Wesley. [§4.5, §4.6]

Fokker, A. D. 1914, *Ann. d. Physik* **43**, 812. [§11.4, §11.6]

Frank, J., King, A. and Raine, D. 1984, *Accretion Power in Astrophysics.* Cambridge University Press. [S. f. r.]

Furth, H. P., Killeen, J. and Rosenbluth, M. N. 1963, *Phys. Fluids* **6**, 459. [§15.2]

Ghosh, P. and Lamb, F. K. 1978, *Astrophys. J. Lett.* **223**, L83. [§14.9]

Giacconi, R., Gursky, H., Paolini, F. R. and Rossi, B. B. 1962, *Phys. Rev. Lett.* **9**, 439. [§5.7]

Gilbert, G. 1600, *De Magnete.* [§16.1]

Gilman, P. 1983, *Astrophys. J. Suppl.* **53**, 243. [§16.7]

Gilman, P. 1986, in *Physics of the Sun, Vol. I* (ed. P. A. Sturrock), p. 95. Reidel. [S. f. r.]

Gold, T. 1968, *Nature* **218**, 731. [§9.4]

Goldreich, P. and Julian, W. H. 1969, *Astrophys. J.* **157**, 869. [§13.7]

Goldstein, H. 1980, *Classical Mechanics*, 2nd edn. Addison-Wesley. [§1.4, §9.2, §10.1, §13.3, §17.1, §17.2, S. f. r.]

Gough, D. O. 1978, *Proc. Workshop on Solar Rotation*, Univ. of Catania, p. 255. [§7.6]

Grad, H. and Rubin, H. 1958, *Proc. 2nd U.N. Conf. on Peaceful Uses of Atomic Energy*, Vol. 31, p. 190. [§14.3]

Grant, H. L., Stewart, R. W. and Moilliet, A. 1962, *J. Fluid Mech.* **12**, 241. [§8.3, Fig. 8.6]

Greenspan, H. P. 1968, *The Theory of Rotating Fluids.* Cambridge University Press. [Fig. 9.2, S. f. r.]

Gull, S. F. 1975, *Mon. Not. Roy. Astron. Soc.* **171**, 263. [§7.4]

Hagen, G. 1839, *Ann. d. Phys. u. Chemie* **46**, 423. [§5.3]

Hale, G. E. 1908, *Astrophys. J.* **28**, 315. [§14.6, §16.1]

Hale, G. E., Ellerman, F., Nicholson, S. B. and Joy, A. H. 1919, *Astrophys. J.* **49**, 153. [§14.7]

Hartree, D. R. 1931, *Proc. Camb. Phil. Soc.* **27**, 143. [§12.5]

Helmholtz, H. 1854, *Lecture at Kant Commemoration, Königsberg.* [§5.7]

Helmholtz, H. 1858, *J. Reine Angew. Math.* **55**, 25. [§4.6]

Helmholtz, H. 1868, *Monatsber. Königl. Preuss. Akad. Wiss. Berlin* **23**, 215. [§7.4]

Herlofson, N. 1950, *Nature* **165**, 1020. [§14.5]

Hess, V. F. 1912, *Sitzungsber. Kaiserl. Akad. Wiss. Wien* **12**, 2001. [§10.5]

Hewish, A. S., Bell, J., Pilkington, J. D. H., Scott, P. F. and Collins, R. A. 1968, *Nature* **217**, 709. [§9.4]

Homann, F. 1936, *Forsch. Ing.-Wes.* **7**, 1. [Fig. 5.3]

Homer Lane, J. 1869, *Amer. J. Sci.* **50**, 57. [Ex. 4.3]

Honsaker, J. L., Karr, H., Osher, J., Phillips, J. A. and Tuck, J. L. 1958, *Nature* **181**, 231. [§15.4]

Hopkins, J. 1980, *Glossary of Astronomy and Astrophysics*, 2nd edn. University of Chicago Press. [S. f. r.]

Huang, K. 1987, *Statistical Mechanics*, 2nd edn. John Wiley. [Pref., §2.1, §3.4, S. f. r.]

Hugoniot, A. 1889, *J. de l'Ecole Polytech.* **58**, 1. [§6.5]

Jackson, J. D. 1975, *Classical Electrodynamics*, 2nd edn. John Wiley. [§4.7, §6.3, §10.1, §10.3, §11.1, §13.6, §13.7, S. f. r.]

Jacobi, C. G. J. 1834, *Ann. d. Phys. u. Chemie* **33**, 229. [§9.3]

Jeans, J. H. 1902, *Phil. Trans. Roy. Soc.* **A 199**, 1. [§7.5]

Jeans, J. H. 1922, *Mon. Not. Roy. Astron. Soc.* **82**, 122. [§3.2]

Jeffreys, H. 1924, *The Earth*. Cambridge University Press. [§5.7]

Jeffreys, H. 1926, *Phil. Mag. (7)* **2**, 833. [§7.3]

Joos, G. 1958, *Theoretical Physics*, 3rd edn. Blackie and Son. Reprinted by Dover. [§14.5]

Joukowski, N. E. 1910, *Zs. Flugt. Motorluftsh.* **1**, 281. [§5.6]

Kelvin, Lord 1861, *Brit. Assoc. Repts.*, Part II, p. 27. [§5.7]

Kelvin, Lord 1869, *Trans. Roy. Soc. Edin.* **25**, 217. [§4.6]

Kelvin, Lord 1871, *Phil. Mag. (4)* **42**, 362. [§7.4]

Kippenhahn, R. 1963, *Astrophys. J.* **137**, 664. [§9.1]

Kippenhahn, R. and Weigert, A. 1990, *Stellar Structure and Evolution*. Springer-Verlag. [§7.6, §8.6, §9.4, S. f. r.]

Kirkwood, J. G. 1946, *J. Chem. Phys.* **14**, 180. [§11.4]

Kittel, C. 1995, *Introduction to Solid State Physics*, 7th edn. John Wiley. [§11.3, §12.2, §13.4, §13.5]

Kolmogorov, A. N. 1941a, *C.R. Acad. Sci. U.S.S.R.* **30**, 301. [§8.3]

Kolmogorov, A. N. 1941b, *C.R. Acad. Sci. U.S.S.R.* **32**, 16. [§8.3]

Korteweg, D. J. and de Vries, G. 1895, *Phil. Mag. (5)* **39**, 422. [§7.7]

Koschmieder, E. L. and Pallas, S. G. 1974, *Int. J. Heat Mass Transfer* **17**, 991. [Fig. 7.4]

Krall, N. A. and Trivelpiece, A. W. 1973, *Principles of Plasma Physics*. McGraw-Hill. [§11.4, S. f. r.]

Krause, F. and Rädler, K.-H. 1980, *Mean-Field Magnetohydrodynamics and Dynamo Theory*. Pergamon. [S. f. r.]

Krymsky, G. F. 1977, *Dokl. Acad. Nauk. U.S.S.R.* **234**, 1306. [§10.5]

Kutta, W. 1911, *Sitzungsber. Bayr. Akad. Wiss., M.-Ph. Kl.* **41**, 65. [§5.6]

Lamb, H. 1908, *Proc. Lond. Math. Soc.* **7**, 122. [§6.2]

Lamb, H. 1932, *Hydrodynamics*, 6th edn. Cambridge University Press. Reprinted by Dover. [S. f. r.]

Landau, L. D. 1944, *C.R. Acad. Sci. U.S.S.R.* **44**, 311. [§7.7]

Landau, L. D. 1946, *J. Phys. (U.S.S.R.)* **10**, 25. [§12.4]

Landau, L. D. and Lifshitz, E. M. 1975, *The Classical Theory of Fields*, 4th edn. Pergamon. [§17.2, S. f. r.]

Landau, L. D. and Lifshitz, E. M. 1976, *Mechanics*, 3rd edn. Pergamon. [S. f. r.]

Landau, L. D. and Lifshitz, E. M. 1987, *Fluid Mechanics*, 2nd edn. Pergamon. [Pref., §4.3, §4.6, §5.5, §6.1, §6.6, §17.2, S. f. r.]

Langmuir, I. 1928, *Proc. Nat. Acad. Sci. U.S.A.* **14**, 627. [§11.1, §12.2]

Laplace, P. S. 1816, *Ann. de Chim. et de Phys.* **3**, 238. [§6.2]

Lawson, J. D. 1957, *Proc. Phys. Soc. (London)* **B70**, 6. [§14.4]

Leighton, R. B., Noyes, R. W. and Simon, G. W. 1962, *Astrophys. J.* **135**, 474. [§7.6]

Lenard, A. 1960, *Ann. Phys. (New York)* **10**, 390. [§11.4]

Liepmann, H. W., Narasimha, R. and Chahine, M. T., 1962 *Phys. Fluids* **5**, 1313. [§3.4]

Liepmann, H. W. and Roshko, A. 1956, *Elements of Gasdynamics*. John Wiley. [§6.1, §6.7, S. f. r.]

Lifshitz, E. M. and Pitaevskii, L. P. 1981, *Physical Kinetics*. Pergamon. [S. f. r.]

Lighthill, J. 1952, *Proc. Roy. Soc.* **A 211**, 564. [§6.3]

Lighthill, J. 1978, *Waves in Fluids*. Cambridge University Press. [§6.3, S. f. r.]

Lin, C. C. 1955, *The Theory of Hydrodynamic Stability*. Cambridge University Press. [S. f. r.]

Lin, C. C. and Shu, F. 1964, *Astrophys. J.* **140**, 646. [§7.5]

Longair, M. S. 1994, *High Energy Astrophysics, Vol. 2*, 2nd edn. Cambridge University Press. [§10.5, S. f. r.]

Lorentz, H. A. 1878, *Verh. Kon. Akad. Wetensch.* **18**. [§12.2]

Lorenz, L. 1872, *Ann. d. Phys. u. Chemie* **147**, 429. [§13.5]

Lovelace, R. V. E., Mehanian, C., Mobarry, C. M. and Sulkanen, M. E. 1986, *Astrophys. J. Suppl.* **62**, 1. [§14.9]

Low, B. C. and Tsinganos, K. 1986, *Astrophys. J.* **302**, 163. [§14.9]

Lundquist, S. 1949, *Phys. Rev.* **76**, 1805. [§14.5]

Lüst, R. 1952, *Z. Naturforsch.* **7a**, 87. [§5.7]

Lüst, R. and Schlüter, A. 1954, *Zs. f. Astrophys.* **34**, 263. [§14.3]

Lynden-Bell, D. 1967, *Mon. Not. Roy. Astron. Soc.* **136**, 101. [§3.7]

Lynden-Bell, D. 1969, *Nature* **223**, 690. [§5.7]

MacDonald, W. M., Rosenbluth, M. N. and Chuck, W. 1957, *Phys. Rev.* **107**, 350. [§11.6]

Maclaurin, C. 1740, *De Causâ Physicâ Fluxus et Refluxus Maris*. [§9.3]

Magnus, G. 1853, *Ann. d. Phys. u. Chemie* [Ex. 4.7] **88**, 1.

Malkus, W. V. R. and Veronis, G. 1958, *J. Fluid Mech.* **4**, 225. [§7.7]

Mathews, P. M. and Venkatesan, K. 1976, *A Textbook of Quantum Mechanics*. Tata McGraw-Hill, New Delhi. [§1.3]

Mathews, J. and Walker, R. L. 1970, *Mathematical Methods of Physics*. W.A. Benjamin. [S. f. r.]

Maunder, E. W. 1904, *Mon. Not. Roy. Astron. Soc.* **64**, 747. [§16.1]

Maxwell, J. C. 1860, *Phil. Mag. (4)* **19**, 19. [§2.3, §3.5]

Maxwell, J. C. 1891, *A Treatise on Electricity and Magnetism*, 3rd edn. Oxford University Press. Reprinted by Dover. [Pref.]

Mayer, J. E. and Mayer, M. G. 1940, *Statistical Mechanics*. John Wiley. [§11.4]

Mestel, L. 1968, *Mon. Not. Roy. Astron. Soc.* **138**, 359. [§14.9]

Mestel, L. and Spitzer, L. 1956, *Mon. Not. Roy. Astron. Soc.* **116**, 583. [§14.10]

Meyer, P. 1969, *Ann. Rev. Astron. Astrophys.* **7**, 18. [Fig. 10.7]

Mihalas, D. 1978, *Stellar Atmospheres*, 2nd edn. Freeman. [§11.1, §17.3]

Mihalas, D. and Binney, J. 1981, *Galactic Astronomy*. Freeman. [§3.7, §9.3, §9.5, S. f. r.]

Mihalas, D. and Mihalas, B. W. 1984, *Foundations of Radiation Hydrodynamics*. Oxford University Press. [§17.2, §17.3]

Milne-Thomson, L. M. 1960, *Theoretical Hydrodynamics*, 4th edn. Macmillan. [S. f. r.]

Misner, C. W., Thorne, K. S. and Wheeler, J. A. 1973, *Gravitation*. Freeman. [§17.2]

Miyoshi, M., Moran, J., Herrnstein, J., Greenhill, L., Nakai, N., Diamond, P. and Inoue, M. 1995, *Nature* **373**, 127. [§5.7]

Moffatt, H. K. 1969, *J. Fluid Mech.* **35**, 117. [§15.3]

Moffatt, H. K. 1978, *Magnetic Field Generation in Electrically Conducting Fluids*. Cambridge University Press. [S. f. r.]

Monin, A. S. and Yaglom, A. M. 1971, *Statistical Fluid Mechanics, Vol. I*. M.I.T. Press. [S. f. r.]

Monin, A. S. and Yaglom, A. M. 1975, *Statistical Fluid Mechanics, Vol. II*. M.I.T. Press. [§8.3, S. f. r.]

Montgomery, D. C. and Tidman, D. A. 1964, *Plasma Kinetic Theory*. McGraw-Hill. [§11.4, §13.5, S. f. r.]

Moreno-Insertis, F. 1986, *Astron. Astrophys.* **166**, 291. [§14.7]

Mouschovias, T. Ch. 1974, *Astrophys. J.* **192**, 37. [§14.8]

Nakagawa, Y. 1955, *Nature* **175**, 417. [§14.6]

Navier, M. 1822, *Mem. de l'Acad. des Sciences* **6**, 389. [§5.2]

Neininger, N. 1992, *Astron. Astrophys.* **263**, 30. [Fig. 16.1]

Newton, I. 1689, *Principia Mathematica Philosophiae Naturalis*. [§6.2, §9.1]

Nicholson, D. R. 1983, *Introduction to Plasma Theory*. John Wiley. [§11.4, §11.6, S. f. r.]

Oort, J. H. 1928, *Bull. Astron. Inst. Netherlands* **4**, 269. [§9.5]

Oort, J. H. 1932, *Bull. Astron. Inst. Netherlands* **6**, 349. [§3.2]

Oort, J. H. 1965, in *Galactic Structure* (ed. A. Blaauw and M. Schmidt), p. 455. University of Chicago Press. [S. f. r.]

Osterbrock, D. E. 1961, *Astrophys. J.* **134**, 347. [§15.5]

Ostriker, J. P. and McKee, C. F. 1988, *Rev. Mod. Phys.* **60**, 1. [S. f. r.]

Panofsky, W. K. H. and Phillips, M. 1962, *Classical Electricity and Magnetism*,

2nd edn. Addison-Wesley. [§4.2, §4.7, §6.3, §10.1, §10.3, §11.1, §11.5, §13.6, §13.7, S. f. r.]

Parker, E. N. 1955a, *Astrophys. J.* **121**, 491. [§14.7]

Parker, E. N. 1955b, *Astrophys. J.* **122**, 293. [§16.3, §16.4, §16.5, §16.6]

Parker, E. N. 1957, *J. Geophys. Res.* **62**, 509. [§15.2]

Parker, E. N. 1958, *Astrophys. J.* **128**, 664. [§4.4, §6.8, §14.9]

Parker, E. N. 1966, *Astrophys. J.* **145**, 811. [§14.8]

Parker, E. N. 1972, *Astrophys. J.* **172**, 499. [§15.5]

Parker, E. N. 1979, *Cosmical Magnetic Fields.* Oxford University Press. [Quot. Part II, S. f. r.]

Parker, E. N. 1994, *Spontaneous Discontinuities in Magnetic Fields.* Oxford University Press. [§15.5]

Parks, G. K. 1991, *Physics of Space Plasmas.* Addison-Wesley. [S. f. r.]

Pedlosky, J. 1987, *Geophysical Fluid Dynamics*, 2nd edn. Springer-Verlag. [§9.2, S. f. r.]

Penning, F. M. 1926, *Nature* **118**, 301. [§12.2]

Petschek, A. 1964, *AAS-NASA Symp. on Solar Flares*, NASA SP-50, p. 425. [§15.2]

Pippard, A. B. 1957, *The Elements of Classical Thermodynamics.* Cambridge University Press. [S. f. r.]

Planck, M. 1917. *Sitz. Preuss. Akad.* 324. [§11.4, §11.6]

Poiseuille, J. L. M. 1840, *Comptes Rendus* **11**, 961 and 1041. [§5.3]

Poisson, S. D. 1829, *J. de l'Ecole Polytech.* **13**, 1. [§5.2]

Prandtl, L. 1905, *Verhandlungen des dritten Internationalen Mathematiker-Kongresses (Heidelberg 1904)*, p. 484. [§5.5]

Prandtl, L. 1925, *Zs. Angew. Math. Mech.* **5**, 136. [§8.6]

Prandtl, L and Tietjens, O. G. 1934a, *Fundamentals of Hydro- and Aeromechanics.* McGraw-Hill. Reprinted by Dover. [Quot. Part I, §5.5, S. f. r.]

Prandtl, L and Tietjens, O. G. 1934b, *Applied Hydro- and Aeromechanics.* McGraw-Hill. Reprinted by Dover. [§5.6, S. f. r.]

Prendergast, K. H. and Burbidge, G. R. 1968, *Ap. J. Lett.* **151**, L83. [§5.7]

Priest, E. R. 1982, *Solar Magnetohydrodynamics.* Reidel. [S. f. r.]

Priest, E. R. and Forbes, T. G. 1986, *J. Geophys. Res.* **91**, 5579. [§15.2]

Pringle, J. E. 1981, *Ann. Rev. Astron. Astrophys.* **19**, 140. [Fig. 5.8, S. f. r.]

Proudman, J. 1916, *Proc. Roy. Soc.* **A 92**, 408. [§9.2]

Rankine, W. J. M. 1870, *Phil. Trans. Roy. Soc.* **160**, 277. [§6.5]

Raychaudhuri, A. K. 1983, *Classical Mechanics.* Oxford University Press. [§1.4, S. f. r.]

Rayleigh, Lord 1881, *Nature* **25**, 52. [§6.2]

Rayleigh, Lord 1883, *Proc. Lond. Math. Soc.* **14**, 170. [§7.4]

Rayleigh, Lord 1916, *Phil. Mag. (6)* **32**, 529. [§7.3]

Rayleigh, Lord 1917, *Proc. Roy. Soc.* **A 93**, 148. [§9.1]

Rees, M. J. 1984, *Ann. Rev. Astron. Astrophys.* **22**, 471. [S. f. r.]

Reif, F. 1965, *Fundamentals of Statistical and Thermal Physics.* McGraw-Hill. [§3.5, §10.5, §11.6, §12.3, S. f. r.]

Reynolds, O. 1883, *Phil. Trans. Roy. Soc.* **174**, 935. [§5.3, §5.4]

Reynolds, O. 1895, *Phil. Trans. Roy. Soc.* **186**, 123. [§8.5]

Riemann, B. 1859, *Abh. Ges. Wiss. Gött.* **8**, 43. [§6.4]

Roberts, B. 1985, in *Solar System Magnetic Fields* (ed. E. R. Priest), p. 37. Reidel. [S. f. r.]

Roberts, P. H. 1967, *An Introduction to Magnetohydrodynamics.* Longmans. [S. f. r.]

Rose, D. J. and Clark, M. 1961, *Plasmas and Controlled Fusion.* M.I.T. Press. [§10.4, S. f. r.]

Rosenbluth, M., MacDonald, W. M. and Judd, D. L. 1957, *Phys. Rev.* **107**, 1. [§13.5]

Rots, A. H. 1975, *Astron. Astrophys.* **45**, 43. [Fig. 14.14]

Rüdiger, G. 1989, *Differential Rotation and Stellar Convection: Sun and Solar-Type Stars.* Akademie-Verlag, Berlin. [S. f. r.]

Ruzmaikin, A. A., Shukurov, A. M. and Sokoloff, D. D. 1988, *Magnetic Fields of Galaxies.* Kluwer. [S. f. r.]

Rybicki, G. B. and Lightman, A. P. 1979, *Radiative Processes in Astrophysics.* John Wiley. [§10.5, §11.1, §17.3]

Saha, M. N. 1920, *Phil. Mag. (6)* **40**, 472. [§11.1]

Saha, M. N. and Srivastava, B. N. 1965, *A Treatise on Heat*, 5th edn. The Indian Press, Allahabad. [§10.5, §12.3, S. f. r.]

Saint-Venant, B. de 1843, *Comptes Rendus* **17**, 1240. [§5.2]

Sakurai, T. 1985, *Astron. Astrophys.* **152**, 121. [§14.9]

Schiff, L. I. 1968, *Quantum Mechanics*, 3rd edn. McGraw-Hill. [§1.3]

Schwarzschild, K. 1906, *Nachr. Ges. Wiss. Gött.*, 41. [§7.2]

Schwarzschild, M. 1958, *Structure and Evolution of the Stars.* Princeton University Press. Reprinted by Dover. [§8.6]

Scott Russell 1844, *Brit. Assoc. Repts.* [§7.7]

Sedov, L. I. 1946, *Prikl. Mat. Mekh.* **10**, 241. [§6.6]

Sedov, L. I. 1959, *Similarity and Dimensional Methods in Mechanics.* Academic Press. [§6.6]

Shafranov, V. D. 1958, *Sov. Phys. JETP* **6**, 545. [§14.3]

Shakura, N. I. and Sunyaev, R. A. 1973, *Astron. Astrophys.* **24**, 337. [§5.7]

Shapiro, S. L. and Teukolsky, S. A. 1983, *Black Holes, White Dwarfs, and Neutron Stars.* John Wiley. [S. f. r.]

Shkarofsky, I. P., Bernstein, I. B. and Robinson, B. B. 1963, *Phys. Fluids* **6**, 40. [§13.5]

Shore, S. N. 1993, *Astrophysical Hydrodynamics.* Academic Press. [S. f. r.]

Shu, F. H. 1982, *The Physical Universe.* University Science Books. [S. f. r.]

Shu, F. H. 1992, *The Physics of Astrophysics, Vol. II: Gas Dynamics.* University Science Books. [Pref., §6.5, S. f. r.]

Shu, F. H., Adams, F. C. and Lizano, S. 1987, *Ann. Rev. Astron. Astrophys.* **25**, 23. [S. f. r.]

Spitzer, L. 1962, *Physics of Fully Ionized Gases.* Interscience. [§13.3, S. f. r.]

Spitzer, L. 1978, *Physical Processes in the Interstellar Medium*. Interscience. [§7.5, S. f. r.]

Spitzer, L. 1987, *Dynamical Evolution of Globular Clusters*. Princeton University Press. [S. f. r.]

Spitzer, L. and Härm, R. 1953, *Phys. Rev.* **89**, 977. [§13.4]

Spruit, H. C. 1981, *Astron. Astrophys.* **98**, 155. [§14.7]

Steenbeck, M. and Krause, F. 1969, *Aston. Nachr.* **291**, 49. [§16.6]

Steenbeck, M., Krause, F. and Rädler, K.-H. 1966, *Z. Naturforsch.* **21a**, 1285. [§16.5]

Stix, M. 1989, *The Sun*. Springer-Verlag. [S. f. r.]

Stix, T. H. 1992, *Waves in Plasmas*. American Institute of Physics. [S. f. r.]

Stokes, G. G. 1845, *Trans. Camb. Phil. Soc.* **8**, 287. [§5.2]

Stokes, G G. 1851, *Trans. Camb. Phil. Soc.* **9**, 8. [§5.5]

Stokes, G. G. 1876, *Smith's Prize Examination*. [§6.2]

Storey, L. R. O. 1953, *Phil. Trans. Roy. Soc.* **A 246**, 113. [§12.5]

Störmer, C. 1907, *Arch. Sci. Phys. Geneve* **24**, 5, 113, 221 and 317. [§10.1]

Sweet, P. A. 1950, *Mon. Not. Roy. Astron. Soc.* **110**, 548. [§9.4]

Sweet, P. A. 1958, *Proc. IAU Symp.* **6**, 123. [§15.2]

Tan, S. M. and Gull, S. F. 1985, *Mon. Not. Roy. Astron. Soc.* **216**, 949. [Fig. 6.3]

Tandberg-Hansen, E. and Emslie, A. G. 1988, *The Physics of Solar Flares*. Cambridge University Press. [S. f. r.]

Tassoul, J.-L. 1978, *Theory of Rotating Stars*. Princeton University Press. [S. f. r.]

Taylor, G. I. 1921a, *Proc. Lond. Math. Soc.* **A 20**, 196. [§8.4]

Taylor, G. I. 1921b, *Proc. Roy. Soc.* **A 100**, 114. [§9.2]

Taylor, G. I. 1935, *Proc. Roy. Soc.* **A 151**, 421. [Quot. Part I, §8.1]

Taylor, G. I. 1950a, *Proc. Roy. Soc.* **A 201**, 159. [§6.6, Fig. 6.4]

Taylor, G. I. 1950b, *Proc. Roy. Soc.* **A 201**, 192. [§7.4]

Taylor, J. B. 1974, *Phys. Rev. Lett.* **33**, 1139. [§15.4]

Taylor, J. B. 1986, *Rev. Mod. Phys.* **58**, 741. [S. f. r.]

Tennekes, H. and Lumley, J. L. 1972, *A First Course in Turbulence*. M.I.T. Press. [§8.5, S. f. r.]

Thomas, J. H. and Weiss, N. O. 1992, in *Sunspots: Theory and Observations* (eds. J. H. Thomas and N. O. Weiss), p. 3. Kluwer. [S. f. r.]

Thomson, J. J. 1897, *Phil. Mag. (5)* **44**, 293. [§12.2, §13.5]

Thomson, W. B. 1951, *Phil. Mag. (7)* **42**, 1417. [§14.6]

Tomczyk, S., Schou, J. and Thomson, M. J. 1996, *Bull. Astron. Soc. India* **24**, 245. [Fig. 7.8]

Tonks, L. and Langmuir, I. 1929, *Phys. Rev.* **33**, 195. [§12.2]

Tritton, D. J. 1988, *Physical Fluid Dynamics*, 2nd edn. Oxford University Press. [S. f. r.]

Unsöld, A. 1977, *The New Cosmos*. Springer-Verlag. [S. f. r.]

Väisälä, V. 1925, *Soc. Sci. Fenn. Commentat. Phys.-Math.* **2**, no. 19, 1. [§7.2]

Van Allen, J. A., Ludwig, G. H., Ray, E. C. and McIlwain, C. E. M. 1958, *Jet Propul.* **28**, 588. [§10.4]

Van Dyke, M. 1982, *An Album of Fluid Motion*. Parabolic Press, Stanford. [S. f. r.]

Vennard, J. K. and Street, R. L. 1976, *Elementary Fluid Mechanics*, 5th edn. John Wiley. [S. f. r.]

Vitense, E. 1953, *Zs. f. Astrophys.* **32**, 135. [§8.6]

Vlasov, A. A. 1945 *J. Phys. (U.S.S.R.)* **9**, 25. [§11.4]

von Kármán, T. 1911, *Nachr. Ges. Wiss. Gött.*, 509. [§5.5]

von Kármaán, T. and Howarth, L. 1938, *Proc. Roy. Soc.* **A 164**, 192. [§8.2]

von Weizsäcker, C. F. 1937, *Phys. Zs.* **38**, 176. [§5.7]

von Weizsäcker, C. F. 1948, *Z. Naturforsch.* **3a**, 524. [§5.7]

Von Zeipel, H. 1924, *Mon. Not. Roy. Astron. Soc.* **84**, 665. [§9.4]

Weber, E. J. and Davis, L. 1967, *Astrophys. J.* **148**, 217. [§14.9]

Weiss, N. O. 1981, *J. Fluid Mech.* **108**, 247. [§14.6, Fig. 14.9]

Wiedemann, G. and Franz, H. 1853, *Ann. d. Phys. u. Chemie* **147**, 429. [§13.5]

Witham, G. B. 1974, *Linear and Nonlinear Waves*. John Wiley. [S. f. r.]

Woltjer, L. 1958, *Proc. Nat. Acad. Sci. U.S.A.* **44**, 489. [§15.3]

Wood, R. W. 1933, *Phys. Rev.* **44**, 353. [§12.2]

Yvon, J. 1935, *La Theorie des Fluides et l'Equation d'Etat*. Hermann et Cie, Paris. [§11.4]

Zeldovich, Ya. B. and Raizer, Yu. P. 1966, *Physics of Shock Waves and High Temperature Hydrodynamic Phenomena*. Academic Press. [S. f. r.]

Zeldovich, Ya. B., Ruzmaikin, A. A. and Sokoloff, D. D. 1983, *Magnetic Fields in Astrophysics*. Gordon and Breach. [S. f. r.]

Zemansky, M. W. and Dittman, R. H. 1981, *Heat and Thermodynamics*, 6th edn. McGraw-Hill. [S. f. r.]

Zener, C. 1933, *Nature* **132**, 968. [§12.2]

Zwaan, C. 1985, *Solar Phys.* **100**, 397. [Fig. 14.11]

Index